ISBN 978-1-334-03377-3
PIBN 10593965

This book is a reproduction of an important historical work. Forgotten Books uses state-of-the-art technology to digitally reconstruct the work, preserving the original format whilst repairing imperfections present in the aged copy. In rare cases, an imperfection in the original, such as a blemish or missing page, may be replicated in our edition. We do, however, repair the vast majority of imperfections successfully; any imperfections that remain are intentionally left to preserve the state of such historical works.

English
Français
Deutsche
Italiano
Español
Português

www.forgottenbooks.com

Mythology Photography **Fiction**
Fishing Christianity **Art** Cooking
Essays Buddhism Freemasonry
Medicine **Biology** Music **Ancient
Egypt** Evolution Carpentry Physics
Dance Geology **Mathematics** Fitness
Shakespeare **Folklore** Yoga Marketing
Confidence Immortality Biographies
Poetry **Psychology** Witchcraft
Electronics Chemistry History **Law**
Accounting **Philosophy** Anthropology
Alchemy Drama Quantum Mechanics
Atheism Sexual Health **Ancient History**
Entrepreneurship Languages Sport
Paleontology Needlework Islam
Metaphysics Investment Archaeology
Parenting Statistics Criminology
Motivational

OBSERVATIONS

ON DAYS OF

UNUSUAL MAGNETIC DISTURBANCE,

MADE AT THE

BRITISH COLONIAL MAGNETIC OBSERVATORIES,

UNDER THE DEPARTMENTS OF THE

ORDNANCE AND ADMIRALTY.

PRINTED BY THE BRITISH GOVERNMENT UNDER THE SUPERINTENDENCE OF

LIEUT.-COLONEL EDWARD SABINE,

OF THE ROYAL ARTILLERY.

Vol. 1, Pt. 1

PART I.—1840–1841.

LONDON:
PUBLISHED FOR HER MAJESTY'S STATIONERY OFFICE,
BY
LONGMAN, BROWN, GREEN, AND LONGMANS.

1843.

LONDON:
PRINTED BY WILLIAM CLOWES AND SONS, STAMFORD STREET,
For Her Majesty's Stationery Office.

PREFACE.

As the principal interest of the Observations, made during periods of magnetic disturbance, appears likely to proceed from their being viewed in connexion with similar observations made simultaneously in other parts of the globe, it has been considered desirable to separate the observations made at such times, from those which are made daily at stated intervals, and to print them in a volume by themselves; and by bringing together the observations of the same date, at the several Observatories, and arranging them in chronological order, to furnish the opportunity of more readily comparing them with each other, and with contemporaneous observations published elsewhere. It is intended that this volume shall comprise the disturbance observations made from 1840 to 1845 inclusive, at the Colonial Observatories instituted by the British Government; the part of the volume which is now published, contains those of 1840-1841.

In addition to the extra observations made at short intervals when the disturbance of the magnets was excessive, the regular observations (which are made hourly at some of the Observatories, and two-hourly at others,) have been extracted from the registers for the days of disturbance, and subjoined; being given for all the Observatories on all days on which extra observations were made at any one. The mean monthly positions of the magnets at the hours of regular observation have also been given, for the purpose of furnishing approximate normal positions, with which the observations on days of disturbance may be compared.

In the hope that an early publication of the observations in 1840-1841, and their comparison with simultaneous observations in other parts of the globe, may lead to the suggestion of more specific points of enquiry than are at present apprehended, and possibly to the substitution of improved instruments and modes of observation,—the printing of this part of the volume has been proceeded with, although all the materials required for the complete reduction of the observations have not yet been received. The coefficients are still wanting for the temperature corrections of the Horizontal and Vertical Force observations of the Antarctic Expedition and of the Van Diemen Island Observatory, and of the Vertical Force observations at St. Helena and the Cape of Good Hope, the necessary experiments for their determination not having yet been made. The scale coefficients

A 2

are also wanting for the Vertical Force Magnetometers of the Antarctic Expedition and of the Van Diemen Island Observatory ; for the present, therefore, the observations with the latter instruments will show only the direction and comparative amount of any changes which may have taken place in the vertical force, leaving the ratio of the changes to the force itself undetermined.

The Observatories were completed, and the instruments moved into them at the following dates: at St. Helena in August, 1840; at Toronto in September, 1840; at Van Diemen Island in October, 1840; and at the Cape of Good Hope in April, 1841. At St. Helena and Toronto, however, the instruments were set up, previous to those dates, in rooms in which temporary accommodation was obtained.

Whilst observations at many stations yet remain to be published, it would be obviously premature to attempt to trace any single disturbance through the various modifications with which it may have manifested itself in different parts of the globe; but it may not be uninstructive to examine in some small degree the general character of these remarkable phenomena, as they have presented themselves at the stations, and during the periods at present under notice ; particularly at Toronto and Van Diemen Island,—as the comparison of stations situated in different magnetic hemispheres, and nearly at the opposite extremities of a diameter of the globe, would seem to offer especial interest.

It has been justly remarked, that in enquiries concerning natural phenomena, which besides their principal cause, have a number of subordinate accidental causes, the course of investigation should be, to separate from the observed march of the phenomena, all that is accidental, for the purpose of drawing forth that which ranges itself under laws, and may thereby be connected with an efficient cause. If such be the course to be pursued when the regular portion of the phenomena forms the subject of investigation, we should adopt the converse proceeding when our attention is designed to be directed to the accidental or irregular portion; we should then endeavour to separate from the general march of the phenomena, all that appears to be subject to laws, leaving the residual to consist entirely of the irregular portion, which we desire to examine.

In our present enquiry, therefore, it is proper, in the first instance, to ascertain and to eliminate the regular diurnal movements of the magnetic elements, or those movements which depend upon the hour of the day, and the season of the year. For this purpose, the mean monthly positions of the magnets, at the regular hours of observation, have been taken as representing the mean diurnal march in the respective months ; omitting in the means, the positions on days when disturbances from irregular causes prevailed to any considerable degree. It is probable that the number of days of observation in each month, may not be sufficient to give the diurnal movement with the precision with which it will hereafter be obtained by the combinations of more than a single year; but the general harmony of the different months, and the correspondence which is shown by the monthly means at Toronto and Van Diemen Island, are an evidence that the approximation is sufficient for the present purpose. In the tables which follow, and which contain the diurnal march derived from the monthly means, the lowest reading at whatever observation hour it may have occurred, has been taken as the zero for the month ; in the case of the Declination, the lowest reading corresponds in both hemispheres to the extreme westerly position of that end of the magnet bar which is directed towards the north ; in the case of the Horizontal and Vertical Force Magnets, the lowest reading implies the weakest force ; the differences at the other hours are expressed, in the Declination, in minutes of Arc, and in the Horizontal and Vertical Force, in parts of those forces respectively.

TORONTO.—Diurnal oscillation of the Declination, 1841.

The actual Times of Observation were more precisely 2¼ minutes before the hours specified.

Mean Time at the Station	WINTER HALF-YEAR.						SUMMER HALF-YEAR.						MEANS.	
	January.	February.	March.	October.	November.	December.	April.	May.	June.	July.	August.	September.	Winter.	Summer.
Hrs.	'	'	'	'	'	'	'	'	'	'	'	'	'	'
0	2·10	2·21	1·69	0·59	0·15	0·84	1·96	1·21	1·16	1·44	0·51	1·01	1·26	1·21
2	0·00	0·00	0·00	0·00	0·00	0·00	0·00	0·00	0·00	0·00	0·00	0·00	0·00	0·00
4	0·63	0·94	2·00	2·56	2·02	2·64	2·49	2·16	2·26	2·46	3·55	3·40	1·80	2·72
6	2·86	2·82	4·25	5·10	4·70	6·45	5·45	5·19	5·62	5·58	5·58	4·79	4·36	5·37
8	5·63	5·00	5·72	6·62	6·08	6·78	7·43	6·74	7·30	7·09	9·30	7·40	5·97	7·54
10	5·02	7·08	6·48	6·65	5·07	5·62	8·76	6·98	6·62	7·82	11·05	6·55	5·99	7·96
12	5·63	5·49	6·62	5·92	5·13	5·27	8·21	5·83	7·17	9·11	7·79	8·31	4·68	7·74
14	5·02	4·21	5·98	6·65	5·05	4·09	8·87	6·19	6·20	7·82	8·11	7·79	5·17	7·50
16	4·92	5·91	7·31	6·64	5·13	4·89	9·25	7·10	7·45	7·89	7·64	7·96	5·80	7·88
18	3·26	5·95	7·66	5·54	3·43	5·54	10·38	10·17	11·56	11·24	12·26	9·24	5·23	10·81
20	5·16	6·95	9·23	7·10	5·70	4·38	11·11	10·40	12·27	12·84	13·38	9·78	6·42	11·63
22	6·09	6·39	7·18	5·80	3·63	4·39	8·02	6·96	6·48	8·28	7·68	5·10	5·58	7·09

VAN DIEMEN ISLAND.—Diurnal oscillation of the Declination, 1841.

The actual Times of Observation were more precisely 10 minutes after the hours specified.

Mean Time at the Station	WINTER HALF-YEAR.						SUMMER HALF-YEAR.						MEANS.	
	April.	May.	June.	July.	August.	October.	January.	February.	March.	October.	November.	December.	Winter.	Summer.
Hrs.	'	'	'	'	'	'	'	'	'	'	'	'	'	'
0	3·74	1·83	1·81	2·27	2·87	3·91	9·01	6·42	7·47	7·26	7·48	7·55	2·74	7·53
2	6·74	4·73	3·58	4·76	5·08	7·00	12·77	8·42	10·07	10·49	11·14	11·36	5·31	10·71
4	5·20	4·25	3·07	3·95	5·03	6·10	10·35	7·31	8·29	9·16	9·73	9·94	4·60	9·13
6	2·84	2·46	1·72	1·97	2·97	3·14	7·72	5·32	5·22	5·89	7·06	7·33	1·52	6·42
8	1·61	0·32	0·72	1·05	1·22	1·16	6·29	4·34	3·86	3·91	4·20	5·66	1·01	4·71
10	0·89	0·17	0·00	0·02	0·00	0·63	4·34	2·67	2·83	2·45	2·09	4·24	0·28	3·10
12	1·10	0·25	0·37	0·55	0·50	1·38	4·59	1·39	2·85	2·32	2·63	2·83	0·69	2·77
14	2·43	1·10	0·83	0·74	0·60	0·75	4·03	1·70	2·91	4·19	4·00	3·15	1·07	3·33
16	2·24	1·82	1·53	2·64	1·64	2·98	4·02	2·02	3·75	4·00	4·22	3·51	2·14	3·59
18	2·51	1·64	1·44	1·82	2·04	2·61	2·00	0·13	2·54	2·69	2·35	2·53	2·01	2·04
20	0·56	1·48	1·70	1·55	1·93	0·55	0·00	0·00	0·00	0·00	0·00	0·00	1·19	0·00
22	0·00	0·00	0·97	0·00	0·84	0·00	1·82	1·64	1·76	1·38	1·40	1·61	0·30	1·53

The observations at Van Diemen Island were made hourly, but the alternate ones only have been taken on this occasion, for the sake of the comparison with Toronto, where during the year 1841 two-hourly observations only were made.

On examining the Table for Toronto, we perceive as follows:—

The march of the regular diurnal oscillation does not consist in a simple movement from one extremity of the range to the other, and back, as has been supposed even at a very recent date,[*] but in an alternate progression and retrogression.

In tracing the course of the oscillation, we may commence with 2 P.M., because in every month of the year, without exception, the north end of the bar is then more to the west, than at any other observation hour. Dating from 2 P.M., the movement is continuous towards the east, until 10 P.M.; when the bar returns towards the west, and reaches, at 2 A.M., a second westerly limit; which, however, is nearer the eastern than the western extremity of the whole diurnal range, in consequence of the retrogression, or westerly movement at this period of the 24 hours, being comparatively of small amount. A second progression towards the east then commences, and continues until 8 A.M., having a much more decided character in the summer than in the winter months, both in its amount, and in the precise hour at which it reaches its limit. From 8 A.M. till 2 P.M., the return is continuous towards the west.

The observation hours which have been named as those at which the alternate movements terminate, viz., the westerly at 2 A.M. and 2 P.M., and the easterly at 10 P.M. and 8 A.M., are indicated alike by the means both of the summer and of the winter half-years; when examined more in detail, there appears a somewhat less regularity in the periods during the nocturnal hours, than during those when the Sun is above the horizon.

The range of the diurnal fluctuation is throughout greater in the summer than in the winter months: this is particularly the case with the easterly movement, which takes place from 2 A.M. to 8 A.M., and the subsequent return. The progressive increase in the extent of this movement may be clearly traced from the mid-winter, when it is barely perceptible, to the mid-summer, when it is very considerable. It is a consequence of this inequality, that in two of the mid-winter months, the *evening* eastern limit becomes the eastern extremity of the whole diurnal range, which is formed in every other month of the year by the *morning* eastern limit.

If we now refer to the Table of the diurnal oscillation of the Declination at Van Diemen Island, we find that its course corresponds with that at Toronto in all the principal features, with one essential distinction, viz., that the hours of easterly movement at Toronto, are those of westerly movement at Van Diemen Island, and *vice versâ*; the north end of the bar being in both cases the one of which the motion is noticed, and the hours being those of mean time at the respective stations. The diurnal range is nearly the same at both places, (rather less at Van Diemen Island), and there is at both

* Arago; Annuaire pour l'an 1836, p. 283.

a similar inequality in its amount in summer and winter. The alternate progression and retrogression are as distinctly marked, and the hours indicated as the turning points are either the same at both stations, or with such slight differences, as cannot be insisted upon, until observation, by its increased frequency and repetition, shall press more closely upon the phenomena.

The next Tables exhibit a similar view of the diurnal oscillation of the Horizontal and Vertical components of the Force at Toronto, derived from the two-hourly observations, corrected for the temperature of the bar, and expressed in terms of the respective forces. An approximate coefficient has been employed in the temperature correction of the Vertical Force observations. Similar Tables for the Van Diemen Island Observatory cannot be prepared, until the observations can be corrected for the temperature of the Magnet bars.

TORONTO.—Diurnal oscillation of the Horizontal Force, 1841.

The actual Times of Observation were at the hours specified.

Mean Time at the Station.	WINTER HALF-YEAR.						SUMMER HALF-YEAR.						MEANS.	
	January.	February.	March.	October.	November.	December.	April.	May.	June.	July.	August.	September.	Winter.	Summer.
Hrs.														
0	·00000	·00000	·00000	·00000	·00001	·00000	·00000	·00052	·00065	·00013	·00070	·00093	·00000	·00049
2	·00051	·00036	·00096	·00092	·00062	·00069	·00147	·00156	·00145	·00139	·00211	·00194	·00068	·00165
4	·00127	·00112	·00150	·00143	·00105	·00109	·00209	·00226	·00188	·00184	·00263	·00236	·00124	·00218
6	·00134	·00125	·00119	·00130	·00096	·00084	·00210	·00204	·00159	·00162	·00205	·00213	·00115	·00192
8	·00115	·00089	·00097	·00106	·00078	·00086	·00141	·00150	·00090	·00076	·00096	·00205	·00095	·00126
10	·00087	·00107	·00084	·00089	·00056	·00087	·00142	·00131	·00072	·00062	·00072	·00160	·00085	·00106
12	·00084	·00085	·00074	·00063	·00033	·00081	·00133	·00124	·00077	·00055	·00077	·00134	·00079	·00103
14	·00100	·00094	·00088	·00103	·00017	·00071	·00155	·00118	·00088	·00064	·00067	·00132	·00079	00104
16	·00097	·00103	·00087	·00113	·00053	·00077	·00156	·00122	·00062	·00067	·00090	·00118	·00088	·00102
18	·00135	·00121	·00099	·00102	·00058	·00129	·00158	·00126	·00103	·00073	·00114	·00179	·00107	·00125
20	·00149	·00104	·00063	·00058	·00046	·00082	·00000	·00072	·00057	·00056	·00093	·00124	·00084	·00092
22	·00103	·00040	·00002	·00021	·00000	·00051	·00003	·00000	·00000	·00000	·00000	·00000	·00000	·00000

Toronto.—Diurnal oscillation of the Vertical Force, 1841.

The actual Times of Observation were more precisely 5 minutes before the hours specified.

Mean Time at the Station	WINTER HALF-YEAR.						SUMMER HALF-YEAR.						MEANS.	
Hrs.	January.	February.	March.	October.	November.	December.	April.	May.	June.	July.	August.	September.	Winter.	Summer.
0	·00012	·00017	·00025	·00038	·00032	·00013	·00018	·00013	·00001	·00020	·00048	·00044	·00023	·00024
2	·00030	·00034	·00040	·00055	·00044	·00029	·00039	·00033	·00018	·00035	·00065	·00063	·00039	·00042
4	·00039	·00047	·00060	·00065	·00052	·00035	·00056	·00057	·00048	·00054	·00081	·00059	·00050	·00059
6	·00041	·00052	·00069	·00064	·00059	·00043	·00069	·00059	·00056	·00062	·00009	·00074	·00055	·00070
8	·00041	·00049	·00067	·00060	·00050	·00034	·00057	·00049	·00047	·00058	·00083	·00064	·00050	·00060
0	·00035	·00043	·00052	·00048	·00020	·00024	·00044	·00040	·00027	·00031	·00044	·00043	·00037	·00038
2	·00018	·00024	·00026	·00016	·00017	·00014	·00015	·00016	·00006	·00008	·00027	·00038	·00019	·00018
4	·00007	·00005	·00000	·00000	·00000	·00000	·00013	·00014	·00003	·00000	·00000	·00010	·00002	·00007
6	·00000	·00005	·00006	·00006	·00018	·00005	·00000	·00001	·00004	·00006	·00012	·00000	·00007	·00004
8	·00001	·00001	·00013	·00021	·00000	·00007	·00006	·00005	·00010	·00009	·00033	·00014	·00007	·00013
0	·00013	·00002	·00018	·00024	·00024	·00009	·00012	·00004	·00009	·00014	·00023	·00042	·00015	·00017
2	·00004	·00000	·00013	·00023	·00023	·00000	·00010	·00000	·00000	·00014	·00034	·00045	·00010	·00017

From these Tables we collect the following particulars:—The diurnal oscillation of the horizontal force consists in an alternate increase and decrease of the force, forming two maxima and two minima in the twenty-four hours. The principal minimum, or the least force at any of the observation hours, occurs at 10 A.M. in the summer half-year, and at noon in the winter half-year. The principal maximum is at 4 or 6 P.M., except in mid-winter, when the afternoon oscillation is so much reduced in amount that the other maximum, which occurs throughout the year at 6 or 8 A.M., becomes, in the months of December and January, the principal maximum. The second minimum takes place during the hours of the night, or from 10 P.M. to 4 A.M., during which period the force is nearly stationary. The whole diurnal oscillation is altogether greater in summer than in winter, and the alternate increase and decrease twice in the twenty-four hours shows itself in the mean of every month of the year with slight variations in the times of the turning points.

The maximum of the vertical force takes place at 6 P.M., and the minimum at 2 or 4 A.M. A second maximum at 8 A.M. and minimum at 10 A.M. are also traceable in some of the months, but a longer period of observation and a more accurate determination of the coefficient in the temperature correction, are requisite for a satisfactory conclusion on this point.

On a review of the accord which shows itself in the diurnal march of each element in the different months, we may feel justified in regarding the mean monthly positions at the several observation hours as approximate normal positions; and we may view the difference between the actual position at any hour during the month, and the normal position at the same hour, as the effect of disturbing causes.

If we represent the Declination observed at the several hours on any particular day by ψ_0, ψ_2, ψ_4 ... ψ_{22}, (the small figures showing the hours of observation in mean Göttingen time,) and the *mean* values at the respective hours in the *month* to which the day belongs, by $\overline{\psi}_0$, $\overline{\psi}_2$, $\overline{\psi}_4$ ... $\overline{\psi}_{22}$; then $\psi_0 - \overline{\psi}_0$ will be the effect of the irregular disturbing force at 0 hours, which we may represent by $\nabla \psi_0$; $\psi_2 - \overline{\psi}_2 = \nabla \psi_2$ is the same at 2 hours; and so forth to $\psi_{22} - \overline{\psi}_{22} = \nabla \psi_{22}$ at 22 hours; the regular or diurnal oscillation of the bar being by this process eliminated. The *fluctuation* of the Declination due to the irregular action between the observations at 0 and at 2 hours will then be $\nabla \psi_2 - \nabla \psi_0$, which we may express by $F(\psi_2)$; the fluctuation between the observations at 2 and 4 hours $\nabla \psi_4 - \nabla \psi_2 = F(\psi_4)$; and so forth: and the *mean* irregular fluctuation for the whole day will be $\overline{F\psi} = \sqrt{\frac{1}{12} \Sigma (F\psi)^2}$ if the number of observation hours have been 12. In like manner the mean irregular fluctuation for the several months in the year will be

$$\overline{F\psi}_{\text{Jan.}} = \sqrt{\frac{1}{n} \Sigma (F\psi_{\text{Jan.}})^2}; \quad \overline{F\psi}_{\text{Feb.}} = \sqrt{\frac{1}{n} \Sigma (F\psi_{\text{Feb.}})^2}; \quad \&c.$$

and the mean irregular fluctuation for the year (1841 for example) will be

$$\overline{F\psi}_{1841} = \sqrt{\frac{1}{n} \Sigma (F\psi_{1841})^2}.$$

In this notation the horizontal bar over the symbols is employed to signify mean values, the particular kind of mean appearing either from the circumstances, or being denoted by the smaller figures; and the horizontal part of the symbol ∇ implies the relation to mean values. The notation is equally applicable to the horizontal and vertical force, of which the absolute values are characterized by X and Z.

The following letters exhibit the values of $\overline{F(\psi)}$ and $\overline{F(X)}$ for every day in 1841 in which observations were made at Toronto and Van Diemen Island; also the mean monthly fluctuation for the several months, and the mean annual fluctuation. All the observations of the horizontal force at Toronto have been reduced to a common temperature: those at Van Diemen Island are necessarily uncorrected.

Table showing the Mean Fluctuation of the Declination and Horizontal Force at
and the Mean

Days of the Month.	JANUARY.		FEBRUARY.		MARCH.		APRIL.		MAY.		JUNE.	
	Decl.	Hor. Force	Decl.	Hor. Force.	Decl.	Hor. Force.	Decl.	Hor. Force.	Decl.	Hor. Force.	Decl.	Hor. Force.
1	2·3	·00036	1·7	·00039	2·1	·00076	1·0	·00055	2·9	·00103	1·9	·00062
2	1·7	·00051	1·3	·00022	1·4	·00055	1·7	·00043	Sunday.		1·9	·00115
3	Sunday.		1·6	·00030	1·7	·00027	2·0	·00095	1·5	·00030	2·6	·00072
4	1·2	·00017	1·6	·00049	1·9	·00026	Sunday.		2·1	·00065	3·2	·00053
5	1·1	·00037	1·1	·00024	1·6	·00045	2·2	·00043	4·8	·00091	3·1	·00070
6	2·6	·00047	2·7	·00031	3·3	·00069	1·9	·00033	5·1	·00079	Sunday.	
7	3·1	·00063	Sunday.		Sunday.		2·4	·00040	1·9	·00076	3·2	·00064
8	2·6	·00095	2·3	·00055	1·1	·00060	4·9	·00144	1·6	·00097	1·2	·00044
9	1·1	·00042	12·9	·00109	1·1	·00035	Good Friday.		Sunday.		4·8	·00066
10	Sunday.		1·7	·00038	1·4	·00043	1·4	·00032	9·7	·00198	2·6	·00045
11	1·9	·00038	2·4	·00059	3·6	·00061	Sunday.		2·2	·00058	4·5	·00063
12	2·2	·00060	5·9	·00084	1·4	·00027	3·9	·00144	2·8	·00077	1·9	·00070
13	5·3	·00047	2·2	·00037	1·7	·00018	1·4	·00037	1·4	·00052	Sunday.	
14	4·2	·00070	Sunday.		Sunday.		3·3	·00049	3·7	·00070	1·9	·00045
15	4·0	·00054	5·8	·00078	8·3	00160	1·3	·00036	1·4	·00056	5·0	·00090
16	4·1	·00045	3·7	·00075	4·5	·00097	5·1	·00042	Sunday.		1·7	·00049
17	Sunday.		1·7	·00023	2·1	·00098	3·8	·00113	3·6	·00081	3·0	·00125
18	1·7	·00032	1·4	·00051	1·9	·00060	Sunday.		2·8	·00110	5·3	·00088
19	2·9	·00073	1·1	·00030	2·6	·00045	2·7	·00102	4·6	·00090	1·7	·00044
20	2·4	·00044	1·3	·00053	4·8	·00036	5·2	·00139	4·0	·00051	Sunday.	
21	2·7	·00060	Sunday.		Sunday.		3·5	·00087	4·7	·00093	1·7	·00067
22	1·3	·00032	4·2	·00084	10·3	·00283	2·4	·00036	2·9	·00109	3·4	·00107
23	1·7	·00039	10·4	·00103	3·4	·00101	2·1	·00033	Sunday.		5·6	·00052
24	Sunday.		3·7	·00064	3·1	·00075	2·0	·00092	1·5	·00048	4·1	·00087
25	7·1	·00077	3·7	·00051	2·9	·00050	Sunday.		*	·00038	4·9	·00178
26	3·2	·00093	7·5	·00134	2·7	·00066	2·9	·00052	*	·00101	1·4	·00036
27	3·5	·00123	3·2	·00064	1·3	·00049	2·0	·00074	4·1	·00033	Sunday.	
28	2·7	·00028	Sunday.		Sunday.		2·4	·00032	2·2	·00075	3·4	·00061
29	1·7	·00036			3·7	·00116	3·8	·00049	4·9	·00055	3·1	·00157
30	2·0	·00047			1·5	·00079	3·5	·00090	Sunday.		2·1	·00152
31	Sunday.				1·4	·00024			6·5	·00060		
Mean Monthly Fluctuations.	3·06	·00058	4·70	·00065	3·52	·00088	3·00	·00071	3·83	·00084	3·35	·00089

Mean Annual Fluctuation of the Declination 3′·39;

*The Declination Magnetometer was not in adjustment on the 25th and 26th of May, in consequence of the suspension thread breaking.

Toronto on the several Days in the Year 1841; also the Mean Monthly Fluctuations, Annual Fluctuation.

JULY.		AUGUST.		SEPTEMBER.		OCTOBER.		NOVEMBER.		DECEMBER.		Days of the Month.
Decl.	Hor. Force.	Decl.	Hor. Force.	Decl.	Hor. Force.	Decl.	Hor. Force.	Decl.	Hor. Force.	Decl.	Hor. Force.	
1·7	·00112	Sunday.		3·5	·00074	1·7	·00036	1·1	·00035	1·4	·00066	1
2·1	·00065	5·4	·00093	3·5	·00084	2·3	·00055	2·9	·00015	3·7	·00101	2
2·7	·00039	2·3	·00069	2·6	·00045	Sunday.		4·2	·00043	5·9	·00157	3
Sunday.		2·4	·00067	2·6	·00057	1·4	·00028	4·5	·00097	3·2	·00096	4
4·6	·00074	2·9	·00095	Sunday.		4·0	·00082	4·1	·00178	Sunday.		5
6·8	·00134	16·4	·00303	1·4	·00035	3·5	·00094	6·7	·00160	4·8	·00062	6
1·6	·00043	2·7	·00141	2·7	·00055	3·7	·00059	Sunday.		2·1	·00063	7
2·2	·00042	Sunday.		1·6	·00037	4·2	·00082	2·8	·00064	4·5	·00133	8
1·4	·00043	3·2	·00039	1·4	·00045	3·2	·00115	1·7	·00049	1·7	·00059	9
1·6	·00074	2·4	·00046	1·1	·00062	Sunday.		1·7	·00039	5·9	·00084	10
Sunday.		5·0	·00110	1·2	·00048	2·2	·00065	2·6	·00194	1·4	·00029	11
2·4	·00048	4·7	·00106	Sunday.		1·5	·00052	1·4	·00043	Sunday.		12
2·1	·00047	2·4	·00062	4·5	·00126	2·8	·00063	2·0	·00039	1·5	·00036	13
3·5	·00083	3·4	·00114	2·8	·00263	3·7	·00066	Sunday.		9·1	·00138	14
2·6	·00070	Sunday.		1·7	·00052	2·0	·00035	1·5	·00048	1·5	·00029	15
1·7	·00122	5·9	·00077	5·6	·00070	4·2	·00049	1·1	·00031	2·4	·00090	16
2·6	·00049	2·8	·00077	2·7	·00042	Sunday.		2·6	·00030	3·2	·00062	17
Sunday.		2·9	·00071	4·9	·00080	2·6	·00095	7·0	·00251	1·2	·00062	18
7·8	·00098	3·0	·00062	Sunday.		3·2	·00064	7·7	·00090	Sunday.		19
4·1	·00141	1·7	·00055	1·7	·00065	7·8	·00122	6·8	·00148	1·9	·00038	20
6·7	·00062	2·8	·00096	2·0	·00040	4·6	·00126	Sunday.		1·2	·00034	21
4·1	·00044	Sunday.		2·6	·00074	1·9	·00048	2·6	·00050	6·1	·00034	22
2·4	·00100	8·6	·00115	1·5	·00033	1·1	·00048	1·3	·00064	3·2	·00027	23
5·0	·00224	3·1	·00057	7·1	·00198	Sunday.		4·5	·00080	3·8	·00042	24
Sunday.		2·7	·00053	21·9	·00515	8·2	·00122	1·3	·00020	Christmas Day.		25
2·9	·00065	11·4	·00065	Sunday.		4·3	·00097	1·6	·00030	Sunday.		26
4·8	·00066	7·8	·00066	5·8	·00156	2·2	·00130	1·6	·00028	1·0	·00036	27
2·4	·00059	5·8	·00081	5·0	·00213	1·3	·00039	Sunday.		1·7	·00033	28
1·6	·00070	Sunday.		4·8	·00164	3·6	·00053	1·3	·00021	1·3	·00025	29
1·9	·00048	†	·00036	4·5	·00072	1·4	·00035	1·4	·00027	4·8	·00160	30
4·6	·00032	6·0	·00109			Sunday.				3·1	·00062	31
3·72	·00083	5·83	·00103	5·18	·00133	3·40	·00079	3·54	·00094	3·73	·00083	Mean Monthly Fluctuations.

of the Horizontal Force ·00086·

† The Declination Magnetometer was not in adjustment on the 30th of August, in consequence of a fibre of the suspension thread having broken; a new suspension thread was fitted.

B 2

Table showing the Mean Fluctuation of the Declination and Horizontal Force at
Fluctuations, and the

Days of the Month.	JANUARY.		FEBRUARY.		MARCH.		APRIL.		MAY.		JUNE.	
	Decl.	Hor. Force.	Decl.	Hor. Force.	Decl.	Hor. Force.	Decl.	Hor. Force.	Decl.	Hor. Force.	Decl.	Hor. Force.
1	2·4	·0006	2·1	·0004	1·6	·0004	1·1	·0004	Sunday.		0·8	·0004
2	Sunday.		1·6	·0004	1·1	·0003	1·1	·0004	0·9	·0002	0·6	·0003
3	2·0	·0006	2·0	·0003	1·4	·0003	Sunday.		2·0	·0002	0·6	·0003
4	1·7	·0003	0·6	·0004	0·9	·0003	0·9	·0004	1·5	·0006	1·1	·0004
5	1·2	·0004	1·3	·0002	0·9	·0002	1·3	·0002	0·8	·0007	Sunday.	
6	2·1	·0009	Sunday.		Sunday.		1·1	·0003	1·7	·0004	0·9	·0003
7	3·1	·0006	3·4	·0009	2·3	·0005	1·1	·0006	1·1	·0003	1·1	·0004
8	1·7	·0004	1·8	·0007	1·3	·0004	1·5	·0003	Sunday.		0·8	·0002
9	Sunday.		5·5	·0005	1·0	·0003	1·1	·0002	3·5	·0013	1·6	·0007
10	1·7	·0004	1·1	·0002	1·1	·0003	Sunday.		4·9	·0013	0·8	·0002
11	1·2	·0004	1·6	·0003	1·2	·0003	1·6	·0006	1·1	·0004	0·5	·0004
12	2·3	·0006	1·7	·0004	1·0	·0004	1·1	·0006	1·4	·0006	Sunday.	
13	1·7	·0007	Sunday.		Sunday.		1·4	·0004	0·6	·0002	0·9	·0003
14	2·7	·0004	1·7	·0006	3·1	·0010	1·2	·0003	0·7	·0002	1·1	·0002
15	1·8	·0004	4·6	·0010	3·8	·0013	1·4	·0001	Sunday.		1·8	·0004
16	Sunday.		3·7	·0006	2·5	·0006	1·3	·0002	2·0	·0005	1·2	·0002
17	1·1	·0004	1·8	·0004	2·0	·0004	Sunday.		3·0	·0004	1·3	·0006
18	1·2	·0006	1·4	·0003	0·8	·0005	1·8	·0009	2·3	·0004	3·0	·0004
19	5·0	·0006	0·5	·0005	2·4	·0004	2·7	·0007	1·1	·0003	Sunday.	
20	1·4	·0005	Sunday.		Sunday.		2·8	·0009	1·3	·0006	0·8	·0003
21	1·3	·0005	0·8	·0003	2·1	·0005	2·2	·0005	3·4	·0006	0·8	·0002
22	2·1	·0005	4·1	·0011	7·6	·0011	1·5	·0008	Sunday.		0·8	·0002
23	Sunday.		3·8	·0007	5·1	·0007	0·7	·0003	0·9	·0003	0·8	·0003
24	0·8	·0004	1·3	·0010	2·5	·0004	Sunday.		0·6	·0001	1·1	·0003
25	1·7	·0008	2·1	·0005	1·6	·0005	1·3	·0004	0·9	·0003	2·8	·0004
26	2·0	·0007	3·8	·0010	0·8	·0004	1·9	·0003	1·3	·0005	Sunday.	
27	2·3	·0008	Sunday.		Sunday.		0·6	·0004	2·1	·0004	0·8	·0002
28	1·5	·0004	3·0	·0010	1·5	·0004	0·7	·0003	1·5	·0002	0·7	·0008
29	1·3	·0004			1·2	·0004	0·8	·0004	Sunday.		1·1	·0007
30	Sunday.				2·3	·0006	2·1	·0006	0·6	·0002	0·7	·0005
31	2·5	·0006			0·8	·0001			0·8	·0004		
Mean Monthly Fluctuations.	2·00	·00055	2·5	·00061	2·53	·00055	1·49	·00048	1·97	·00054	1·31	·00038

Mean Annual Fluctuation of the Declination 2′·02 ;

Van Diemen Island, on the several Days in the Year 1841; also the Mean Monthly Mean Annual Fluctuation.

JULY.		AUGUST.		SEPTEMBER.		OCTOBER.		NOVEMBER.		DECEMBER.		Days of the Month.
Decl.	Hor. Force.	Decl.	Hor. Force.	Decl.	Hor. Force.	Decl.	Hor. Force.	Decl.	Hor. Force.	Decl.	Hor. Force.	
0·8	·0002	1·0	·0002	2·3	·0004	1·3	·0004	0·9	·0004	1·9	·0003	1
0·6	·0003	1·7	·0004	2·0	·0005	Sunday.		0·9	·0003	2·6	·0006	2
Sunday.		2·1	·0010	1·3	·0004	1·5	·0006	3·1	·0007	4·8	·0005	3
0·9	·0003	0·8	·0004	Sunday.		0·9	·0002	2·4	·0011	Sunday.		4
2·3	·0006	1·8	·0004	0·8	·0002	1·3	·0007	2·5	·0010	1·3	0003	5
2·1	·0005	6·1	·0013	1·1	·0003	1·6	·0004	Sunday.		1·6	·0005	6
0·8	·0002	Sunday.		1·4	·0004	1·5	·0005	1·3	·0005	1·6	·0005	7
0.6	·0003	0·5	·0004	1·2	·0003	3·8	·0013	1·6	·0005	4·5	·0007	8
0·8	·0002	0·8	·0002	1·0	·0003	Sunday.		1·6	·0002	2·2	·0004	9
Sunday.		1·0	·0003	0·6	·0003	1·5	·0003	1·2	·0005	1·5	·0003	10
1·6	·0002	2·3	·0006	Sunday.		1·1	·0004	2·6	·0008	Sunday.		11
1·3	·0003	1·5	·0010	1·7	·0005	1·0	·0003	2·1	·0003	1·4	·0004	12
1·2	·0002	0·9	·0003	2·8	·0013	1·3	·0004	Sunday.		1·5	·0005	13
3·1	·0003	Sunday.		1·7	·0004	0·8	·0003	1·4	·0004	2·7	·0008	14
2·3	·0004	1·8	·0006	0·8	·0004	0·8	·0003	1·3	·0004	0·6	·0005	15
0·8	·0002	1·1	·0006	1·8	·0005	Sunday.		1·4	·0005	1·9	·0004	16
Sunday.		1·0	·0003	0·6	·0004	1·1	·0005	2·5	·0013	2·7	·0006	17
0·8	·0003	1·1	·0003	Sunday.		1·3	·0007	2·3	·0014	Sunday.		18
1·5	·0007	0·9	·0002	1·1	·0003	1·1	·0004	5·2	·0010	1·6	·0004	19
4·5	·0003	0·9	·0002	1·7	·0005	1·9	·0007	Sunday.		1·4	·0002	20
0·8	·0007	Sunday.		0·8	·0004	1·7	·0009	1·7	·0003	1·3	·0002	21
2·1	·0006	1·6	·0008	0·6	·0004	0·9	·0001	3·3	·0003	1.2	·0006	22
1·8	·0006	3·3	·0004	1·0	·0003	Sunday.		1·5	·0004	1·3	·0005	23
Sunday.		1·8	·0005	4·3	·0011	2·8	·0007	1·9	·0005	1·4	·0006	24
0·6	·0004	1·2	·0002	Sunday.		2·7	·0007	2·1	0003	Sunday.		25
1·3	·0004	2·6	·0006	1·4	·0004	2·8	·0008	1·6	·0007	1·1	·0003	26
1·5	·0003	2·8	·0005	2·0	·0011	1·4	·0005	Sunday.		0·8	·0002	27
0·8	·0003	Sunday.		3·0	·0012	3·3	·0004	1·1	·0007	2·3	·0003	28
0·9	·0002	1·3	·0007	4·0	·0005	1·4	·0007	1·7	·0003	1·8	·0004	29
0·8	·0002	1·9	·0002	2·0	·0002	Sunday.		1·3	·0004	3·5	·0010	30
Sunday.		2·0	·0007			1·3	·0005			1·3	·0005	31
1·71	·00039	2·03	·00058	1·92	·00060	1·83	·00059	2·21	·00069	2·20	·00050	Mean Monthly Fluctuations.

of the Horizontal Force ·00054.

A cursory examination of these tables is sufficient to show that some connexion exists between the disturbances of principal magnitude at Toronto and those at Van Diemen Island. If twenty or thirty of the most disturbed days in the year be selected from each of the stations, the days will be found to be for the most part the same at both; and the three days of most remarkable disturbance at Van Diemen Island, *viz.*, the 22nd March, 10th May, and 6th of August, are also the most disturbed days at Toronto, (excepting the 25th September, which, falling on Sunday at Van Diemen Island, is excluded from the comparison).

Another circumstance which cannot fail to be noticed, even on a mere inspection, is the general inferiority in amount of the disturbances at Van Diemen Island when compared with those at Toronto; an inferiority which shows itself both in the Horizontal Force and in the Declination. The fluctuation from one hour of observation to the next, on the average of the whole year, is, at Toronto, of the Declination $3' \cdot 99$, and of the Horizontal Force $\cdot 00086$; at Van Diemen Island, Declination $2' \cdot 02$, Horizontal Force $\cdot 00054$; and if portions of the year taken in corresponding seasons be compared, the inequality will be found to be generally nearly in the same proportion. The extra observations which are made at both Observatories, at short intervals, on days of unusual disturbance, afford also consistent evidence of the lesser effect which the disturbing influences produce at Van Diemen Island. The Magnetic relations of the two stations,— those at least which might be supposed to have the principal influence on the comparative effect of disturbing forces,—do not so greatly differ; the terrestrial magnetic intensity is nearly the same at both stations; the inclination is—70° $40'$ at Van Diemen Island, 75° $10'$ at Toronto; the component of the force which acts on the bar in its horizontal position is $4 \cdot 5$ at Van Diemen Island, $3 \cdot 5$ at Toronto. The range of the regular diurnal oscillation, though somewhat less at Van Diemen Island than at Toronto, approaches much nearer an equality than the irregular fluctuation. Though in nearly corresponding latitudes, the two stations differ considerably in other geographical circumstances which materially affect climatological relations: Toronto is situated in the interior of a great continent, and Van Diemen Island, though in the neighbourhood of land of great extent, is surrounded in all other directions by an expanse of ocean. These circumstances may not be undeserving of notice, though no connexion has yet been established between climatological and magnetic influences. That there should be a fluctuation of nearly four minutes of arc in the Declination, and of nearly $\frac{1}{1156}$th part of the whole horizontal force, on the average of the year, between one two-hourly observation and the succeeding one (as appears to be the case at Toronto),—arising from causes, and subject to laws, of which we have at present no knowledge whatsoever,—is a circumstance which cannot fail to draw the attention of Magneticians even more strongly than hitherto to the subject of the irregular disturbances. Refinements in instruments, particularly the precautions which have been adopted since 1841 to prevent the generation of currents of air within the external magnetometer case, may possibly in succeeding years somewhat diminish the .value of the mean fluctuations which have been derived from the observations of 1841 ; but the correspondence which has been stated to exist, and which may be pursued in detail in the Observations contained in this volume, in the times of the occurrence of the disturbances (of those of principal magnitude at least), at stations so widely remote from each other as Toronto and Van Diemen Island, and so opposed in season and horary angle, appear fully to justify the persuasion, that the fluctuations, of which we have sought to obtain at least approximate mean values, are actual natural phenomena, and not, as some have imagined, mere instrumental deceptions.

Further evidence of the generality of these phenomena may be obtained by comparing the days of principal disturbance at Prague with those at Toronto and Van Diemen Island. At Prague, as at the British Colonial Observatories, observations at short intervals are commenced whenever the attention of the Director is arrested by the occurrence of a change of more than usual magnitude between one regular observation and the next; and are continued until the fluctuations assume again their ordinary character. The disturbance observations at Prague are printed in detail, and in a form very similar to that which has been adopted in this volume, in a separate section of the "Magnetische und Meteorologische Beobachtungen zu Prag" now in course of publication, by M. Kreil, Director of the Observatory. This portion of the work has not yet advanced beyond November, 1840; but the volume in which it is comprised contains also a table, in which the relative magnitude of the disturbance at the observation hours on every day from July, 1839, to June, 1841, inclusive, is given both in Declination and Horizontal Intensity ; and M. Kreil has been so obliging as to furnish me with a MS. continuation to the end of December, 1841. In this table the days of disturbance are marked by an asterisk($*$), and the day of greatest disturbance in each month is marked thus ($*$ ☽). The days which M. Kreil characterizes as days of disturbance are those in which the sum of the fluctuations from one observation hour to the next throughout the twenty-four hours, including both the irregular and diurnal changes, equals or exceeds the double of the sum of the corresponding fluctuations derived from the mean monthly positions at the same hours. This mode of comparison serves to show, equally with the one which has been adopted in this volume, the days in each month when the irregular fluctuation deduced from the usual observation hours has been greatest, and will answer therefore the present purpose ; though inasmuch as in M. Kreil's method the unit of comparison varies in different months, an amount of irregular fluctuation which may constitute a day of disturbance in one month might not do so in another.

The days on which the extra observations recorded in this volume were made at Toronto and Van Diemen Island are inserted in the following table ; and the final column shows the character attached to the same day in M. Kreil's Table. The comparison is limited to 1841 because the Van Diemen Island Observatory can scarcely be considered to have been in regular action before the commencement of that year.

1841.	Days of Extra Observation at Toronto.	Days of Extra Observation at Van Diemen Island.	Days of Disturbance at Prague.
Jan. 13	13	None	13 ✳ Decl. and ✳ ☽ H. F.
18 & 19	18	19	18 ✳ H. F.; 19 ✳ Decl., and ✳ H. F.
25, 26, & 27	25, 26, & 27	None	25 ✳ H. F.; 26 ✳ Decl., and ✳ H. F.; 27 ✳ H. F.
Feb. 7	(Sunday)	7	7 ✳ Decl. and ✳ H. F.
„ 9	9	9	9 ✳ ☽ Decl. and ✳ H. F.
„ 15	15	15	15 ✳ Decl. and ✳ H. F.
„ 22 & 23	23	22	22 not dist.; 23 ✳ Decl., and ✳ H. F.
„ 26	26	26	26 ✳ Decl. and ✳ ☽ H. F.
March 14, 15, & 16	14, 15	15, 16	14 ✳ H. F.; 15 ✳ Decl. and ✳ H. F.; 16 ✳ H. F.
„ 22 & 23	22	22	22 ✳ ☽ Decl. and ✳ H. F.; 23 ✳ Decl. and ✳ ☽ H. F.
April 18, 19, & 20	None	18, 19, 20	18 ✳ H. F.; 19 not dist.; 20 ✳ ☽ H. F.
May 10	10	10	10 ✳ ☽ Decl. and ✳ ☽ H. F.
July 19 & 20	19	20	19 not dist.; 20 ✳ ☽ Decl., and ✳ H. F
Aug. 2 & 3	2	3	2 not dist.; 3 ✳ H. F.
„ 6	6	6	6 ✳ ☽ H. F.
„ 23	23	23	23 ✳ H. F.
„ 27	27	27	27 ✳ H. F.
„ 31, & Sept. 1	31	1	31 not dist.; 1 ✳ H. F.
Sept. 12 & 13	13	12, 13	12 ✳ H. F.; 13 ✳ Decl., and ✳ H. F.
„ 24, 25, & 26	24, 25, 26	24, 25, 26	24 not dist.; 25 ✳ ☽ Decl. and ✳ ☽ H. F.; 26 ✳ H. F.
„ 27 & 28	27, 28	27, 28	27 ✳ Dec. aud ✳ H F.; 28 ✳ H. F.
Oct. 8 & 9	9	8	8 and 9 not dist.
„ 24 & 25	24, 25	25	24 not dist.; 25 ✳ ☽ Dec., and ✳ H. F.
Nov. 6	6	None	6 ✳ Decl., and ✳ H. F.
„ 18 & 19	18	18, 19	18 not dist.; 19 ✳ Dec., and ✳ H. F
Dec. 3	None	3	3 ✳ ☽ Decl. and ✳ H. F.
„ 8	None	8	8 ✳ Decl. and ✳ H. F.
., 14	14	None	14 ✳ Decl. and ✳ H. F.
„ 30	None	30	30 ✳ Decl. and ✳ ☽ H. F.

In the following table the day in each month is shown which is characterized by M. Kreil, in the manner already described, as the most disturbed day in the month at Prague. On a reference to the tables in pages x, xi, xii, xiii, the amount can be examined of the disturbance in Declination and Horizontal intensity on the same days at Toronto and Van Diemen Island: the most disturbed day in each month at those stations has been inserted in the present table with a view of showing how frequently the days are the same.

1841.	Prague.		Toronto.		Van Diemen Island.		Remarks.
	Decl.	Hor. Force.	Decl.	Hor. Force.	Decl.	Hor. Force.	
January .	31	13	25	27	19	6	Jan. 31 fell on Sunday at Toronto.
February .	9	26	9	26	9	22	
March . .	22	23	22	22	22	15	
April .	No day of marked dist.	20	20	12	20	20	
May .	10	10	10	10	10	10	
June .	No day of marked dist.	25	23	25	18	28	
July .	20	24	19	24	20	19	
August . .	No day of marked dist.	6	6	6	23	6	
September .	25	25	25	25	24	13	Sept. 25 fell on Sunday at Van Diemen Island.
October .	25	21	25	27	8	8	
November .	5	4	19	18	19	18	
December .	3	30	14	30	3	30	

We have thus three stations, one in the interior of Europe, one in the interior of America, and a third near the mid latitude in the southern hemisphere and in a meridian very distant from the other two. Of the twenty-nine principal disturbances recorded in the table,—some confined to a single day, others running through two or three successive days, and comprehending altogether forty-nine days,—by far the greater part are shown to have manifested themselves at the three stations, though variously modified in the intensity of the effect, in the particular time in which the action was greatest, and in the element most affected. Nor is it by any means to be understood that on the very few occasions when no extra observations were made at one of the two British stations, no disturbance existed there; the irregularity may not in all cases have been of sufficiently decided character to have appeared to require extra observations; or circumstances may not always have been suitable for the effort of supporting observations at short intervals with three instruments. The practice, however, which has been adopted in this volume, of giving the positions of the Magnets at the usual hours of observation at all the Observatories, on every day when extra observations have been made at any one of them, furnishes a ready means of examining, in the very few cases referred to, whether the disturbance be traceable, although no extra observations were made: in most of the instances it will be found to be distinctly traceable.

Of the twenty-nine periods of principal disturbance above noticed, fifteen were also marked by extra observations at St. Helena, showing a still more general extension. There is, however, a difference in the character of the phenomenon as it manifests itself at St. Helena (and possibly elsewhere in the low latitudes) from that which it bears in the higher latitudes, which renders it less fitted to arrest the attention of the observer at the moment, and may, therefore, in such localities, occasion a disturbance to pass more frequently without extra observation. In the higher latitudes great and rapid fluctuations

c

both in direction and force, which even a casual observer could scarcely overlook, appear
the ordinary and leading characteristic; whilst at St. Helena this feature is far less obvious,
and the more usual form of a disturbance is that of a sustained deviation, in one direction
or the other, from the normal position at the same hours. In the Returns transmitted
from Toronto and Van Diemen Island, the extra observations are generally introduced
by a remark, "commenced in consequence of considerable or rapid changes occurring in
the position of the Magnets." The corresponding notice at St. Helena is "commenced
in consequence of the readings of the Declination or Horizontal Force Magnetometer
being unusually high or low" as the case may be.*

It has been inferred by M. Kreil from the Prague observations (Mag. und Met. Beob.
zu Prag Zwnt. Jahr, page 28) that the disturbances have little or no effect on the total
magnetic intensity, and consequently that the variations which take place at such times
in the horizontal force must be regarded as due to corresponding changes in the inclination. Dr. Lamont is also of opinion that "it is not improbable that the total force of
terrestrial Magnetism remains unchanged, and that the variations perceived by us relate
to direction only, so that the changes of inclination are given by changes of horizontal
intensity," (Res. der. Mag. Beob. in München, 1840, 1841, 1842, page 678). The disturbances recorded in this volume by no means confirm the view, that the disturbing forces
act at all times only in the direction in which they would produce no effect on the total
intensity. On the contrary, many instances will be found on examination in which the
total intensity was obviously and considerably influenced. A change in the total intensity
shows itself by an alteration in the readings of the Horizontal and Vertical Force Magnetometers, the alteration being in both in the same sense, i. e., either increased or decreased
readings in both. A change of inclination on the other hand shows itself by the readings of the two force Magnetometers being simultaneously affected in opposite directions,
i. e., the reading of the one increasing when the other decreases. Instances of both
kinds are frequent (as well as of the more complicated effect arising from both elements
being affected). The very remarkable disturbance which took place about 16ʰ. on the
29th May, 1840, at Toronto, is a striking instance of change in the total intensity, both
the force Magnetometers having been deflected beyond the scale at the same time in the
same direction.†

For the purpose of exhibiting the proportionate amount of disturbance in the different

---

* In much the greater number of instances it is the low reading of the Horizontal Force Magnet, indicating a
diminution of horizontal intensity, which occasioned the commencement of extra observations at St. Helena.

† The remarkable disturbance referred to, on the 29th of May, 1840, took place during the occurrence of an
Aurora, and appears to have been connected with a peculiar phase of that meteor, as described by the Director
of the Observatory in the note printed in page 3. Taking the positions of the Horizontal and Vertical Force
Magnets when reported off the scale as if they had been at the zero of the scale, and regarding the mean positions in the month as normal positions, we have the disturbance of the Magnets at 16ʰ. 25ᵐ. in scale divisions
respectively as follows :—

months of the year, we may take the mean annual fluctuation in declination and horizontal intensity at Toronto and Van Diemen Island respectively as unity, and express the mean fluctuations of either element in each month in terms of its own mean annual fluctuation. Toronto and Van Diemen Island having opposite seasons, (*i. e.*, the winter of the one being the summer of the other, and *vice versâ*), any effect of season would

---

Declin., 129·4 westerly; Hor. Force, 145 decrease; Vert. Force, 61·7 decrease: equivalent to—

$$\delta\psi = 1° \ 33'·3$$

$$\frac{\delta X}{X} = -·0304$$

$$\frac{\delta Z}{Z} = -·0092$$

If we suppose these disturbances to be occasioned by a small force f in a direction of which the Declination is ψ' and the Dip θ' we have the equations—

$$f \cos\theta' \cos(\psi' - \psi) = \delta X \quad \text{- - - - - - - - - - - -} \quad (1)$$

$$f \cos\theta' \sin(\psi' - \psi) = (X + \delta X)\tan\delta\psi \quad \text{- - - - - - -} \quad (2)$$

$$f \sin\theta' \qquad\qquad = \delta Z \quad \text{- - - - - - - - - - -} \quad (3)$$

$$\tan(\psi' - \psi) = \tan\delta\psi \cdot \frac{X + \delta X}{\delta X} \quad \text{- - - -} \quad (4)$$

$$f \sin\theta' = \phi\sin\theta \frac{\delta Z}{Z} = A \quad \text{- - - - -} \quad (5)$$

$$f \cos\theta' = \phi\cos\theta \sec(\psi' - \psi)\frac{\delta X}{X} = B \quad (6)$$

$$f \cos\theta' = \phi\cos\theta \csc(\psi' - \psi)\left(1 + \frac{\delta X}{X}\right)\tan\delta\psi = B \quad \text{- -} \quad (7)$$

$$\tan\theta' = \frac{A}{B}.$$

$$f = \frac{A}{\sin\theta'} = \frac{B}{\cos\theta'}$$

By means of these formulæ we obtain the direction and magnitude of the small disturbing force on May 29th, 16ʰ. 25ᵐ., as follows :—

$$\psi' - \psi = 130° \ 10'$$

$$\theta' = -40 \ 50$$

$$f = ·0136 \phi$$

October 24th, 1841, between 18ʰ. and 19ʰ., and November 18th, 1841, between 16ʰ. and 17ʰ., at Toronto, are other instances of remarkable disturbances apparently connected with peculiar phases of the Aurora. If we compute in a similar manner the direction and magnitude of the disturbing force at 18ʰ. 40ᵐ. on the 24th October, and at 16ʰ. 25ᵐ. on the 18th November, we obtain as follows :—

Oct. 24.	Nov. 18.
$\psi' - \psi = 135° \ 0'$;	$\psi' - \psi = 139° \ 08'$
$\theta' = -39 \ 6$;	$\theta' = -31 \ 15$
$f = ·0054 \phi$;	$f = ·0093 \phi$

It is deserving of notice that the directions should be so nearly the same in the three examples which have been cited.

tend to produce dissimilarity in the proportions, when the amount of disturbance in the same month at the two stations is compared; but as we have already seen the correspondence which exists in regard to the days of principal disturbance at both, it may be proper to examine in the first instance the degree in which an accordance may further be traced in the mean monthly fluctuations.

1841.	Toronto.		Van Diemen Island.	
	Declin.	Hor. Force.	Declin.	Hor. Force.
January	·77	·68	1·00	1·00
February	1·18	·75	1·33	1·12
March	·88	1·02	1·27	1·00
April	·75	·82	·75	·88
May	·94	·97	·99	1·00
June	·84	1·03	·66	·70
July	·91	·96	·86	·71
August	1·46	1·19	1·02	1·08
September . . .	1·30	1·55	·97	1·09
October	·87	·91	·92	1·09
November . . .	·89	1·09	1·11	1·25
December. . . .	·93	·96	1·10	·92

The influence of the general disturbances which took place in more than ordinary measure in August, September, and February, is here visible; April, on the other hand, shows itself to have been a remarkably tranquil month at both stations.* We may still remark, however, even on general inspection, an indication of the apparent influence of season, which becomes more manifest when the months are arranged according to the respective seasons.

	Toronto.		Van Diemen Island.		Mean.
	Declin.	Hor. Force.	Declin.	Hor. Force.	
4 Summer Months	1·05	1·04	1·13	1·07	1·07
4 Winter Months	0·94	0·87	0·88	0·87	0·89
4 Months (2 Spring & 2 Autumn) .	0·94	1·07	0·98	1·01	1·00

The excess of the proportionate fluctuation in the summer over the winter months, in both elements, and at both stations, is too consistent and considerable to be altogether regarded as accidental, and is more deserving of attention when it is remembered that it must be viewed as a residual quantity after the counter-action of the causes, the syn-

---

* At Prague, in April, no day occurred in which the sum of the fluctuations of the Declination equalled twice the sum of the fluctuations of the mean monthly positions at the same hours; consequently, according to M. Kreil's mode of characterizing disturbed days, there was no day of disturbance of the Declination at Prague in the month of April.

chronous influence of which must now be regarded as an established fact. It must be admitted, however, that the evidence of a single year does not afford a sufficient basis on which to found the important deduction of an annual period in these hitherto apparently irregular phenomena; more especially as M. Kreil has drawn from the observations at Prague a contrary inference, namely, that the disturbances are more considerable in winter than in summer. (Mag. und Met. Bcob. zu Prag Zwnt. Jahr, pages 14 and 18).

We may adopt a similar mode of examining the proportionate fluctuation, on the average of the whole year, at the different observation hours of the day and night. The mean fluctuation in the year of the Declination (for example) at 0 hour is

$$\overline{F\,(\psi_0)} = \frac{1}{n}\sqrt{\overline{\Sigma\,(F\,\psi_0)^2}}\,; \text{ at 2 hours } \overline{F\,(\psi_2)} = \frac{1}{n}\sqrt{\overline{\Sigma\,(F\,\psi_2)^2}}\,; \&c., \&c.,$$

and $\dfrac{\overline{F\,(\psi_0)}}{\overline{F\,(\psi_{1841})}}, \; \dfrac{\overline{F\,(\psi_2)}}{\overline{F\,(\psi_{1841})}}$, &c., exhibit the proportionate amount of the mean fluctuation

at the several hours of observation to the mean annual fluctuation regarded as unity.

Mean fluctuation of the Declination at the several hours of observation in 1841.

Toronto.		Van Diemen Island.		Remarks.
18 to 20	1·16	17 to 19	0·82	
20 ,, 22	0·94	19 ,, 21	0 84	Mean of the hours of the day.
22 ,, 0	0·74	21 ,, 23	0·91	
0 ,, 2	0·67	23 ,, 1	0·70	Toronto, V. D. Island,
2 ,, 4	0·55	1 ,, 3	0·65	0·82 0·78
4 ,, 6	0·84	3 ,, 5	0·77	
6 to 8	1·20	5 to 7	1·05	
8 ,, 10	1·39	7 ,, 9	1·12	Mean of the hours of the night.
10 ,, 12	1·12	9 ,, 11	1·31	
12 ,, 14	0·87	11 ,, 13	1·36	Toronto, V. D. Island,
14 ,, 16	1·10	13 ,, 15	1·25	1·13 1·18
16 ,, 18	1·09	15 ,, 17	1·02	

The fluctuation of the Declination appears from this table to be considerably greater during the hours of the night than in those of the day, both at Toronto and Van Diemen Island.

In regard to the horizontal force we can obtain no satisfactory deduction from the Observations at Van Diemen Island, until we are enabled to correct the results for the influence of variations of temperature on the magnetism of the bar.

The following are the mean fluctuations of the Horizontal Force at Toronto at the several hours of observation in 1841, from which no very obvious inference appears to be deducible :—

Day .			Mean, 1·02		Night.			Mean, 1·00
	18 to 20	1·00				6 to . 8	1·20	
	20 ,, 22	1·02				8 ,, 10	0·94	
	22 ,, 0	1·17				10 ,, 12	0·87	
	0 ,, 2	0·91				12 ,, 14	0·89	
	2 ,, 4	0·98				14 ,, 16	0·89	
	4 ,, 6	1·02				16 ,, 18	1·20	

On examining the meteorological registers at the Toronto Observatory, with reference to the appearance of Aurora, on the twenty-four days of principal magnetic disturbance at that station, it appears that on thirteen days of the twenty-four the Aurora was visible, and that on the remaining eleven days the sky was either densely overcast or heavily clouded, so that the Aurora, though it might exist, could not be seen.

January 18 Faint Auroral light.
  „   27 Auroral light in North.
February 9 Densely overcast.
  „   23 Faint Auroral light.
March 15 Bright Aurora.
  „   22 Generally overcast.
May   10 Densely clouded with rain.
July   19 Bright Aurora.
  „   24 Heavily overcast, with thunder.
August 6 Brilliant Aurora.
  „   23 Faint Aurora.
  „   26 Overcast with haze.

September 13 Aurora, with streamers and pulsations.
  „   24 Heavily clouded with rain.
  „   25 Bright Aurora.
October 9 Brilliant Aurora.
  „   25 Bright Aurora.
November 5 Faint Auroral light.
  „   6 Densely clouded.
  „   11 Densely clouded, with rain.
  „   18 Brilliant Aurora.
December 3 Densely clouded, with rain.
  „   14 Densely clouded.
  „   30 Overcast with dense haze.

The connexion between Aurora and magnetic disturbance, each viewed as a local phenomenon, has often been remarked : the days in the above list on which both occurred together were however days of disturbance at Prague and Van Diemen Island, as well as at Toronto; it would seem therefore that we may view the occurrence of Aurora at Toronto on those occasions as a local manifestation connected with magnetic effects, which, whatever may have been their origin, probably prevailed on the same day over the whole surface of the globe.

The observations with the Vertical Force Magnetometer at St. Helena prior to July 1841, have been omitted. The magnet, which had been much rusted on the passage out, was replaced in October, 1841, by one of improved construction, and in the observations made subsequently the connexion appears to have been generally preserved for short intervals.

The scale coefficients (k) of the Horizontal and Vertical Force Magnetometers, being the values respectively of one division of the scale in parts of the force at the station, are given in the heading of each day at each of the Observatories; except for the Vertical

Force Magnetometers of the Antarctic Expedition and of the Van Diemen Island Observatory, from which the data for computing the coefficients have not yet been received.

The temperature coefficients (q) of the Horizontal and Vertical Force Magnets, or the changes of the magnetic moment of the several bars, produced by one degree of Fahrenheit, are also given in the several headings, as far as they are yet known.

EDWARD SABINE.

Woolwich, August 1, 1843.

OBSERVATIONS WITH THE MAGNETOMETERS

ON DAYS OF

UNUSUAL MAGNETIC DISTURBANCE,

1840-1841.

MARCH 21, 1840.

TORONTO { Decl. 1 Scale Division = 0'·72 ; H. F. k = ·00017; q = ·00034 [a] ; V. F. [b]

Extra observations.

The H. F. was observed at 1m. 40s. before the times specified.

M. Gött. Time.				Decl.	Hor. Force.		Vert. Force.	
d.	h.	m.	s.	Sc.-Div^{ns}.	Sc.-Div^{ns}.	Ther.	Sc.-Div^{ns}.	Ther.
21	3	45	0	63·4	37·6	46		
		47	30	57·8				
		50	0	58·6	38·2			
		52	30	56·9				
		55	0	55·1	39·9			
		57	30	54·3				
	4	0	0	54·3	40·4	47		
		5	0	47·8	39·7			
		10	0	46·1	39·9			
		15	0	46·1	40·9			
		20	0	46·8	41·0			
		25	0	48·1	42·2			
		30	0	48·9	43·0			
		35	0	51·7	44·5			
		40	0	54·3	44·8			
		45	0	56·3	43·9			
		50	0	55·7	43·1			
		55	0	54·2	41·6			
	5	0	0	53·6	40·7			
		5	0	53·1	40·0			
		10	0	51·6	38·8			
		15	0	51·6	37·5			
		20	0	49·1	37·2			
		25	0	47·0	37·2			
		30	0	47·4	37·2			
		35	0	46·7	36·6			
		40	0		37·3	50		
	15	55	0	60·9	54·8			
	16	0	0	60·9	55·2	48		
		5	0		54·0			
		15	0	61·6	54·2			
		20	0	63·2				
		25	0	61·0	53·7			

Positions at the usual hours of observation, March 21st.

21	0	0	0	63·8	61·7	44		
	2	0	0	61·8	47·3	46		
	4	0	0	54·3	39·7	47		
	6	0	0	49·6	38·8	50		
	8	0	0	44·6	48·6	49		

MARCH 21, 1840.

M. Gött. Time.			Decl.	Hor. Force.		Vert. Force.	
d.	h.	m.	s.	Sc.-Div^{ns}.	Sc.-Div^{ns}.	Ther.	Sc.-Div^{ns}.	Ther.
21	10	0	0	51·7	54·9	49		
	12	0	0	58·8	56·4	50		
	14	0	0	59·8	55·2	49		
	16	0	0 [c]	60·9	54·0	48		

Mean Positions at the same hours during the Month. [d]

			63·4	59·9	45		
0	0	0	63·4	59·9	45		
2	0	0	67·2	58·2	46		
4	0	0	63·4	48·9	47		
6	0	0	53·3	47·3	49		
8	0	0	48·9	53·7	50		
10	0	0	52·4	66·9	51		
12	0	0	56·7	56·2	52		
14	0	0	59·3	56·4	51		
16	0	0	60·1	57·0	49		

APRIL 25, 1840.

TORONTO { Decl. 1 Scale Division = 0'·72 ; H. F. k = ·0002; q = ·00034 . ; V. F. [e]

Extra observations made during a violent hail-storm, occurring at the close of a light thunder-storm. [f]

25	9	50	0	48·0		59
		53	0	52·2		
		56	0	52·5		
		57	25		46·7	
		58	40		46·8	
		59	0	52·4		
		59	55		47·6	
	10	0	0	51·9	47·9	59
		1	0	52·1		
		1	10		47·9	
		2	0	52·2		
		2	25		47·7	
		3	0	53·0		
		3	40		47·5	
		4	55		48·1	
		6	10		48·6	
		7	25		48·3	
		8	0	53·3		
		9	0	53·5		
		10	0	53·3		
		19	0	54·0		

APRIL 25, 1840.

Positions at the usual hours of observation, April 25th.

M. Gött. Time.				Decl.	Hor. Force.		Vert. Force.	
d.	h.	m.	s.	Sc.-Div^{hs}.	Sc.-Div^{ns}.	Ther.	Sc.-Div^{ns}.	Ther.
25	0	0	0	67·1	55·0	54		
	2	0	0	64·7	52·0	55		
	4	0	0	55·0	36·5	56		
	6	0	0	53·0	42·1	57		
	8	0	0	48·8	46·1	59		
	10	0	0	51·9	47·9	59		
	12	0	0	57·5	49·2	62		
	14	0	0	59·7	43·5	63		
	16	0	0 [g]	58·8	44·6	62		

Mean Positions at the usual hours of observation, April 18th to April 30th.

			66·8	58·0	52		
0	0	0	66·8	58·0	52		
2	0	0	68·5	55·9	52		
4	0	0	61·2	49·0	53		
6	0	0	51·8	49·0	54		
8	0	0	46·9	55·0	55		
10	0	0	51·2	57·5	56		
12	0	0	56·7	56·0	57		
14	0	0	57·5	52·8	56		
16	0	0	61·0	52·0	56		
18	0	0	61·2	54·6	53		
20	0	0	62·2	56·8	53		
22	0	0	62·8	58·5	52		

MAY 29 and 30, 1840.

TORONTO { Decl. 1 Scale Division = 0'·72 ; H. F. k = ·00021; q = ·00034 ; V. F. k = ·00015; q = ·0002 [h]

Regular, extra, and term observations, May 28th, 18h., to May 30th.

The V. F. was observed at 2m. 30s. before, and the H. F. at 2m. 30s. after the times specified. [i]

				Decl.	Hor. Force.		Vert. Force.	
28	18	0	0	158·9	142·7	71	55·2	70
	20	0	0	161·8	140·5	71	53·1	71
	22	0	0	169·4	142·7	70	59·1	69
29	0	0	0	191·2	162·7	69	58·7	69
	25	0		170·1	152·8		57·5	
	30	0		169·3				
	35	0		177·5	154·7			
	40	0		179·4				

---

[a] The value of q for this bar, which was used from March to August, is approximate.
[b] The V. F. magnet was not in adjustment.   [c] Saturday midnight at Toronto.
[d] The mean positions of the H. F. magnet are from March 19th to March 31st inclusive; they cannot be considered as strictly comparable with the daily positions on the 21st, in consequence of the possible stretching of the suspension wires for some weeks after the first adjustment.
[e] The V. F. magnet was observed, but the readings were not as yet to be relied on.
[f] At 25. 9h. 40m. a dense mass of black clouds, covering five-eighths of the sky in N. and W.; frequent peals of distant thunder commenced at about 9h.
55   Clouds spreading upwards from the West, covering nearly all the sky; loud thunder, and vivid flashes of forked lightning. Lightning proceeding from W.N.W., where the sky was broken into bright masses of clouds, with rounded and well-defined upper edges; sky clear to the Southward.
58   A tremendous shower of hail fell very suddenly; and about five minutes after, a heavy peal of thunder, accompanied with rain, and a heavy squall from the Westward; some of the hailstones were fully *one inch in diameter;* the whole of them were very large. The fall lasted about ten minutes, when the clouds broke to the Northward and Westward, and the whole sky became covered with masses of cumuli passing over from the West. The wind drew again to the East.

25 10 45   Sky clear to the W. of North; storm passing over to the East; heavy and lowering in the North-east; wind drawing back to the West; clouds passing from the West. The hail-storm appears to have extended in a westerly direction, the breadth within which hail fell being about four or five miles; the weight of the squall was not more than a mile and a half, or two miles in width.

[g] Saturday midnight at Toronto.   [h] This value is approximate.

Description of an Aurora observed on the night of the 29th and 30th May 1840, and of its effect on the Magnets in the Observatory at Toronto.

The disturbance of the magnets commenced between 28d. 22h. and 29d. 0h. 0m.; the magnetometer readings at those hours showing a decrease in the declination, amounting to 15'·1, and an increase in the H. F. equal to 0·004 of the whole force.

At 29d. 0h. 25m., the disturbance of the declination was so great *that in one minute only,* the change of angle amounted to about 10'; the disturbance subsided rapidly, and little more than the usual changes took place until 29d. 4h., at which time the observations at short intervals were discontinued.

MAY 29 and 30, 1840.				MAY 29 and 30, 1840.				MAY 29 and 30, 1840.			
M. Gött. Time.	Decl.	Hor. Force.	Vert. Force.	M. Gött. Time.	Decl.	Hor. Force.	Vert. Force.	M. Gött. Time.	Decl.	Hor. Force.	Vert. Force.
d. h. m. s.	Sc.-Div^m.	Sc.-Div^m. Ther.	Sc.-Div^m. Ther.	d. h. m. s.	Sc.-Div^m.	Sc.-Div^m. Ther.	Sc.-Div^m. Ther.	d. h. m. s.	Sc.-Div^m.	Sc.-Div^m. Ther.	Sc.-Div^m. Ther.
29 0 45 0	175·7	153·1	57·9	29 1 50 0	175·7			29 3 30 0	164·9		
50 0	174·2			2 0 0	170·9	144·1 69	56·5 68	35 0	162·3	149·4	55·3
55 0	174·9	152·3	57·5	35 0	169·6	144·8		40 0	160·0		
1 0 0	173·0		59·4 69	50 0	167·2			45 0	160·3	148·8	54·7
10 0	172·8	148·6		55 0	163·2	148·4	55·2	50 0	161·0		
15 0	170·4	145·7		3 0 0	165·5		69	55 0	157·9	147·5	54·2
20 0	168·2			5 0	163·3	145·9	53·3	4 0 0	157·3		68 68
26 0	156·8	137·5		10 0	164·6		54·4	5 0	158·0	150·2	54·5
31 0	167·2			15 0	166·3	144·5		10 0	155·8		
34 0	172·2	138·5		20 0	163·9		54·5	15 0	155·5	147·7	54·8
45 0	176·9	142·3		25 0	167·2	147·5	54·5	20 0	156·6		

The term day observations commenced at 29ᵈ.10ʰ. The weather continued fair and calm, and the sky perfectly clear, except a light haze round the horizon, until 29ᵈ.13ʰ., when a few cirro-strati appeared in the North and West horizons. The sun set with a glow which extended to the N.E.

A faint light appeared in the North about 9 o'clock, followed soon after by faint streamers to the East of North; at 9ʰ. 22ᵐ. the sky to the North had become one bright sheet of light; at 9ʰ. 30ᵐ. a dark cloud lay along the horizon, arching upwards; brilliant streamers shot up along its extent from N.W. to N.E., reaching nearly to the zenith, and occasionally sending out flashes like a gleam of lightning, having a most beautiful effect.

The remarkable feature of this Aurora consisted in a continued succession of most brilliant flashes or pulsations, at times covering the whole upper portion of the sky from W.N.W. to E.N.E., but never lower than an altitude of 15° or 20°; the streamers were faint at their first appearance, but soon brightened up, and would occasionally shoot upwards, darting across the flashes; the latter resembled large waves of light thrown upwards in successive layers, appearing usually at different points in detached masses, but occasionally each layer or wave extending round the whole Northern semi-circle and being thrown upwards with a regular although incessant motion. They were only once observed to flash across the zenith, but frequently round it.

The greatest disturbance of the declination magnetometer commenced about 15ʰ. 35ᵐ., at which time or a few minutes later, two remarkably brilliant streamers, shaded off with lesser ones, were seen to rise from a patch of light in N.W. at an alt. of about 15°, converging, as appeared to be the case with all the others, to a point about 5° West of the zenith, and, for a few seconds, forming a very distinct corona about 10° from the point of convergence; the sky within the circle or corona being perfectly clear.

The most beautiful and remarkable appearance during the night occurred a few minutes after 16ʰ.; a perfect arch extended from N.W. to N.E., splendid streamers shot up from it to the point of convergence already mentioned, and most brilliant and incessant flashes were thrown upwards to the South of the zenith in a series of concentric rings, the centre of rings and arch being about 10° East of North. The flashes extended only within about 15° or 20° to the North of the zenith, but closed nearer to the South, and formed for a few seconds almost a perfect ring of light; the Northern part of the circle remained clear for some time, the dimensions and other particulars were unfortunately not noted as above.

It was while this appearance lasted that the H. F. and V. F. magnets were drawn out of the range of their scales. The pulsations and streamers became fainter, and the extreme disturbance of the magnets ceased after 16ʰ. 30ᵐ. A faint light and streamers were visible at intervals until 20ᵈ. 8ʰ.; towards 19ʰ. the flashes and streamers again appeared with nearly their former brilliancy; a very light air had sprung up from the North, and in lieu of their former uniform pulsations the flashes appeared to have an incessant dancing motion, downwards as well as sideways. On its again becoming calm the motion ceased.

All traces of the Aurora vanished between 20ʰ. and 21ʰ., the magnets became comparatively steady, and continued without disturbance until 30ᵈ. 10ʰ., when very sudden changes took place for a short time.

Flashes of such magnitude and brilliancy did not appear to have been commonly seen in this part of the country.

The disturbance of the magnets, more particularly those of the H. F. and V. F., was far greater than could have been anticipated from the observations of the preceding months. The greatest changes that had been observed between any succeeding pair of two hourly observations between the 15th February and 29th May were of the declination about 16°·5; of the H. F., 0·0044; and of the V. F. from 20th April to 29th May, 0·00088 of the whole; while the entire range of the declination during the preceding months amounted only to an angle of 28°·0, and the maximum monthly changes of the H. F. and V. F. to ·0163 and ·0024 of the respective forces.

The extreme change of declination during this disturbance amounted to 1° 59', the readings of the scale varying from 25·4 to 190·1, the mean reading being about 160·0; the Aurora caused an increase to the mean Westerly declination of about 1° 37'.

The highest reading of the H. F. magnetometer was 214·0, the change to 0 of the scale corresponds to a decrease of 0·044, or about 1-23rd of the whole force. The extreme disturbance must have exceeded this very considerably.

No colour could be distinguished in the Aurora except the usual pale yellowish tinge.

Corporal Johnston, one of the assistants at the Observatory, heard a distinct hissing noise like the escape of steam at about 29ᵈ. 16ʰ., when the flashes were most brilliant and the force magnets were beyond the range of the scales; particular attention was paid throughout the night, but there were few intervals before midnight when the neighbourhood was sufficiently quiet to allow of a slight sound being heard.

Detailed account of the Aurora.

d. h. m.	
29 14 0	Haze round horizon, remainder clear.
15 0	Haze round horizon; faint light on either side of North to an altitude of about 5°.
15	A very faint streamer shot up in N. by E. to an alt. of about 60°.
19	Sky to North a bright sheet of light.
24	Bright arch of light from N.N.W. to N.N.E., alt. in North about 15°; faint streamers shooting up in North by East to an alt. of 45°.
27	Brilliant streamers in close arrangement shooting up from extreme end of arch at N.W.
27	Streamers rising from N.W. to N.E. very bright at intervals, and occasionally appearing to flash like a gleam of lightning; height nearly to zenith; dark clouds along the horizon, arching upwards in North; greatest alt. 15°.
37	Streamers fainter in the N.E. drawing more to the Westward, converging to a point about 5° West of zenith.
42	Two brilliant streamers shaded off with lesser ones darting up from a patch of light in N.W. Streamers at intervals from N.W. to N. converging to the point above named; for a few seconds a distinct corona within 10° of point of convergence.
45	Streamers fainter; incessant pulsations like faint flashes of lightning through the whole sky to the Northward; dark in the North except when lighted up by the pulsations increased in brilliancy; stars visible through them; incessant flashes appearing to circle round the zenith to Westward.
52	Patches of light at different alts from W.N.W. to S.E.; no streamers. Flashes from the whole semi-circle across and round zenith, appearing bent in their course upwards.
16 0	Brilliant streamers and pulsations
2	Perfect arch from N.W. to N.E., splendid streamers from it and most beautiful and incessant flashes—almost one continued stream of light.
7	Same continuing, streamers converging as before, pulsations thrown upwards in a series of concentric rings to the South of zenith, round point of convergence, centre of arch and rings about 10° East of North.
9	A bright and broad streamer rising in magnetic North and extending nearly to zenith; pulsations fainter and less frequent; streamers ascending in the West from a patch of light, circling outwards, and returning towards the point of convergence. Scale of H. F. magnet re-appearing for a few seconds.
12	Flashes brighter—rapid as could be conceived; two small but very brilliant streamers in North, one very bright shot up in North and disappeared instantly. Patches of light and faint streamers from N.W. to N.
17	Bright streamers from W. to S.E. converging to Westward of zenith and forming a corona.
20	North clear except occasional bright flashes and streamers.
27	Faint streamers; pulsations most brilliant from N.W. to E.
28	Aurora much fainter; streamers scarcely visible, faint pulsations at intervals.
32	Most brilliant streamers in N.N.W., and flashes converging, &c., as before.
42	Faint streamers and flashes from N. by W. to N.E.; none to Westward.
53	Streamers disappeared; faint light above horizon, stratus clouds visible in it; flashes continuing near West and East.
17 7	No streamers or flashes; faint patches of light remaining in N.N.W.
13	Faint light in N.W. brightening to Eastward of North.
27	Streamers shooting up from a bright patch extending from N.E. to N.W.; flashes following each other in quick succession.
37	Light fainter; a few streamers and flashes at intervals; several bright patches above bank of light from N.E. to N.W.
42	Light increasing; bright flashes extending to zenith.
47	Light very faint; brightest in the Eastward.

OBSERVATIONS WITH THE MAGNETOMETERS ON DAYS OF UNUSUAL MAGNETIC DISTURBANCE, 1840-1841.

MAY 29 and 30, 1840.

d.	h.	m.	s.	Decl. Sc.-Div	Hor. Force Sc.-Div	Ther.	Vert. Force Sc.-Div	Ther.
29	4	25	0	156·7	146·8		54·6	
		30	0	161·0				
		35	0	155·9	151·0		54·6	
		40	0	156·9				
		45	0	159·7	152·8		53·6	
		50	0	162·1				
		55	0	158·3	150·1		53·9	
	5	0	0	158·8		68		68
		5	0	154·4	147·6		53·4	
		10	0	154·3				
		15	0	156·1	145·6		54·2	
		20	0	158·0				
		25	0	158·9	147·9		54·7	
		30	0	157·9				
		35	0	157·7	145·8		55·3	
		40	0	156·8				
		45	0	155·7	150·1		55·1	
		50	0	153·5				
		55	0	151·0	151·4		55·1	
	6	0	0	150·1		68		
		5	0	150·5	148·2		55·0	67
		10	0	153·1				
		15	0	152·6	147·5		54·9	
		20	0	153·3				
		25	0	149·8	155·4		56·5	
		30	0	146·9				
		35	0	147·8	151·1		56·5	
		40	0	150·4				
		45	0	151·4	151·2		56·4	
		50	0	151·6				
		55	0	159·8	151·9		56·9	
	7	0	0	154·9				
	8	0	0	151·2	155·6	68	56·9	68
	10	0	0	153·8	160·3	69	70·8	68
		5	0	154·8				
		10	0	152·5	158·5		70·8	
		15	0	150·8				
		20	0	154·2	160·1		71·3	
		25	0	156·1				
		30	0	156·2	165·1			
		35	0	153·6				
		40	0	147·6	159·0		75·2	
		45	0	150·3				
		50	0	154·9	156·7		79·0	
		55	0	163·8				
	11	0	0	161·6	184·8	69	80·4	69
		5	0	155·0				
		10	0	152·6	185·0		84·2	
		15	0	166·9				
		20	0	159·4	214·0		91·6	
		25	0	150·8				
		30	0	164·4	203·3		101·2	
		35	0	175·9				

MAY 29 and 30, 1840.

d.	h.	m.	s.	Decl. Sc.-Div	Hor. Force Sc.-Div	Ther.	Vert. Force Sc.-Div	Ther.
29	11	40	0	141·6	207·7		89·5	
		45	0	145·1				
		50	0	159·5	198·8		93·3	
		55	0	138·4				
	12	0	0	123·3	214·8	69	87·3	69
		5	0	161·1				
		10	0	151·3	160·9		87·5	
		15	0	164·8				
		20	0	160·0	152·9		78·5	
		25	0	148·7				
		30	0	143·5	143·2		75·2	
		35	0	142·1				
		40	0	146·5	135·9		75·4	
		45	0	153·7				
		50	0	157·0	133·8		73·0	
		55	0	157·1				
	13	0	0	159·6	137·3	70	74·7	
		5	0	158·1				
		10	0	171·1	137·7		77·2	
		15	0	168·3				
		20	0	171·7	127·4		72·1	
		25	0	164·4				
		30	0	177·0	119·8		74·5	
		35	0	180·1				
		40	0	182·3			74·8	
		42	30*	185·5	125·8			
		45	0	184·6				
		47	30	180·4			72·9	
		50	0	175·3				
		52	30	171·8	118·4			
		55	0	173·5				
		57	30	176·0			70·5	69
	14	0	0	177·4		70		
		2	30	176·6	111·6			
		5	0	176·8				
		7	30	180·2			59·8	
		10	0	181·2				
		12	30	182·4	99·2			
		15	0	185·0				
		17	30	186·7			54·6	
		20	0	185·2				
		22	30	184·6	98·8			
		25	0	183·3				
		27	30	179·8			55·1	
		30	0	174·5	101·5			
		32	30	172·8				
		35	0	171·3				
		37	30	171·3			56·5	
		40	0	173·9				
		42	30	175·9	110·3			
		45	0	176·1				
		47	30	172·1			53·3	
		50	0	168·9				

MAY 29 and 30, 1840.

d.	h.	m.	s.	Decl. Sc.-Div	Hor. Force Sc.-Div	Ther.	Vert. Force Sc.-Div	Ther.
29	14	52	30	162·1	99·7			
		55	0	157·8				
		57	30	152·7			51·2	
	15	0	0	151·3				69
		2	30	153·3	84·9	70		
		5	0	155·9			40·1	
		7	30	157·1	82·0			
		10	0	159·3			41·8	
		12	30	161·7	80·8			
		15	0	164·5			44·2	
		17	30	168·4	91·6			
		20	0	168·1			52·3	
		22	30	169·7	97·3			
		25	0	170·4			53·4	
		27	30	168·8	93·7			
		30	0	170·5			48·1	
		32	30	171·7	84·6			
		35	0	164·9			27·5	
		37	30	170·8				
		40	0	179·6			27·4	
		42	30	160·6	67·2			
		45	0	138·8			b	
		47	30	133·1	41·2			
		50	0	133·7				
		52	30	143·3	36·6			
		55	0	148·4				
		57	30	147·6	22·1			
	16	0	0	136·5		70	c	69
		2	30	109·0				
		5	0	181·9				
		7	30	91·0				
		10	0	114·2				
		12	30	83·4				
		15	0	82·4				
		17	30	59·9				
		20	0	31·0				
		22	30	31·0				
		25	0	25·4				
		27	30	37·0				
		30	0	56·0			19·4	
		32	30	115·1				
		35	0	162·7			22·4	
		37	30	158·5				
		40	0	159·0				
		42	30	140·3	9·6			
		45	0	123·0			24·0	
		47	30	136·2	31·2			
		50	0	154·1			29·2	
		52	30	163·7	87·5			
		55	0	175·3			30·7	
		57	30	186·9	89·5			
	17	0	0	185·4			19·3	69
		2	30	176·1	75·8	70		

d. h. m.
29 17 57   Faint arch of light; dark clouds rising from North horizon, remainder clear.

   18 42   Light increasing; numerous streamers, and flashes of a yellowish hue brightest in N.E.

      57   An arch of dark clouds along North horizon; greatest alt. about 15° (in North); streamers rising from it from N.W. to N.E.; pulsations very vivid, striking upwards and in a horizontal direction, having a waving motion.

   19 0   Light air from North; pulsations appearing affected by it; the flashes having a dancing motion, sideways as well as downwards, instead of their former uniform motion upwards.

      5   Pulsations and streamers fainter; faint patches of light from N.W. to N.E.

     37   Light very faint; a few streamers.

     57   Two very clear streamers, one to the East the other to the West of North, with flashes between them, following each other in quick succession like waves.

d. h. m.
29 20 27   Light almost extinct; one very faint streamer to N.E.

     57   Auroral light gone; sky clear until nearly noon, when it clouded over with cumuli and cumu-strati mixed with strati. Calm and partially clouded throughout the day; rugged and broken masses of cumu-strati covering the whole sky at about 7h. 30m. very black and dense in the N.W. The night and morning of the 31st were very clear; clouded again towards sunset, and a heavy thunderstorm with rain occurred during the night.

* The H. F. and V. F. were observed at the times specified, during the continuance of the extra observations, viz., from 29d. 13h. 42m. 30s. to 29d. 19h. 47m. 30s.

b The V. F. magnet was out of the field from 15d. 45h. to 16d. 25h.

c The scale of the H. F. magnet was out of the field from 16h. 0m. to 16h. 40m. During the periods of extreme disturbance the movements of the magnets were so rapid and irregular, that it was impossible to observe with the usual accuracy, especially with the V. F. magnet.

MAY 29 and 30, 1840.

M. Gött. Time (d. h. m. s.)	Decl. Sc.-Div.	Hor. Force Sc.-Div.	Ther.	Vert. Force Sc.-Div.	Ther.
29 17 5 0	183·1			10·4	
7 30	171·2	81·2			
10 0	163·6	79·8		14·6	
12 30	161·2				
15 0	157·3			17·5	
17 30	166·5	72·3			
20 0	176·7			25·9	
22 30	190·1	84·9			
25 0	187·7			25·1	
27 30	181·7				
30 0	166·0			20·5	
32 30	156·0	99·6			
35 0	153·6			32·9	
37 30	153·7	08·6			
40 0	157·0			38·4	
42 30	155·3	99·7			
45 0	153·4			34·3	
47 30	157·4	95·1			
50 0	164·0			39·2	
52 30	164·0	98·6			
55 0	165·5			36·9	
57 30	167·4	06·7			
18 0 0	173·2			53·5	69
2 30	175·6	15·3	70		
5 0	175·0			53·1	
7 30	174·5	17·8			
10 0	171·6			53·3	
12 30	171·1	19·3			
15 0	166·8			51·6	
17 30	166·9	14·6			
20 0	168·0	99·6		52·5	
25 0	157·3				
30 0	156·0				
32 30	144·3	45·5			
35 0	137·8			28·9	
37 30	130·0	29·6			
40 0	127·3			24·5	
42 30	129·6	21·1			
45 0	121·6			20·6	
47 30	111·3	12·6			
50 0	102·3	21·4		12·4	
52 30					
55 0	99·9			23·0	
57 30	110·8	58·4			
19 0 0	122·1			28·3	68
2 30	114·1	72·4	69		
5 0				30·3	
7 30	122·2				
10 0				26·8	
12 30	131·4	78·9			
15 0	136·0			27·6	
17 30	143·6	80·8			
20 0	147·4			28·1	
22 30	149·4	93·4			
25 0	150·4			28·5	
27 30	150·8	102·9			
30 0	149·2			32·9	
32 30	151·5	98·2			
35 0	152·0			31·4	
37 30	157·7	101·6			
40 0	157·9	102·6		36·2	
45 0	164·6			38·4	
47 30		103·8			
50 0	169·3	106·8		41·4	
55 0	172·3				
20 0 0	174·9	114·0	69		68
5 0	172·2				

MAY 29 and 30, 1840.

M. Gött. Time (d. h. m. s.)	Decl. Sc.-Div.	Hor. Force Sc.-Div.	Ther.	Vert. Force Sc.-Div.	Ther.
29 20 10 0	173·6	118·2		50·6	
15 0	174·4				
20 0	175·2	115·7		48·9	
25 0	180·0				
30 0	181·8	123·3		49·9	
35 0	182·9				
40 0	178·7	125·1		55·6	
45 0	178·2				
50 0	176·2	129·6		56·5	
55 0	174·7				
21 0 0	169·6	133·4	68	57·4	69
5 0	165·3				
10 0	163·8	130·8		58·1	
15 0	163·5				
20 0	163·2	129·3		59·3	
25 0	162·9				
30 0	161·1	126·6		60·1	
35 0	160·4				
40 0	160·5	127·1		60·0	
45 0	159·4				
50 0	157·3	125·2		59·5	
55 0	156·0				
22 0 0	154·7	125·3	68		67
5 0	156·5				
10 0				58·4	
15 0	150·4				
20 0	161·3	130·7		59·8	
25 0	162·0				
30 0	161·2	130·0		60·6	
35 0	161·7				
40 0	161·7	126·9		60·3	
45 0	160·2				
50 0	159·9	127·1		60·2	
55 0	158·8				
23 0 0	156·5	123·6	67	60·3	67
5 0	156·6				
10 0	155·6	126·3		58·1	
15 0	153·9				
20 0	155·2	125·2		57·4	
25 0	155·9				
30 0	154·9	127·7		57·2	
35 0	155·6				
40 0	155·4	124·4		57·0	
45 0	157·8				
50 0	158·7	125·4		57·1	
55 0	160·0				
30 0 0 0	158·0	124·0	66	56·8	66
5 0	157·4				
10 0	159·6	129·1		57·1	
15 0	158·2				
20 0	160·7	131·4		57·0	
25 0	161·6				
30 0	163·4	128·5		57·5	
35 0	163·7				
40 0	162·6	121·8		58·1	
45 0	163·7				
50 0	165·9	119·6		57·9	
55 0	165·2				
1 0 0	165·0	117·6	66	58·1	66
5 0	163·0				
10 0	163·0	115·7		58·4	
15 0	161·8				
20 0	160·7	112·3		58·6	
25 0	159·5				
30 0	158·7	115·4		58·5	
35 0	157·3				
40 0	157·7	117·3		58·5	

MAY 29 and 30, 1840.

M. Gött. Time (d. h. m. s.)	Decl. Sc.-Div.	Hor. Force Sc.-Div.	Ther.	Vert. Force Sc.-Div.	Ther.
30 1 45 0	156·7				
50 0	155·1	117·4		58·9	
55 0	154·9				
2 0 0	153·1	115·0	66	59·5	66
5 0	150·9				
10 0	151·0	119·4		59·0	
15 0	153·6				
20 0	158·3	111·4		59·2	
25 0	155·6				
30 0	153·1	117·7		59·3	
35 0	148·5				
40 0	147·6	114·6		59·6	
45 0	147·1				
50 0	144·1	109·4		57·1	
55 0	145·4				
3 0 0	145·1	109·4	67	56·3	66
5 0	143·1				
10 0	140·0	110·8		57·2	
15 0	140·9				
20 0	139·8	111·0		58·1	
25 0	140·3				
30 0	140·7	111·0		57·7	
35 0	141·6				
40 0	142·6	113·2		58·6	
45 0	143·7				
50 0	143·5	116·7		59·3	
55 0	143·9				
4 0 0	143·6	119·4	67	60·1	66
5 0	143·3				
10 0	142·7	121·6		60·8	
15 0	142·6				
20 0	142·5	121·1		60·9	
25 0	143·8				
30 0	143·7	124·9		62·9	
35 0	143·7				
40 0	143·6	128·1		63·9	
45 0	144·3				
50 0	145·1	131·8		64·4	
55 0	145·5				
5 0 0	145·1	135·0	67	64·4	67
5 0	145·0				
10 0	144·7	136·7		63·9	
15 0	144·8				
20 0	144·4	137·2		63·5	
25 0	144·1				
30 0	144·8	137·6		63·5	
35 0	145·2				
40 0	144·9	138·3		64·0	
45 0	144·3				
50 0	143·5	138·2		64·0	
55 0	143·8				
6 0 0	143·9	138·3	72	63·5	71
5 0	143·5				
10 0	143·8	139·7		63·2	
15 0	143·9				
20 0	144·1	142·2		62·9	
25 0	144·2				
30 0	144·7	143·7		62·9	
35 0	145·3				
40 0	145·5	144·9		62·8	
45 0	145·7				
50 0	146·0	145·8		62·9	
55 0	146·2				
7 0 0	146·7	146·5	67	63·3	66
5 0	146·7				
10 0	147·0	145·7		63·2	
15 0	147·3				

MAY 29 and 30, 1840.

M. Gött. Time. d. h. m. s.	Decl. Sc.-Div	Hor. Force Sc.-Div	Ther. °	Vert. Force Sc.-Div	Ther. °
30 7 20 0	147·3	145·0		62·9	
25 0	147·1				
30 0	147·1	144·0		60·0	
35 0	147·2				
40 0	147·6	144·5		58·7	
45 0	148·0				
50 0	148·2	146·3		60·0	
55 0	148·7				
8 0 0	149·0	148·0	67	59·1	67
5 0	149·9				
10 0	150·1	148·7		64·7	
15 0	151·2				
20 0	151·3	149·4		64·9	
25 0	150·8				
30 0	150·8	148·0		65·3	
35 0	151·1				
40 0	150·7	150·8		65·9	
45 0	151·2				
50 0	153·1	156·5		67·1	
55 0	155·1				
9 0 0	154·5	159·7	67	68·0	67
5 0	155·1				
10 0	154·6	159·7		68·6	
15 0	153·8				
20 0	154·8	157·2		69·5	
25 0	155·7				
30 0	161·4	159·1			
35 0	155·6				
40 0	156·9	149·1		71·0	
45 0	158·6				
50 0	161·9	155·9		72·0	
55 0	166·6				
10 0 0	167·7	192·3	68	72·4	68
5 0					
10 0	165·6	166·4			
15 0	161·5				
20 0	154·5	159·6		69·9	
25 0	151·0				
30 0	150·6	152·1		71·0	
35 0	155·7				
40 0	163·4	155·5		74·6	
45 0	162·3				
50 0	160·4	150·3		71·6	
55 0	154·6				
11 0 0	150·6	146·3	68	69·7	68
5 0	150·6				
10 0	150·5	142·0			
15 0	151·0				
20 0	151·9	139·4		65·3	
25 0	153·1				
30 0	152·5	142·0		63·8	
35 0	151·7				
40 0	151·1	140·2		62·7	
45 0	151·4				
50 0	151·2	140·1		66·7	
55 0	151·0				
12 0 0	151·2	143·8	68	67·0	68
14 0 0	152·1	134·6	69	63·7	68
16 0 0	153·8	136·5	68	60·2	68

Mean Positions at the usual hours of observation, between the 25th of May and the 4th June inclusive, omitting May 28 and 29.

M. Gött. Time. d. h. m. s.	Decl. Sc.-Div	Hor. Force Sc.-Div	Ther. °	Vert. Force Sc.-Div	Ther. °
0 0 0	162·3	146·0	60	60·1	60
2 0 0	164·5	145·1	60	60·3	60
4 0 0	158·4	140·1	61	60·1	60

MAY 29 and 30, 1840.

M. Gött. Time. d. h. m. s.	Decl. Sc.-Div	Hor. Force Sc.-Div	Ther. °	Vert. Force Sc.-Div	Ther. °
6 0 0	150·1	145·3	62	60·0	61
8 0 0	147·1	150·8	62	60·3	61
10 0 0	153·4	158·0	63	62·7	62
12 0 0	154·9	150·4	63	62·5	63
14 0 0	154·3	147·1	64	61·7	63
16 0 0	154·8	145·1	64	61·7	64
18 0 0	156·8	148·6	62	61·4	62
20 0 0	156·7	146·0	61	61·0	61
22 0 0	161·3	148·3	61	60·7	61

ST. HELENA { Decl. 1 Scale Division = $0'\cdot71$
H. F. $k = \cdot00012$; $q =$
V. F. $k = \cdot0004$; $q =$

Regular and term observations, May 28th, 18^h, to May 29th, 12^h.
The V. F. was observed at $2^m\,30^s$ before, and the H. F. at $2^m\,30^s$ after the times specified.

M. Gött. Time. d. h. m. s.	Decl. Sc.-Div	Hor. Force Sc.-Div	Ther. °
28 18 0 0	39·0	38·3	63
20 0 0	44·2	42·5	63
22 0 0	41·5	38·5	63
29 0 0 0	41·6	59·4	63
2 0 0	39·0	41·2	63
4 0 0	36·4	36·5	65
6 0 0	34·4	31·0	64
8 0 0	34·8	49·6	64
10 0 0		14·1	64
5 0	32·4		
10 0	31·4	13·0	
15 0			
20 0	33·0	14·0	
25 0	33·0		
30 0	33·1	16·9	
35 0	33·0		
40 0	34·1	16·0	
45 0	35·0		
50 0	33·7	17·3	
55 0	33·3		
11 0 0	31·5	20·7	65
5 0	30·6		
10 0	30·6	16·5	
15 0	30·6		
20 0	30·7	17·8	
25 0	30·5		
30 0	29·9	14·4	
35 0	29·3		
40 0	29·4	14·6	
45 0	29·1		
50 0	29·1	14·4	
55 0	28·4		
12 0 0	28·1	15·4	65
5 0	29·1		
10 0	29·8	10·6	
15 0	30·4		
20 0	28·1	106·5	
25 0	30·2		
30 0	30·5	9·0	
35· 0	30·0		
40 0	28·6	10·2	
45 0	29·2		
50· 0	28·0	9·6	
55 0	28·3		
13 0 0	28·7	24·0	65
5 0	20·0		
10 0	29·4	14·0	
15 0	30·2		
20 0	30·5	18·6	

MAY 29 and 30, 1840.

M. Gött. Time. d. h. m. s.	Decl. Sc.-Div	Hor. Force Sc.-Div	Ther. °	Vert. Force Sc.-Div	Ther. °
29 13 25 0	28·8				
30 0	20·0			10·4	
35 0	28·3				
40 0	28·2			9·4	
45 0	29·2				
50 0	29·0			20·2	
55 0	28·5				
14 0 0	28·2			19·5	65
5 0	28·4				
10 0	28·3				
15 0	27·5				
20 0	26·9			24·2	
25 0	27·5				
30 0	28·0			20·2	
35 0	27·8				
40 0	25·4			19·2	
45 0	31·5				
50 0	29·7			18·5	
55 0	28·6				
15 0 0	28·5			17·6	65
5 0	29·1				
10 0	28·2			18·2	
15 0	28·1				
20 0	27·9			20·6	
25 0	26·6				
30 0	27·5			18·4	
35 0	28·4				
40 0	27·3			19·5	
45 0	28·2				
50 0	28·5			19·9	
55 0	29·6				
16 0 0	30·1			19·4	65
5 0	30·1				
10 0	29·8			17·6	
15 0	29·1				
20 0	29·0			17·7	
25 0	29·9				
30 0	31·5			20·0	
35 0	31·5				
40 0	32·9			21·5	
45 0	31·4				
50 0	33·0			25·0	
55 0	32·8				
17 0 0	31·4			28·5	65
5 0	22·0				
10 0	32·7			29·0	
15 0	32·5				
20 0	31·0			29·0	
25 0	31·6				
30 0	31·2			28·9	
35 0	32·0				
40 0	32·8			28·0	
45 0	32·6				
50 0	33·6			28·0	
55 0	33·1				
18 0 0	33·4			24·4	64
5 0	32·0				
10 0	32·6			22·0	
15 0	32·5				
20 0	33·1			22·2	
25 0	32·6				
30 0	32·2			22·0	
35 0	32·5				
40 0	31·1			21·3	
45 0	32·4				
50 0	32·5			21·0	
55 0	32·6				

MAY 29 and 30, 1840.

M. Gött. Time d. h. m. s.	Decl. Sc.-Div	Hor. Force Sc.-Div	Ther.	Vert. Force Sc.-Div	Ther.
29 19 0 0	32·9	20·0	65		
5 0	33·0				
10 0	33·5	18·3			
15 0	34·2				
20 0	34·4	17·5			
25 0	34·9				
30 0	35·9	17·5			
35 0	36·4				
40 0	36·5	18·4			
45 0	36·9				
50 0	37·4	19·6			
55 0	37·6				
20 0 0	39·2	20·8	65		
5 0	39·1				
10 0	41·7	22·1			
15 0	40·0				
20 0	38·5	21·3			
25 0	40·0				
30 0	40·1	21·8			
35 0	40·4				
40 0	40·5	21·3			
45 0	37·6				
50 0	39·9	21·2			
55 0	39·6				
21 0 0	40·0	20·8	65		
5 0	39·2				
10 0	38·8	21·8			
15 0	40·8				
20 0	38·0	22·3			
25 0	38·2				
30 0	40·5	23·4			
35 0	38·3				
40 0	38·5	24·5			
45 0	38·2				
50 0	37·2	25·8			
55 0	37·6				
22 0 0	37·7	27·1	64		
5 0	36·8				
10 0	35·8	27·3			
15 0	38·3				
20 0	35·5	26·7			
25 0	35·5				
30 0	34·9	27·0			
35 0	34·7				
40 0	34·9	28·0			
45 0	34·3				
50 0	34·4	28·4			
55 0	34·7				
23 0 0	34·7	29·5	64		
5 0	35·0				
10 0	35·1	29·6			
15 0	35·1				
20 0	35·3	30·1			
25 0	35·3				
30 0	35·3	30·4			
35 0	35·5				
40 0	35·9	30·2			
45 0	35·7				
50 0	35·5	30·8			
55 0	35·7				
30 0 0	35·4	30·5	64		
5 0	36·1				
10 0	34·5	30·0			
15 0	33·9				
20 0	33·7	30·2			
25 0	33·7				
30 0	33·9	30·3			

MAY 29 and 30, 1840.

M. Gött. Time d. h. m. s.	Decl. Sc.-Div	Hor. Force Sc.-Div	Ther.	Vert. Force Sc.-Div	Ther.
30 0 35 0	33·9	31·0			
40 0	33·7				
45 0	33·6				
50 0	33·7	31·6			
55 0	32·8				
1 0 0	32·9	32·6	65		
5 0	32·5				
10 0	31·6	31·2			
15 0	31·7				
20 0	31·7	31·5			
25 0	31·7				
30 0	32·2	31·2			
35 0	31·6				
40 0	30·5	30·5			
45 0	30·4				
50 0	30·8	30·2			
55 0	30·4				
2 0 0	29·5	28·7	65		
5 0	29·6				
10 0	29·5	27·1			
15 0	28·9				
20 0	28·4	26·1			
25 0	27·9				
30 0	28·6	26·0			
35 0	29·4				
40 0	27·6	24·1			
45 0	29·4				
50 0	29·0	22·8			
55 0	29·4				
3 0 0	29·5	23·5	65		
5 0	31·5				
10 0	31·0	23·7			
15 0	32·5				
20 0	32·7	22·0			
25 0	30·0				
30 0	34·2	25·0			
35 0	32·4				
40 0	27·6	23·8			
45 0	30·2				
50 0	30·5	16·5			
55 0	33·5				
4 0 0	33·0	20·5	64		
5 0	33·3				
10 0	31·5	20·0			
15 0	31·2				
20 0	31·4	19·5			
25 0	22·3				
30 0	32·0	19·5			
35 0	34·7				
40 0	21·5	19·0			
45 0	29·5				
50 0	18·5	17·5			
55 0	29·2				
5 0 0	29·2	19·5	64		
5 0	29·5				
10 0	29·2	19·5			
15 0	29·2				
20 0	29·2	19·8			
25 0	29·2				
30 0	29·3	20·1			
35 0	29·3				
40 0	29·2	20·5			
45 0	29·6				
50 0	29·9	19·4			
55 0	29·7				
6 0 0	30·0	20·1	64		
5 0	29·7				

MAY 29 and 30, 1840.

M. Gött. Time d. h. m. s.	Decl. Sc.-Div	Hor. Force Sc.-Div	Ther.	Vert. Force Sc.-Div	Ther.
30 6 10 0	31·4	19·8			
15 0	31·1				
20 0	29·4	19·9			
25 0	30·8				
30 0	31·5	19·7			
35 0	32·9				
40 0	32·9	19·3			
45 0	33·0				
50 0	33·0	18·9			
55 0	33·4				
7 0 0	33·1	18·6	64		
5 0	33·4				
10 0	33·5	18·4			
15 0	33·6				
20 0	33·1	18·6			
25 0	33·2				
30 0	33·3	18·5			
35 0	33·3				
40 0	32·6	18·2			
45 0	32·0				
50 0	32·0	18·4			
55 0	32·0				
8 0 0	31·6	18·5	66		
5 0	31·4				
10 0	30·4	18·5			
15 0	31·4				
20 0	31·8	18·5			
25 0	31·9				
30 0	32·0	18·1			
35 0	32·2				
40 0	31·7	18·5			
45 0	31·9				
50 0	31·8	18·5			
55 0	32·4				
9 0 0	31·5	17·9	65		
5 0	31·5				
10 0	28·7	17·5			
15 0	29·1				
20 0	29·3	17·5			
25 0	28·5				
30 0	28·9	18·7			
35 0	29·2				
40 0	29·0	20·0			
45 0	29·0				
50 0	29·1	20·0			
55 0	30·0				
10 0 0	29·5	22·0	65		
12 0 0	32·0	21·5	65		

The Magnetometers were not in adjustment at this period during a sufficient number of consecutive days, to give satisfactory mean positions comparable with those of the 29th and 30th of May.

ANTARCTIC EXPEDITION AT KERGUELEN ISLAND.

Decl. 1 Scale Division = 0'·73

H. F. $k = ·00019$; $q =$

The Magnetometers were observed simultaneously at the times specified.

M. Gött. Time d. h. m. s.	Decl. Sc.-Div	Hor. Force Sc.-Div	Ther.	Vert. Force Sc.-Div	Ther.
28 18 0 0	56·8	50·1	38		
19 0 0	62·4	51·6			
20 0 0	61·2	49·8	39		
21 0 0	62·2	52·9			
22 0 0	63·7	55·5	39		
23 0 0	66·1	52·9			
29 0 0 0	68·5	50·2	39		
1 0 0	64·7	50·6			

MAY 29 and 30, 1840.

M. Gött. Time (d. h. m. s)	Decl. Sc.-Div.ns	Hor. Force Sc.-Div.ns	Ther.	Vert. Force Sc.-Div.ns	Ther.
29 2 0 0	65·6	51·3	38		
3 0 0	67·1	50·0			
4 0 0	63·2	51·4	38		
5 0 0	62·5	48·5			
6 0 0	61·6	52·5	39		
7 0 0	63·4	48·2			
8 0 0	56·1	49·2	39		
9 0 0	53·6	45·0			
10 0 0	55·2	58·8	39		
2 30	55·2	58·3			
5 0	54·7	57·9			
7 30	53·7	58·0			
10 0	52·2	57·8			
12 30	57·2	57·2			
15 0	58·3	59·9			
17 30	58·8	60·4			
20 0	58·5	60·4			
22 30	60·9	57·4			
25 0	63·5	57·2			
27 30	64·9	58·1			
30 0	58·5	57·1			
32 30	56·7	57·3			
35 0	54·9	55·3			
37 30	52·5	50·2			
40 0	59·1	47·6			
42 30	66·1	47·4			
45 0	68·7	46·4			
47 30	69·3	44·6			
50 0	65·6	48·1			
52 30	63·2	53·2			
55 0	60·7	53·7			
57 30	52·5	57·7			
11 0 0	43·9	59·2	41		
2 30	40·7	51·3			
5 0	47·3	50·5			
7 30	53·0	52·5			
10 0	61·2	57·7			
12 30	67·2	65·1			
15 0	69·3	72·8			
17 30	64·4	78·7			
20 0	51·3	76·8			
22 30	44·6	75·8			
25 0	41·6	73·2			
27 30	41·8	70·9			
30 0	38·7	71·1			
32 30	39·2	70·3			
35 0	38·6	70·5			
37 30	34·7	70·8			
40 0	37·1	70·1			
42 30	40·2	70·4			
45 0	43·3	72·3			
47 30	47·5	74·9			
50 0	49·2	74·9			
52 30	50·4	75·3			
55 0	49·2	74·9			
57 30	46·6	75·7			
12 0 0	45·8	75·2	41		
2 30	42·1	77·2			
5 0	39·8	77·6			
7 30	37·6	75·8			
10 0	37·4	74·4			
12 30	35·7	75·7			
15 0	34·9	76·3			
17 30	36·6	76·6			
20 0	39·1	76·1			
22 30	38·8	73·5			
25 0	39·1	72·6			

MAY 29 and 30, 1840.

M. Gött. Time (d. h. m. s)	Decl. Sc.-Div.ns	Hor. Force Sc.-Div.ns	Ther.	Vert. Force Sc.-Div.ns	Ther.
29 12 27 30	39·4	70·5			
30 0	40·7	68·0			
32 30	41·2	66·2			
35 0	42·5	64·5			
37 30	42·6	64·1			
40 0	43·4	64·4			
42 30	42·5	65·5			
45 0	40·3	66·6			
47 30	39·7	67·0			
50 0	40·6	65·9			
52 30	41·5	65·6			
55 0	42·8	65·6			
57 30	45·2	65·5			
13 0 0	46·2	66·8	41		
2 30	48·8	68·1			
5 0	48·9	70·0			
7 30	50·2	72·0			
10 0	52·5	73·5			
12 30	47·7	71·5			
15 0	42·1	73·7			
17 30	36·9	73·1			
20 0	33·7	71·8			
22 30	33·4	66·5			
25 0	35·7	61·7			
27 30	36·5	59·7			
30 0	41·0	58·6			
32 30	41·9	61·2			
35 0	44·7	62·4			
37 30	47·4	63·1			
40 0	50·6	64·7			
42 30	52·9	67·1			
45 0	53·8	70·3			
47 30	54·0	75·2			
50 0	55·6	75·3			
52 30	56·1	71·8			
55 0	58·1	69·1			
57 30	60·1	70·7			
14 0 0	61·3	70·2	41		
2 30	64·3	71·3			
5 0	67·8	73·3			
7 30	71·4	76·0			
10 0	74·8	76·4			
12 30	74·3	78·0			
15 0	74·6	79·7			
17 30	75·4	78·2			
20 0	76·3	76·6			
22 30	75·3	74·0			
25 0	73·8	72·4			
27 30	68·5	73·2			
30 0	63·5	74·0			
32 30	61·8	72·7			
35 0	62·6	72·3			
37 30	63·0	73·5			
40 0	65·2	73·7			
42 30	68·1	73·2			
45 0	73·3	71·7			
47 30	75·6	70·9			
50 0	75·7	74·0			
52 30	75·4	76·7			
55 0	74·4	76·0			
57 30	74·0	76·2			
15 0 0	74·1	76·5	41		
2 30	71·2	74·1			
5 0	68·8	81·0			
7 30	69·7	80·0			
10 0	75·5	80·8			
12 30	71·8	78·4			

MAY 29 and 30, 1840.

M. Gött. Time (d. h. m. s)	Decl. Sc.-Div.ns	Hor. Force Sc.-Div.ns	Ther.	Vert. Force Sc.-Div.ns	Ther.
29 15 15 0	78·1	75·9			
17 30	75·5	76·5			
20 0	75·9	80·7			
22 30	78·0	80·4			
25 0	79·4	78·4			
27 30	83·3	75·9			
30 0	85·1	76·8			
32 30	92·2	78·3			
35 0	95·4	79·7			
37 30	97·6	81·8			
40 0	96·0	83·3			
42 30	93·9	83·2			
45 0	99·9	90·5			
47 30	92·8	93·8			
50 0	78·4	97·7			
52 30	80·5	96·7			
55 0	81·7	95·9			
57 30	84·5	95·3			
16 0 0	81·4	96·0	41		
2 30	80·8	95·6			
5 0	72·0	94·9			
7 30	52·4	97·3			
10 0	47·1	100·9			
12 30	56·8	98·9			
15 0	60·8	95·1			
17 30	62·4	92·6			
20 0	58·6	92·1			
22 30	70·3	94·9			
25 0	73·4	99·0			
27 30	71·8	97·0			
30 0	74·3	97·8			
32 30	74·1	101·2			
35 0	69·9	100·1			
37 30	71·5	97·6			
40 0	67·5	102·1			
42 30	80·1	109·6			
45 0	91·9	114·6			
47 30	84·0	117·6			
50 0	110·2	120·0			
52 30	110·5	110·2			
55 0	101·9	110·6			
57 30	102·9	111·9			
17 0 0	99·2	109·1	41		
2 30	88·6	113·9			
5 0	97·0	109·2			
7 30	102·5	106·9			
10 0	103·2	95·7			
12 30	101·7	92·7			
15 0	100·5	89·3			
17 30	101·9	83·5			
20 0	101·6	80·8			
22 30	99·3	81·4			
25 0	99·6	76·0			
27 30	98·0	76·6			
30 0	93·0	77·2			
32 30	92·1	70·5			
35 0	90 7	70·3			
37 30	90·1	70·1			
40 0	84·6	71·5			
42 30	86·4	67·0			
45 0	85·0	69·0			
47 30	83·7	67·9			
50 0	82·4	64·7			
52 30	82·2	64·7			
55 0	80·2	63·4			
57 30	80·3	64·3			
18 0 0	82·1	63·4	42		

MAY 29 and 30, 1840.

M. Gött. Time (d. h. m. s.)	Decl. Sc.-Div.	Hor. Force Sc.-Div.	Ther. °	Vert. Force Sc.-Div.	Ther. °
29 18 2 30	83·6	61·4			
5 0	81·1	61·3			
7 30	77·6	61·6			
10 0	74·7	61·5			
12 30	71·6	61·9			
15 0	69·7	60·2			
17 30	67·2	60·3			
20 0	70·9	60·1			
22 30	68·7	61·5			
25 0	65·6	62·5			
27 30	70·1	60·4			
30 0	68·2	61·8			
32 30	68·9	62·7			
35 0	68·6	64·5			
37 30	75·5	65·0			
40 0	66·3	65·9			
42 30	64·1	67·1			
45 0	62·8	67·4			
47 30	60·8	68·2			
50 0	60·3	67·5			
52 30	61·3	67·9			
55 0	61·0	67·2			
57 30	61·0	67·0			
19 0 0	60·6	66·6	43		
2 30	61·0	66·2			
5 0	60·9	66·2			
7 30	58·0	66·5			
10 0	57·1	66·5			
12 30	57·0	66·1			
15 0	57·5	65·5			
17 30	58·0	65·3			
20 0	56·4	65·2			
22 30	56·2	64·4			
25 0	58·2	64·0			
27 30	59·5	65·3			
30 0	61·3	65·3			
32 30	61·0	65·5			
35 0	60·7	65·8			
37 30	59·6	66·5			
40 0	59·7	66·2			
42 30	60·5	65·9			
45 0	60·8	66·1			
47 30	62·5	66·6			
50 0	63·2	67·1			
52 30	63·5	67·8			
55 0	62·9	67·7			
57 30	63·4	67·6			
20 0 0	61·8	67·5	43		
2 30	59·4	67·3			
5 0	59·1	66·7			
7 30	60·6	65·1			
10 0	60·5	65·4			
12 30	59·7	64·8			
15 0	59·3	64·8			
17 30	60·5	63·8			
20 0	61·7	64·1			
22 30	62·0	65·3			
25 0	60·6	65·4			
27 30	60·8	64·9			
30 0	60·8	64·4			
32 30	59·8	65·2			
35 0	64·8	63·8			
37 30	60·8	62·0			
40 0	64·9	61·9			
42 30	60·8	63·1			
45 0	61·5	64·1			
47 30	63·9	64·0			

MAY 29 and 30, 1840.

M. Gött. Time (d. h. m. s.)	Decl. Sc.-Div.	Hor. Force Sc.-Div.	Ther. °	Vert. Force Sc.-Div.	Ther. °
29 20 50 0	64·1	64·0			
52 30	60·0	63·8			
55 0	60·5	64·0			
57 30	60·7	63·7			
21 0 0	61·4	62·9	44		
2 30	58·4	63·2			
5 0	62·1	62·9			
7 30	63·6	62·9			
10 0	62·8	62·7			
12 30	63·4	63·0			
15 0	62·6	62·3			
17 30	64·6	62·9			
20 0	65·1	63·0			
22 30	63·5	62·7			
25 0	62·2	61·9			
27 30	62·4	62·5			
30 0	62·6	62·3			
32 30	62·7	62·7			
35 0	62·2	62·8			
37 30	62·6	63·4			
40 0	63·8	62·9			
42 30	68·6	63·0			
45 0	62·9	63·1			
47 30	63·0	63·2			
50 0	62·6	63·5			
52 30	62·8	63·2			
55 0	62·9	63·1			
57 30	63·2	63·2			
22 0 0	62·4	62·9	44		
2 30	62·1	62·9			
5 0	62·1	62·6			
7 30	62·4	62·6			
10 0	63·5	63·2			
12 30	64·3	63·6			
15 0	62·2	63·8			
17 30	61·6	63·0			
20 0	61·9	62·6			
22 30	59·9	63·2			
25 0	59·7	63·0			
27 30	59·8	62·6			
30 0	59·5	62·4			
32 30	59·8	61·7			
35 0	60·1	61·5			
37 30	59·9	61·4			
40 0	60·0	61·5			
42 30	60·8	61·6			
45 0	61·6	61·6			
47 30	60·5	61·9			
50 0	62·0	61·9			
52 30	65·5	61·8			
55 0	60·9	61·8			
57 30	60·4	61·8			
23 0 0	61·6	62·1	43		
2 30	60·7	60·6			
5 0	61·3	62·3			
7 30	60·8	62·5			
10 0	61·2	62·6			
12 30	62·3	62·8			
15 0	62·5	62·9			
17 30	62·2	62·9			
20 0	61·9	63·7			
22 30	61·8	63·7			
25 0	61·8	63·4			
27 30	61·7	63·6			
30 0	61·4	63·6			
32 30	61·3	63·8			
35 0	60·9	64·1			

MAY 29 and 30, 1840.

M. Gött. Time (d. h. m. s.)	Decl. Sc.-Div.	Hor. Force Sc.-Div.	Ther. °	Vert. Force Sc.-Div.	Ther. °
29 23 37 30	60·7	64·0			
40 0	61·6	63·9			
42 30	62·0	63·8			
45 0	63·6	63·7			
47 30	62·7	64·0			
50 0	62·4	63·8			
52 30	63·5	63·3			
55 0	64·3	63·1			
57 30	64·5	63·8			
30 0 0	64·6	64·0	44		
2 30	64·5	64·4			
5 0	63·8	64·8			
7 30	62·8	64·5			
10 0	62·4	64·5			
12 30	62·5	64·7			
15 0	61·1	64·8			
17 30	61·0	64·5			
20 0	61·5	64·0			
22 30	61·1	63·6			
25 0	59·1	63·2			
27 30	58·8	62·7			
30 0	59·2	61·8			
32 30	59·7	61·3			
35 0	61·5	61·4			
37 30	60·6	61·5			
40 0	60·8	61·8			
42 30	61·2	61·9			
45 0	60·5	62·5			
47 30	60·3	62·4			
50 0	60·2	62·2			
52 30	60·6	61·7			
55 0	60·2	61·9			
57 30	60·5	61·7			
1 0 0	61·0	62·2	43		
2 30	60·5	63·0			
5 0	60·7	63·7			
7 30	60·2	64·0			
10 0	58·6	64·3			
12 30	57·8	64·0			
15 0	54·8	64·1			
17 30	52·7	63·7			
20 0	52·1	62·9			
22 30	52·5	61·0			
25 0	53·7	60·9			
27 30	55·3	60·9			
30 0	55·7	61·0			
32 30	57·5	60·4			
35 0	61·9	60·3			
37 30	61·1	61·3			
40 0	62·3	62·4			
42 30	60·7	62·9			
45 0	59·4	61·1			
47 30	59·7	59·5			
50 0	60·9	59·9			
52 30	61·6	60·8			
55 0	61·5	62·5			
57 30	60·3	63·0			
2 0 0	60·5	62·6	43		
2 30	61·6	63·1			
5 0	61·6	63·3			
7 30	59·9	67·6			
10 0	59·9	66·5			
12 30	60·2	66·2			
15 0	60·7	66·5			
17 30	59·0	67·2			
20 0	58·5	67·2			
22 30	57·5	67·0			

MAY 29 and 30, 1840.

M. Gött. Time.			Decl.	Hor. Force.		Vert. Force.	
d.	h. m. s		Sc.-Div⁹⁸.	Sc.-Div⁹⁸.	Ther.	Sc.-Div⁹⁸.	Ther.
30	2 25 0		55·9	67·5			
	27 30		55·1	65·2			
	30 0		55·5	65·1			
	32 30		53·3	66·2			
	35 0		53·1	66·1			
	37 30		52·7	65·5			
	40 0		50·7	66·7			
	42 30		43·6	67·1			
	45 0		42·2	58·1			
	47 30		46·1	56·8			
	50 0		49·1	57·3			
	52 30		50·7	59·0			
	55 0		51·1	61·0			
	57 30		50·9	60·5			
3	0 0		49·3	59·6	43		
	2 30		48·8	56·6			
	5 0		50·5	56·5			
	7 30		51·5	56·9			
	10 0		51·5	57·0			
	12 30		54·0	57·3			
	15 0		56·4	60·4			
	17 30		56·5	61·4			
	20 0		56·5	63·9			
	22 30		55·5	63·8			
	25 0		55·0	61·5			
	27 30		56·1	61·0			
	30 0		57·8	62·2			
	32 30		57·4	61·6			
	35 0		57·6	61·8			
	37 30		56·3	62·0			
	40 0		56·3	61·5			
	42 30		56·8	61·5			
	45 0		56·8	61·7			
	47 30		56·4	61·1			
	50 0		57·4	61·0			
	52 30		57·7	61·6			
	55 0		57·6	61·5			
	57 30		57·8	61·4			
4	0 0		58·1	61·6	43		
	2 30		58·1	62·0			
	5 0		57·1	61·9			
	7 30		57·7	61·6			
	10 0		57·9	62·0			
	12 30		56·3	62·1			
	15 0		56·2	62·2			
	17 30		55·7	62·1			
	20 0		55·1	62·0			
	22 30		55·5	61·2			
	25 0		56·1	61·1			
	27 30		56·5	61·2			
	30 0		56·9	61·5			
	32 30		57·1	61·7			
	35 0		57·3	61·8			
	37 30		56·7	61·8			
	40 0		56·4	61·6			
	42 30		56·4	61·7			
	45 0		56·7	61·6			
	47 30		57·0	61·6			
	50 0		57·8	61·1			
	52 30		58·1	61·4			
	55 0		58·2	61·5			
	57 30		58·1	61·5			
5	0 0		57·9	61·5	43		
	2 30		57·5	61·6			
	5 0		57·3	61·6			
	7 30		57·1	61·4			
	10 0		56·8	61·6			

MAY 29 and 30, 1840.

M. Gött. Time.			Decl.	Hor. Force.		Vert. Force.	
d.	h. m. s		Sc.-Div⁹⁸.	Sc.-Div⁹⁸.	Ther.	Sc.-Div⁹⁸.	Ther.
30	5 12 30		56·5	61·2			
	15 0		56·6	61·0			
	17 30		56·4	60·7			
	20 0		56·8	60·7			
	22 30		56·6	60·7			
	25 0		56·5	60·8			
	27 30		56·5	60·5			
	30 0		56·1	60·5			
	32 30		55·9	60·2			
	35 0		55·9	60·2			
	37 30		56·1	60·2			
	40 0		56·2	60·1			
	42 30		57·4	60·0			
	45 0		57·1	60·4			
	47 30		57·2	60·4			
	50 0		57·2	60·6			
	52 30		57·2	60·6			
	55 0		56·8	60·6			
	57 30		56·8	60·4			
6	0 0		56·7	60·4	43		
	2 30		56·9	60·4			
	5 0		56·8	60·1			
	7 30		56·8	60·1			
	10 0		56·9	59·9			
	12 30		56·9	59·8			
	15 0		56·9	60·0			
	17 30		57·6	60·3			
	20 0		57·0	60·4			
	22 30		56·9	60·4			
	25 0		56·9	60·2			
	27 30		56·9	60·3			
	30 0		56·9	60·3			
	32 30		57·1	60·1			
	35 0		57·5	60·2			
	37 30		57·6	60·2			
	40 0		57·6	60·1			
	42 30		57·7	60·0			
	45 0		58·1	60·0			
	47 30		58·2	60·1			
	50 0		57·8	60·1			
	52 30		57·6	60·0			
	55 0		57·5	60·0			
	57 30		57·5	60·0			
7	0 0		57·4	60·0	43		
	2 30		57·5	59·9			
	5 0		57·7	59·8			
	7 30		57·8	59·9			
	10 0		57·9	60·0			
	12 30		57·8	60·1			
	15 0		57·8	60·1			
	17 30		57·8	60·1			
	20 0		57·7	60·1			
	22 30		57·6	59·9			
	25 0		57·9	60·0			
	27 30		57·8	60·0			
	30 0		58·1	60·0			
	32 30		58·2	59·9			
	35 0		58·2	60·0			
	37 30		58·5	60·0			
	40 0		58·6	59·9			
	42 30		58·6	59·9			
	45 0		58·5	59·7			
	47 30		58·5	59·5			
	50 0		58·7	59·5			
	52 30		58·6	59·4			
	55 0		58·8	59·1			
	57 30		59·0	59·0			

MAY 29 and 30, 1840.

M. Gött. Time.			Decl.	Hor. Force.		Vert. Force.	
d.	h. m. s		Sc.-Div⁹⁸.	Sc.-Div⁹⁸.	Ther.	Sc.-Div⁹⁸.	Ther.
30	8 0 0		59·0	59·0	43		
	2 30		59·0	59·0			
	5 0		58·8	59·0			
	7 30		57·6	59·0			
	10 0		58·2	59·0			
	12 30		58·2	59·0			
	15 0		58·3	58·6			
	17 30		58·6	58·7			
	20 0		58·5	58·6			
	22 30		58·5	58·6			
	25 0		58·4	58·8			
	27 30		58·3	58·8			
	30 0		58·0	59·1			
	32 30		58·0	59·2			
	35 0		58·0	59·1			
	37 30		58·1	59·1			
	40 0		58·2	58·8			
	42 30		58·2	58·4			
	45 0		58·3	59·2			
	47 30		58·0	59·2			
	50 0		58·0	59·1			
	52 30		57·6	59·3			
	55 0		56·6	59·3			
	57 30		55·9	59·5			
9	0 0		54·8	59·5	44		
	2 30		54·1	59·6			
	5 0		53·3	59·2			
	7 30		53·4	59·2			
	10 0		53·3	59·5			
	12 30		54·1	59·8			
	15 0		53·8	60·1			
	17 30		53·4	59·7			
	20 0		53·6	59·4			
	22 30		53·7	59·4			
	25 0		54·2	59·0			
	27 30		55·8	59·8			
	30 0		57·8	60·6			
	32 30		58·3	61·5			
	35 0		58·5	62·5			
	37 30		58·5	63·2			
	40 0		58·1	64·0			
	42 30		58·1	65·7			
	45 0		52·6	65·7			
	47 30		51·1	65·2			
	50 0		55·0	64·6			
	52 30		54·9	64·1			
	55 0		54·8	62·6			
	57 30		55·8	59·7			

Mean Positions at the usual hours of observation, from May 27th to June 27th inclusive.

d.	h. m. s		Decl.	Hor. Force.	Ther.
	0 0 0		60·9	54·2	38
	1 0 0		59·9	52·9	
	2 0 0		59·2	52·1	37
	3 0 0		58·9	52·4	
	4 0 0		57·9	52·5	37
	5 0 0		57·9	52·5	
	6 0 0		57·1	52·9	37
	7 0 0		57·4	52·8	
	8 0 0		57·0	52·2	37
	9 0 0		57·1	51·8	
	10 0 0		59·1	52·4	37
	11 0 0		58·7	52·6	
	12 0 0		58·3	53·4	36
	13 0 0		59·5	52·2	
	14 0 0		60·1	51·9	36

MAY 29 and 30, 1840.

M. Gött. Time.				Decl.	Hor. Force.		Vert. Force.	
d.	h.	m.	s.	Sc.-Div^m.	Sc.-Div^m.	Ther.	Sc.-Div^m.	Ther.
15	0	0		59·3	52·1			
16	0	0		58·9	51·8	36		
17	0	0		59·1	51·6			
18	0	0		56·7	51·8	36		
19	0	0		56·3	52·8			
20	0	0		57·0	54·5	37		
21	0	0		58·1	55·6			
22	0	0		59·9	55·8	37		
23	0	0		60·0	55·7			

JULY 5 and 6, 1840.

TORONTO { Decl. 1 Scale Division = 0'·72
H. F. k = ·00021; q = ·00034
V. F. k = ·0002; q = ·0002

Extra and regular observations.

The V. F. was observed at 1m. 30s. before, and the H. F. 2m. 0s. after the times specified.

M. Gött. Time.				Decl.	Hor. Force.		Vert. Force.	
d.	h.	m.	s.	Sc.-Div^m.	Sc.-Div^m.	Ther.	Sc.-Div^m.	Ther.
5	18	0	0	65·9	41·6	67	38·0	67
		24	18	31·4	52·9		30·0	
		29	18	30·0	52·3		30·1	
		34	18	33·2	50·9		31·6	
		39	18	33·0	48·7		31·6	
		44	18	32·8	46·1		31·9	
		49	18	38·0	45·1		32·9	
		54	18	39·0	39·2		33·6	
		59	18	40·0	39·5		32·7	
	19	4	18	40·2	41·2	67	32·3	67
		9	18	38·0	38·9		32·3	
		14	18	35·7	37·4		32·1	
		24	18	45·5	41·6		31·9	
		29	18	47·2	43·3		31·8	
		34	18	48·6	44·3		31·8	
		39	18	48·9	45·5		34·1	
		44	18	51·0	49·2		34·8	
	20	0	0	43·2	37·5	67	34·3	67
		14	18	36·6	43·0		34·8	
		19	18	42·0	52·3		39·6	
		24	18	42·3	50·9		38·8	
		29	18	40·8	49·7		39·7	
		34	18	47·0	49·1		38·2	
		39	18	54·9	51·8		30·5	
		44	18	58·5	50·6		35·1	
		49	18	59·4	51·5		35·1	
		59	18	52·0	50·0		40·7	
	21	4	18	44·5	51·6	67	40·7	68
		9	18	42·0	51·7		40·0	
		14	18	39·6	51·9		40·0	
		19	18	37·5	47·1		36·4	
		24	18	38·1	48·8		36·4	
		29	18	41·4	49·4		39·6	
		34	18	41·6	47·8		39·6	
		44	18	39·0	49·2		37·4	
		49	18	38·7	48·6		37·7	
	22	0	0	41·4	49·5	68	38·9	68
6	0	0	0	28·3	64·6	67	46·2	67
	2	0	0	36·0	56·2	67	48·8	67
	4	0	0	17·0	65·3	69	48·5	68
	6	0	0	7·0	67·2	69	49·6	68
	8	0	0	7·5	76·5	70	49·1	69

JULY 5 and 6, 1840.

M. Gött. Time.				Decl.	Hor. Force.		Vert. Force.	
d.	h.	m.	s.	Sc.-Div^m.	Sc.-Div^m.	Ther.	Sc.-Div^m.	Ther.
6	10	0	0	12·6	86·7	71	52·0	70
	12	0	0	17·7	74·9	71	51·7	70
	14	0	0	12·2	67·3	71	50·7	71
	16	0	0	12·5	66·8	71	48·6	70
	18	0	0	14·4	70·8	70	48·7	70
	20	0	0	15·4	74·7	70	49·3	69
	22	0	0	19·2	72·2	69	45·4	69

Mean Positions at the usual hours of observation during the Month.^a

	0	0	0	22·6	75·3	68	47·3	66
	2	0	0	24·2	70·6	68	47·2	66
	4	0	0	19·3	66·8	69	47·5	67
	6	0	0	9·2	71·8	69	47·8	68
	8	0	0	7·5	79·3	70	48·8	68
	10	0	0	11·6	80·2	71	49·1	69
	12	0	0	15·3	74·6	71	49·5	70
	14	0	0	16·0	73·5	71	49·6	70
	16	0	0	17·2	72·7	71	49·6	70
	18	0	0	20·2	71·0	70	46·7	68
	20	0	0	21·5	71·2	69	46·7	68
	22	0	0	20·8	73·1	68	46·3	67

ST. HELENA { Decl. 1 Scale Division = 0'·71
H. F. k = ·00013; q = ·0003

Positions at the usual hours of observation, from July 5, 14h. to July 6, 22h.

5	14	0	0	73·4	63·2		61
	16	0	0	72·1	57·9		61
	18	0	0	71·7	62·9		61
	20	0	0	77·3	60·5		61
	22	0	0	77·5	58·8		61
6	0	0	0	73·0	59·5		62
	2	0	0	67·1	62·1		62
	4	0	0	68·2	56·4		63
	6	0	0	67·1	51·7		63
	8	0	0	63·8	48·6		63
	10	0	0	68·0	54·8		62
	12	0	0	69·5	55·2		62
	14	0	0	69·4	57·6		62
	16	0	0	71·0	55·7		62
	18	0	0	73·1	59·0		62
	20	0	0	73·5	62·0		61
	22	0	0	71·6	71·0		62

Mean Positions at the same hours during the Month.^b

	0	0	0	71·5	72·4		61
	2	0	0	71·8	71·2		62
	4	0	0	72·0	64·9		63
	6	0	0	70·4	61·2		62
	8	0	0	71·5	60·5		62
	10	0	0	71·9	61·2		62
	12	0	0	72·2	61·5		62
	14	0	0	72·5	62·8		61
	16	0	0	72·3	63·1		61
	18	0	0	72·8	64·5		61
	20	0	0	75·6	66·1		61
	22	0	0	72·5	70·6		61

JULY 17 and 18, 1840.

TORONTO { Decl. 1 Scale Division = 0'·72
H. F. k = ·00021; q = ·00034
V. F. k = ·0002; q = ·0002

Extra observations.

The V. F. was observed at 1m. 30s. before, and the H. F. 2m. 30s. after the times specified.

M. Gött. Time.				Decl.	Hor. Force.		Vert. Force.	
d.	h.	m.	s.	Sc.-Div^m.	Sc.-Div^m.	Ther.	Sc.-Div^m.	Ther.
17	5	29	43	12·2	65·8		61·9	
		34	43	11·2	68·0		62·6	
		39	43	10·2	68·0		62·7	
		44	43	9·3	69·7		62·7	
		49	43	10·4	67·0		63·1	
		59	43	7·9	69·6		63·3	
	6	4	43	5·5	69·0	74	63·4	74
		9	43	7·1	66·4		63·5	
		14	43	7·8	74·7		63·6	
		19	43	6·5	75·9		63·9	
		24	43	6·4	74·2		63·8	
		29	43	6·0	74·0		63·7	
		34	43	6·3	73·9		63·6	
		39	43	5·6	75·7		63·7	
		44	43	5·0	76·0		63·7	
		49	43	5·0	76·5		63·8	
	7	29	43	3·8	80·6		60·6	
		34	43	4·7	83·1		64·8	
		39	43	5·9	83·9		63·3	
		44	43	6·9	83·0		63·0	
		49	43	6·9	84·3		64·7	
		59	43	7·0	83·6		64·9	
	8	9	43	7·7	83·1	74	63·5	74
		14	43	7·9	81·0		63·5	
		19	43	9·9	82·3		63·7	
		24	43	10·9	84·1		64·0	
		29	43	11·1	86·3		64·4	
		34	43	11·2	88·1		64·6	
		39	43	10·0	88·3		64·8	
		44	43	9·0	86·0		64·8	
		49	43	9·1	84·4		64·7	
		54	43	8·1	80·1		65·2	
		59	43	8·7	80·6		65·0	
	9	4	43	8·0	80·1	74	64·9	74
		9	43	8·0	80·3		65·3	

Positions at the usual hours of observation, from July 17d. 0h. to July 18d. 16h.

17	0	0	0	10·8	69·1	75	50·5	75
	2	0	0	15·1	54·9	74	55·9	74
	4	0	0	19·1	57·0	74	58·3	74
	6	0	0	7·9	69·6	74	63·3	74
	8	0	0	7·0	83·6	74	64·9	74
	10	0	0	13·0	84·1	74	66·9	74
	12	0	0	10·5	73·8	74	65·7	74
	14	0	0	17·9	67·3	73	64·3	73
	16	0	0	17·6	72·0	73	62·9	73
	18	0	0	12·0	68·4	72	60·7	73
	20	0	0	31·9	58·7	71	51·5	72
	22	0	0	29·2	69·7	70	47·2	71
18	0	0	0	9·6	69·5	69	44·8	69
	2	0	0	17·8	59·8	70	50·7	70
	4	0	0	16·5	58·2	70	54·6	70
	6	0	0	15·1	68·1	71	58·5	70

---

^a The mean positions of the Declination magnet are from the 3rd to the 25th July inclusive; those of the H. F. magnet from the 1st to the 18th inclusive; and those of the V. F. magnet from the 1st to the 14th inclusive.

^b The mean positions of the H. F. magnet are from the 1st to the 11th July inclusive.

JULY 17 and 18, 1840.

M. Gött. Time. d. h. m. s.	Decl. Sc.-Div.	Hor. Force. Sc.-Div.	Ther.	Vert. Force. Sc.-Div.	Ther.
18 8 0 0	5·0	74·1	71	63·1	71
10 0 0	7·0	87·8	72	62·8	72
12 0 0	13·9	77·2	74	60·4	73
14 0 0	13·5	76·4	74	60·5	74
16 0 0*	11·2	72·7	73	60·8	73

The Mean Positions of the Decl. and H. F. magnetometers are given in page 11.[b]

St. Helena { Decl. 1 Scale Division = 0'·71; H. F. k = ·00018; q = ·0003

Positions at the usual hours of observation, from July 17d. 0h. to July 18d. 12h.

M. Gött. Time. d. h. m. s.	Decl. Sc.-Div.	Hor. Force. Sc.-Div.	Ther.
17 0 0 0	71·6	56·5	62
2 0 0	71·0	56·0	62
4 0 0	69·8	49·2	63
6 0 0	69·9	46·1	63
8 0 0	71·4	45·0	62
10 0 0	73·1	47·6	62
12 0 0	73·0	44·6	62
14 0 0	73·5	46·5	
16 0 0	74·6	51·1	
18 0 0	73·5	50·9	61
20 0 0	76·5	55·0	61
22 0 0	73·5	52·6	61
18 0 0 0	69·7	50·8	62
2 0 0	71·3	51·6	62
4 0 0	71·2	45·5	62
6 0 0	70·0	47·5	63
8 0 0	71·5	43·5	63
10 0 0	71·8	47·8	63
12 0 0	72·5	49·0	63

Mean Positions at the same hours during the Month.[c]

0 0 0	71·5	67·8	61
2 0 0	71·8	65·7	62
4 0 0	72·0	62·0	62
6 0 0	70·4	59·9	62

JULY 17 and 18, 1840.

M. Gött. Time. t. h. m. s.	Decl. Sc.-Div.	Hor. Force. Sc.-Div.	Ther.	Vert. Force. Sc.-Div.	Ther.
8 0 0	71·5	58·1	61		
10 0 0	71·9	57·8	61		
12 0 0	72·2	57·2	61		
14 0 0	72·5	59·2	61		
16 0 0	72·3	59·3	61		
18 0 0	72·8	60·4	61		
20 0 0	75·6	61·7	60		
22 0 0	72·5	64·9	60		

AUGUST 19, 20, and 21, 1840.

Toronto { Decl. 1 Scale Division = 0'·72; H. F. k = ·00022; q = ·00034; V. F. k = ·00016; q = ·0002

Regular and extra observations.

The V. F. was observed at 1m. 30s. before, and the H. F. at 1m. 30s. after the times specified.

M. Gött. Time. d. h. m. s.	Decl. Sc.-Div.	Hor. Force. Sc.-Div.	Ther.	Vert. Force. Sc.-Div.	Ther.
19 0 0 0	59·3	152·4	70	49·0	69
2 0 0	55·7	133·1	71	46·6	68
4 0 0	41·7	145·3	72	44·1	69
6 0 0	33·3	146·1	73	45·6	70
8 0 0	40·4	146·2	74	48·2	71
10 0 0	34·2	145·4	75	57·3	72
12 0 0	47·6	142·5	76	58·4	73
14 0 0*	70·3	132·1	76	66·6	72
52 30	63·0	129·1		53·4	
57 30	62·6	129·8		53·2	
15 2 30	59·2	127·4	76	52·2	76
7 30	60·0	127·8		52·2	
17 30	62·5	122·5		52·3	
22 30	61·0	119·1		51·8	
27 30	64·6	122·0		51·9	
32 30	68·0	120·2		53·6	
37 30	85·5	112·2		57·9	
42 30	94·4	116·9		56·8	
47 30	94·0	114·5		56·7	
52 30	91·2	118·8		56·0	
16 0 0	87·3	119·0	76	54·8	76

AUGUST 19, 20, and 21, 1840.

M. Gött. Time. d. h. m. s.	Decl. Sc.-Div.	Hor. Force. Sc.-Div.	Ther.	Vert. Force. Sc.-Div.	Ther.
19 16 7 30	77·4	123·8		54·4	
12 30	69·2	123·0		51·5	
17 30	66·8	123·7		50·5	
22 30	66·0	126·7		49·4	
32 30	63·2	128·5		48·4	
37 30	60·3	127·5		46·6	
47 30	65·2	123·3		46·5	
52 30	71·9	122·5		46·5	
57 30	70·5	125·3		46·7	
17 2 30	61·7	127·3	76	47·0	76
7 30	59·0	126·9		45·5	
12 30	60·1	126·0		46·7	
17 30	60·0	126·3		46·0	
27 30	59·7	126·8		45·9	
47 30	56·8	126·6		47·7	
18 0 0	56·0	128·4	75	48·8	74
20 0 0	56·4	137·5	74	47·6	74
22 0 0	52·4	143·7	73	51·8	74
20 0 0 0	55·5	144·8	72	50·8	72
2 0 0	59·3	138·9	72	50·7	71
4 0 0	43·1	128·3	72	49·5	72
6 0 0	43·1	139·2	73	51·4	72
8 0 0	41·4	145·0	74	52·4	73
10 0 0	45·5	142·8	75	53·0	74
12 0 0	48·2	146·2	76	54·2	75
14 0 0	50·4	141·5	76	52·8	75
16 0 0	52·5	139·3	75	53·4	76
18 0 0	49·2	140·1	75	50·6	75
20 0 0	47·8	138·4	74	42·7	74
22 0 0	51·0	139·3	74	46·0	74
21 0 0 0	52·5	138·6	73	46·9	73
2 0 0	59·6	138·2	73	47·3	73
4 0 0	42·0	137·0	74	49·1	74
6 0 0	39·7	137·0	74	49·1	74
8 0 0	42·2	142·4	76	52·2	75
10 0 0	46·1	146·9	77	52·8	76
12 0 0	46·0	138·1	77	52·7	76
14 0 0	47·0	136·0	77	51·5	77
16 0 0	65·6	133·9	76	47·7	76
18 0 0*	45·4	124·0	76	37·4	76

a Saturday midnight at Toronto.

b In consequence of breaks in the series, there are no mean positions of the V. F. magnet comparable with the positions of July 18th and 19th.

c The mean positions of the H. F. magnet are from the 15th to the 31st July inclusive.

d. h. m.	Wind.	
19 14 0	Calm	Arch of light in north horizon extending from N.E. to N.W.; centre about 5° East of North, alt. 20°; remainder perfectly clear.
15 0	Calm	A double arch of auroral light extending from N. E. to N.W.; greatest alt. of interior arch about 15°, of exterior 20°, centre of arches about 5° E. of N.; remainder of sky clear.
2	Calm	Faint pulsations or waves of light alternately advancing and receding from principal arch; no streamers. Two meteors, second magnitude, fell from zenith towards N.W.
9	Calm	A bright patch forming in North and progressing Westward, shooting up streamers during its course to an alt. of 45°.
17	Calm	Three bright concentric arches of light; alt. of exterior arch 35°; stars visible between second and third arches. Streamer like a pillar, rising from a bright patch and progressing Westward, became indistinct in N. N. W. Another meteor from alt. 70° to 30°, direction N.W.
27	Calm	Two arches increasing in extent from E.N.E. to W.N.W., alt. 40°, with strongly defined upper surfaces, although occasionally breaking into irregular surface-like waves, and shooting up a series of small streamers from the more prominent parts. Stars visible below the interior arch.
43	Calm	Arch-like character of Aurora nearly gone, leaving a continuous chain of bright patches in its stead, each shooting up streamers of considerable brilliancy. One in West of great brilliancy, and of an orange tint, alt. 60°.
47	Calm	Aurora decreasing in brilliancy; nothing remaining except a bank of yellowish light from N.W. to N.E., alt. about 20°.

d. h. m.	Wind.	
19 15 57	Calm	Aurora decreasing; faint light alone remaining.
16 27	Calm	A few cirri and cirro-strati appearing in North horizon; light of the moon (which rose about 16h.) obscuring Eastern portion of auroral light, faint light remaining in N.W. and North.
52	Calm	Light increasing and forming an arch from N.N.E. to N.N.W., alt. 15°; faint pulsations and waving motion in upper surface of arch.
57	Calm	A bright patch of light in North, with streamers from its Western portion.
17 7	Calm	Haze and cirri rising in North horizon; the Aurora nearly indistinct; no feature visible except a low bank of light with faint pulsations, but evidently in full operation, obscured only by the haze and cirri.
32	Calm	Nothing remarkable except a faint bank of light in North.
42	Calm	Light becoming fainter.
18 0	Calm	Sky clear, faint Aurora in N.; light haze round the moon.
*21 18 0	Calm	Auroral light in the N.; lightning in N.W., near horizon.
40		Light growing fainter.
40		Disappearing rapidly.
19 5		Scarcely visible, clouds rising in N.W.; sheet lightning in N.W., near horizon; distant thunder.
40		Sky partially clouded in the North; light cirri. Auroral light appearing through the clouds in West and North.
18 to 19 40 }	Calm	Constant lightning in N.W. Distant thunder during the latter part of the period.
20 0	Calm	Three-quarters clouded, with light cirri and haze; faint light in North horizon.
21 35	Calm	A low bank of strati in N. and N.W. horizon; haze round remainder of horizon; zenith clear.

August 19, 20, and 21, 1840.

M. Gött. Time.			Decl.	Hor. Force.		Vert. Force.	
d.	h.	m.	s.	Sc.-Div^m.	Sc.-Div^m.	Ther.	Sc.-Div^m.	Ther.
21 18	32	0	48·9	128·5		38·0	
	37	0	48·8	128·6		39·3	
	42	0	47·2	129·0		39·7	
	47	0	48·8	132·7		41·1	
	52	0	50·8	134·4		42·8	
	57	0	52·3	136·3		44·2	
19	2	0	55·1	137·5	75	45·6	75
	7	0	57·0	138·4		46·2	
	12	0	56·0	136·6		46·7	
	17	0	54·0	135·7		46·8	
	22	0	51·0	134·8		47·3	
	27	0	50·4	133·9		48·1	
	32	0	45·4	129·9		46·6	
	37	0	41·7	128·9		43·0	
	42	0	39·1	127·3		41·6	
20	0	0	29·5	118·4	77	37·1	77
	37	0	45·9	138·7		41·8	
	42	0	47·4	139·2		42·8	
	47	0	46·1	139·0		44·0	
	52	0	46·9	142·0		45·1	
	57	0	47·5	139·1		46·3	
21	2	0	49·1	138·2	77	46·4	77
	7	0	50·7	139·7		46·6	
	12	0	55·0	142·9		47·8	
	17	0	54·5	142·3		48·6	
	22	0	52·9	143·0		48·7	
	27	0	53·4	142·5		49·5	
22	0	0	55·1	141·1	75	50·5	76
22	0	0	60·7	140·6	74	50·4	75

Mean Positions at the usual hours of observation, during the Month.[b]

0	0	0	55·7	149·4	69	·49·0	69
2	0	0	57·6	142·0	69	49·0	69
4	0	0	45·9	140·1	70	49·0	70

August 19, 20, and 21, 1840.

M. Gött. Time.			Decl.	Hor. Force.		Vert. Force.	
d.	h.	m.	s.	Sc.-Div^m.	Sc.-Div^m.	Ther.	Sc.-Div^m.	Ther.
	6	0	0	39·1	142·1	70	50·1	70
	8	0	0	39·9	150·0	71	51·4	71
	10	0	0	44·4	152·2	72	52·8	71
	12	0	0	47·1	148·0	72	52·6	71
	14	0	0	54·1	148·5	72	52·1	72
	16	0	0	56·1	144·6	71	48·7	72
	18	0	0	49·6	141·6	71	45·2	70
	20	0	0	45·4	144·9	70	44·8	70
	22	0	0	51·0	147·0	69	47·0	70

St. Helena { The Magnetometers were removing from the temporary to the permanent Observatory.

August 28 and 29, 1840.

Toronto {
Decl. 1 Scale Division = 0'·72
H. F. k = ·00022 ; q = ·00034
V. F. k = ·00016 ; q = ·0002

Regular, term, and extra observations.

The V. F. was observed at 2^m. 30^s. before, and the H. F. at 2^m. 30^s. after the times specified.

				Decl.	Hor. Force.		Vert. Force.	
28	0	0	0	156·4	154·0	68	53·5	68
	2	0	0	161·2	145·1	67	52·8	68
	4	0	0	152·1	144·9	68	53·1	68
	6	0	0	144·9	145·9	69	53·4	69
	8	0	0	142·4	156·2	71	55·0	70
	10	0	0	146·7	163·5	72	57·7	72
	5	0		146·2				
	10	0		146·7	164·5		55·8	
	15	0		146·5				
	20	0		146·9	165·2		55·4	
	25	0		146·7				
	30	0		153·7	165·4		63·8	
	35	0		157·1				

August 28 and 29, 1840.

M. Gött. Time.			Decl.	Hor. Force.		Vert. Force.	
d.	h.	m.	s.	Sc.-Div^m.	Sc.-Div^m.	Ther.	Sc.-Div^m.	Ther.
28 10	40	0	165·0	154·0		61·3	
	45	0	177·0				
	50	0	177·0	158·6		59·2	
	55	0	172·0				
11	0	0	166·5	153·6	73	56·2	73
	5	0	161·4				
	10	0	158·0	154·2		55·2	
	15	0	155·6				
	20	0	152·5	146·7		53·8	
	25	0	148·8				
	30	0	143·2	142·5		54·2	
	35	0	141·8				
	40	0	143·1	146·2		54·1	
	45	0	145·3				
	50	0	148·0	155·2		55·5	
	55	0	146·3				
12	0	0	144·6	146·0	73	55·0	72
	5	0	143·6				
	10	0	141·0	144·7		54·8	
	15	0	139·9				
	20	0	136·1	144·9		54·2	
	25	0	135·4				
	30	0	135·6	146·5		53·3	
	35	0	137·3				
	40	0	138·7	146·8		53·1	
	45	0	140·2				
	50	0	141·3	148·9		53·6	
	55	0	141·0				
13	0	0	142·7	151·0	73	52·9	73
	5	0	142·1				
	10	0	141·9	153·1		53·5	
	15	0	144·6				
	20	0	141·7	145·8		52·3	
	25	0	139·0				
	30	0	141·1	158·1		52·7	
	35	0	144·7				

a The H. F. was observed at 2^m. after the times specified from 21^d. 18^h. 32^m. to 21^d. 21^h. 27^m.

b The mean positions are from August 17th to 31st inclusive.

d.	h.	m.	Wind.	
28 12	0	Calm	Six-eighths clouded, with light cirri and cirro-cumuli, clearing in N.W. and zenith.
14	0	Calm	A few strati in East horizon, remainder clear. Bank of light in North horizon.
	25	Calm	Arch of light in North, extending from N.W. to N.E., alt. in centre about 60°.
	35	Calm	A number of concentric arches, with patches of light in different places.
	40	Calm	Arches breaking up, four or five streamers shooting up in N.W. to alt. of 40° or 50°.
	45	Calm	Light in North remaining; a number of streamers in North.
15 30	Calm	Clouds rising in N.W., gradually obscuring the whole sky.
16	0	Calm	Densely clouded.
17	0	Calm	Densely clouded; air close and oppressive.
	30	Calm	Streamers proceeding from East to West across zenith; much obscured by cirrus; haze covering the whole sky.
	32	Calm	Pulsations of light, and streamers darting from the horizon on every side, and meeting in a circle about 5° in diameter in zenith; halo of auroral light, composed of waves and pulsations, proceeding from every quarter towards it; stars visible in the centre of the ring of light, but nowhere else; centre of circle about 10° S. W. of zenith.
	35		
	36	Calm	Bright gleam of light darting in wavy pulsations from East to West, across zenith. North and South portions of the sky densely clouded.
	40	Calm	Brilliant belt or streamer darting upwards in pulsations, and progressing from East to West faster than the eye could follow, smaller streamers from North and South entering it at right angles throughout its whole course.
	45	Calm	Clouded all over, except a bright patch of light in East, emitting faint streamers.

d.	h.	m.	Wind.	
28 17 50			Clearing in E. and S.E.; faint streamers. Large meteor (above first magnitude) fell from alt. 60° to 40°; brilliant tail, time of flight about 2·; direction from zenith to S.E.
	55	Calm	Clouds rising in N.W., and totally obscuring the sky.
18	0	Calm	Densely clouded; auroral light almost entirely obscured.
	24		Sky almost clear; dark cumulous masses in N. and N.E. thrown into strong relief by auroral light, and pale flashes occurring at intervals; flashes and pulsations in N.W., extending upwards to zenith. ;
	30		Clear; bright streamers from W. to N.E.; faint pulsations reaching nearly to zenith.
	42		A few strati in the North; streamers; bright patch in N.N.W.; flashes to zenith.
	52		Overcast to within 10° or 15° of North horizon; bright light and a few very faint streamers still visible.
19 4			Overcast, cumuli and cumulous haze, except near horizon. No streamers or pulsations.
	26		Overcast, with broken cumulous masses; flashes visible in zenith only.
	35		Clear, except heavy bank of cumuli in North; brilliant and incessant flashes converging from all sides towards an oval space about 10° E. and W. of zenith.
	37		Sudden start of H. F. magnet.
	46		Nearly overcast; a few streamers to N.W., in clear intervals.
	51		Clouded, bright to Northward.
20 3			Bright light in breaks of clouds.
	38		Clear, bright flashes and pulsations.
	56		Light bank of clouds in N. and W. horizons, alt. 10° or 15°; remainder clear; faint auroral light and low streamers in N.W.; none in any other part of the sky.
21 31			Clouded along North horizon to alt. of 15° or 20°; very faint auroral light at edges only.

The general character of this Aurora, as far as could be judged by the casual observations above given, was the same as that of the 28th and 29th May; flashes or pulsations, extending to the zenith, formed in both cases the remarkable features. On both days, the magnets were considerably disturbed during the earlier part of the day.

AUGUST 28 and 29, 1840.

d.	h.	m.	s.	Decl. Sc.-Div.	Hor. Force Sc.-Div.	Ther.	Vert. Force Sc.-Div.	Ther.
28	13	40	0	148·3	155·3		54·4	
		45	0	147·6				
		50	0	146·7	149·9		53·1	
		55	0	147·0				
	14	0	0	149·1	146·8	73	54·5	73
		5	0	150·4				
		10	0	155·0	144·9		60·1	
		15	0	166·1				
		20	0	187·9	149·2		60·7	
		25	0	207·6				
		30	0	219·2	157·4		69·7	
		35	0	242·9				
		40	0	258·5	162·8		63·5	
		45	0	243·9				
		50	0	223·1	150·7		46·0	
		55	0	227·4				
	15	0	0	207·0	142·0	73	41·6	73
		5	0	199·6				
		10	0	202·8	131·6		36·7	
		15	0	211·5				
		20	0	205·4	148·2		36·3	
		25	0	203·0				
		30	0	199·6	157·6		34·6	
		35	0	175·0				
		40	0	163·9	141·1		31·0	
		45	0	169·3				
		50	0	183·4	140·0		38·2	
		55	0	180·7				
	16	0	0	183·8	143·7	73	36·3	73
		5	0	174·0				
		10	0	171·0	131·8		33·3	
		15	0	163·2				
		20	0	151·6	136·1		28·6	
		25	0	152·0				
		30	0	162·4	121·0		34·5	
		35	0	163·9				
		40	0	155·2	116·4		27·9	
		45	0	151·8				
		50	0	150·1	103·8		30·0	
		55	0	155·1				
	17	0	0	161·3	114·7	73	34·2	73
		5	0	163·5				
		10	0	152·0	84·9		38·3	
		15	0	133·6	59·8†			
		20	0	114·6	91·3		35·2	
		25	0	136·1	97·3†		57·7‡	
		30	0	154·0	119·8		59·1	
		31	50	135·4				
		35	0	172·6	140·5†		59·4‡	
		37	40	162·9	160·5†			
		40	0	181·2	222·8		49·3	
		42	40	187·9	180·5†		49·2‡	
		45	0	153·1			48·3‡	
		47	40	143·3	116·4†			
		50	0	171·6	134·8		35·8	
		52	40	193·1			44·9‡	
		55	0	180·5	124·3†		40·7‡	
		57	40	165·8				
	18	0	0	153·1	110·2	73	36·5	73
		2	40	140·2	100·4†			
		5	0	136·4			35·3‡	
		7	40	165·7	123·4†			
		10	0	184·1	124·3		45·6	
		12	40	190·7			46·6‡	

AUGUST 28 and 29, 1840.

d.	h.	m.	s.	Decl. Sc.-Div.	Hor. Force Sc.-Div.	Ther.	Vert. Force Sc.-Div.	Ther.
28	18	15	0	189·8				
		20	0	180·3	113·9		29·0	
		22	40	170·2				
		25	0	163·5	103·2		19·6	
		27	40	161·3				
		30	0	160·0	90·9			
		32	40	168·9				
		35	0	174·2	95·9		18·4	
		37	40	172·9				
		40	0	172·0	99·7			
		42	40	162·1			27·4	
		45	0	147·3	85·7		35·0	
		47	40	138·8				
		50	0	135·2	96·6		26·5	
		52	40	137·5				
		55	0	144·2	83·4		30·8	
		57	40	152·0				
	19	0	0	149·7	74·9†	73	25·5	73
		2	40	154·7				
		5	0	160·6	88·2†		31·1	
		7	40	167·3				
		10	0	170·6				
		12	40	171·8	92·1†			
		15	0	168·9	82·6		32·6	
		17	40	169·8				
		20	0	171·6	83·8			
		22	40	170·5			32·5	
		25	0	168·0			36·4	
		30	0	162·3	79·5		37·4	
		32	40	165·3				
		35	0	157·0	82·1		32·2	
		37	40	150·2				
		40	0	153·5	15·4		36·5	
		42	40	167·0				
		45	0	167·5	59·1		44·1	
		47	40	194·5			47·6	
		50	0	194·5	73·0			
		52	40	180·4				
		55	0	170·5	69·6		41·2	
		57	40	148·1				
	20	0	0	140·5	28·0	73	41·1	73
		2	40	147·9				
		5	0	143·4	42·2		42·1	
		7	40	137·8				
		10	0	145·1	31·0		34·3	
		12	40	131·5				
		15	0	128·9			31·1	
		17	40	137·0				
		20	0	132·5	09·0		32·9	
		22	40	130·7				
		25	0	140·8	00·0		33·6	
		27	40	147·4				
		30	0	137·4	00·4		34·3	
		32	40	154·3				
		35	0	149·3	06·9		37·0	
		37	40	155·7				
		40	0	157·6	08·2		36·8	
		42	40	159·1				
		45	0	157·0	14·9		36·5	
		47	40	163·4				
		50	0	161·4	20·7		38·6	
		52	40	160·4				
		55	0	164·4	27·8		39·7	
		57	40	157·9				

AUGUST 28 and 29, 1840.

d.	h.	m.	s.	Decl. Sc.-Div.	Hor. Force Sc.-Div.	Ther.	Vert. Force Sc.-Div.	Ther.
28	21	0	0	158·6	129·9	73	41·1	73
		2	40	154·8				
		5	0	154·7	125·8		41·5	
		7	40	154·6				
		10	0	153·9	127·4		43·4	
		15	0	154·9				
		20	0	157·2	132·8		47·9	
		30	0	159·1	135·8		49·5	
		35	0	158·7				
		40	0	159·5	138·1		50·5	
		45	0	158·9				
		55	0	156·9	135·0		51·3	
		55	0	155·8				
	22	0	0	160·2	128·4	73	48·9	73
		5	0	161·1				
		10	0	155·9	123·7		48·9	
		15	0	155·6				
		20	0	162·3	133·0		50·0	
		25	0	159·2				
		30	0	156·7	134·9		52·1	
		35	0	154·7				
		40	0	155·9	130·8		49·3	
		45	0	153·4				
		50	0	153·5	128·4		46·8	
		55	0	154·1				
	23	0	0	155·5	132·4	73	49·3	73
		5	0	157·7				
		10	0	156·5	132·9		47·8	
		15	0	159·4				
		20	0	160·6	134·1		48·5	
		25	0	165·7				
		30	0	164·6	137·0		48·9	
		35	0	158·9				
		40	0	159·3	131·7		46·8	
		45	0	162·4				
		50	0	162·6	135·8		48·3	
		55	0	166·0				
29	0	0	0	165·8	136·1	73	48·2	73
		5	0	167·9				
		10	0	167·0	168·6		47·2	
		15	0	168·6				
		20	0	167·5	136·7		50·9	
		25	0	168·4				
		30	0	169·3	136·1		54·0	
		35	0	171·7				
		40	0	167·7	137·9		53·3	
		45	0	166·6				
		50	0	168·1	138·6		53·0	
		55	0	169·8				
	1	0	0	177·0	141·3	73	53·4	73
		5	0	165·5				
		10	0	174·1	140·6		53·4	
		15	0	174·2				
		20	0	175·9	141·0		53·3	
		25	0	173·7				
		30	0	171·9	141·0		53·5	
		35	0	170·1				
		40	0	168·1	138·8		53·5	
		45	0	167·3				
		50	0	168·9	137·7			
		55	0	168·9				
	2	0	0	168·0	138·7	73	50·3	73
		5	0	165·3				
		10	0	166·3	145·3		50·7	

The H. F. observations marked † were taken respectively at the times 17^h. 19^m., 25^m. 50^s., 35^m. 30^s., 39^m., 30^s., 45^m. 30^s., 56^m., 18^h. 5^m., 9^m, 30^s., 19^h. 2^m. 51^s., 8^s., 13^m.; the others as usual, 2^m. 30^s. after the times specified.

The V. F. observations marked ‡ were taken at the times 17^h. 24^m. 30^s., 34^m., 40^m. 30^s., 44^m.,51^m., 56^m., 18^h. 3^m. 30^s., 11^m. ; and the observations from 18^h. 25^m. to 21^h. 10^m. at the times specified; all others as usual, 2^m. 30^s. before the times specified.

August 28 and 29, 1840.

M. Gött. Time.				Decl.	Hor. Force.		Vert. Force.	
d.	h.	m.	s.	Sc.-Div.	Sc.-Div.	Ther.	Sc.-Div.	Ther.
29	2	15	0	167·2				
		20	0	164·5			51·5	
		25	0	163·5				
		30	0	161·2	135·9		51·5	
		25	0	159·7				
		40	0	159·9	134·2		48·2	
		45	0	159·1				
		50	0	157·0	133·6		48·6	
		55	0	156·1				
	3	0	0	155·3	133·1	73	48·4	73
		5	0	154·1				
		10	0	153·1	132·5		48·3	
		15	0	152·8				
		20	0	151·2	133·2		48·1	
		25	0	150·9				
		30	0	151·3	133·6		47·8	
		35	0	149·7				
		40	0	150·2	134·6		47·7	
		45	0	148·3				
		50	0	145·7	134·7		48·0	
		55	0	145·1				
	4	0	0	145·0	135·4	72	47·9	72
		5	0	143·9				
		10	0	143·9	135·3		47·5	
		15	0	143·7				
		20	0	143·0	136·5		48·2	
		25	0	142·8				
		30	0	141·6	137·2		48·7	
		35	0	141·1				
		40	0	141·9	137·4		48·7	
		45	0	141·4				
		50	0	140·0	139·7		48·7	
		55	0	139·0				
	5	0	0	138·4	138·6	72	48·7	73
		5	0	138·6				
		10	0	138·8	139·3		48·5	
		15	0	139·2				
		20	0	137·9	138·9		49·2	
		25	0	137·0				
		30	0	138·0	141·2		50·4	
		35	0	138·0				
		40	0	137·8	141·7		50·7	
		45	0	137·5				
		50	0	137·0	142·9		50·6	
		55	0	137·0				
	6	0	0	136·5	143·4	73	50·6	73
		5	0	136·3				
		10	0	136·8	144·7		50·9	
		15	0	137·1				
		20	0	137·1	144·8		51·3	
		25	0	137·4				
		30	0	136·3	146·1		51·4	
		35	0	137·1				
		40	0	137·0	147·0		51·8	
		45	0	137·2				
		50	0	137·2	142·4		51·5	
		55	0	137·6				
	7	0	0	137·2	142·1	73	51·6	73
		5	0	137·5				
		10	0	138·0	148·0		51·3	
		15	0	138·1				
		20	0	138·3	149·4		51·3	
		25	0	139·0				
		30	0	138·8	150·8		51·2	
		35	0	138·8				
		40	0	139·2	150·2		51·2	
		45	0	138·9				

August 28 and 29, 1840.

M. Gött. Time.				Decl.	Hor. Force.		Vert. Force.	
d.	h.	m.	s.	Sc.-Div.	Sc.-Div.	Ther.	Sc.-Div.	Ther.
29	7	50	0	139·0	150·7		51·2	
		55	0	139·1				
	8	0	0	139·5	150·4	73	51·2	73
		5	0	140·0				
		10	0	140·3	150·1		51·4	
		15	0	140·9				
		20	0	141·0	150·6		51·2	
		25	0	141·7				
		30	0	141·6	159·5		53·7	
		35	0	144·0				
		40	0	141·6	162·4		53·5	
		45	0	140·5				
		50	0	141·8	157·9		52·5	
		55	0	141·8				
	9	0	0	142·0	155·3	73	52·1	73
		5	0	142·2				
		10	0	142·0	161·4		52·0	
		15	0	142·2				
		20	0	143·7	149·1		51·9	
		25	0	143·7				
		30	0	143·9	147·7		50·7	
		35	0	143·8				
		40	0	144·0	153·2		51·1	
		45	0	143·9				
		50	0	143·7	150·7		50·9	
		55	0	143·9				
	10	0	0	144·0	151·1	73	50·8	73
	12	0	0	147·0	143·2	73	50·0	73
	14	0	0	145·8	147·0	72	50·5	72
	16	0	0	144·7	147·0	72	50·5	72

The Mean Positions at the usual hours of observation, are given in page 13.

St. Helena { Decl. 1 Scale Division = 0'·71 ; H. F. k = ·00019 ; q = ·0003

Regular and term observations.

The H. F. was observed at 2m. 30s. after the times specified.

M. Gött. Time.				Decl.	Hor. Force.		Vert. Force.	
d.	h.	m.	s.	Sc.-Div.	Sc.-Div.	Ther.	Sc.-Div.	Ther.
28	0	0	0	67·0	222·4	61		
	2	0	0	67·1	222·0	62		
	4	0	0	67·0	216·0	62		
	6	0	0	66·1	214·6	61		
	8	0	0	66·1	212·0	60		
	10	0	0	67·1	207·5	60		
		5	0	66·8				
		10	0	66·6	207·7			
		15	0	66·3				
		20	0	66·6	210·5			
		25	0	66·8				
		30	0	66·6	213·2			
		35	0	66·7				
		40	0	66·7	214·3			
		45	0	66·1				
		50	0	66·7	213·4			
		55	0	66·8				
	11	0	0	67·0	211·4			
		5	0	67·2				
		10	0	67·3	211·4			
		15	0	67·3				
		20	0	67·3	209·1			
		25	0	67·1				
		30	0	66·9	209·1			
		35	0	66·9				
		40	0	67·2	209·2			
		45	0	67·7				

August 28 and 29, 1840.

M. Gött. Time.				Decl.	Hor. Force.		Vert. Force.	
d.	h.	m.	s.	Sc.-Div.	Sc.-Div.	Ther.	Sc.-Div.	Ther.
28	11	50	0	67·2	210·3			
		55	0	67·9				
	12	0	0	66·4	208·2	60		
		5	0	66·4				
		10	0	66·5	207·2			
		15	0	66·3				
		20	0	66·4	207·5			
		25	0	66·4				
		30	0	67·0	208·5			
		35	0	67·4				
		40	0	67·2	209·5			
		45	0	67·5				
		50	0	67·5	209·4			
		55	0	67·4				
	13	0	0	67·5	210·2			
		5	0	67·5				
		10	0	67·6	212·2			
		15	0	67·9				
		20	0	67·6	213·2			
		25	0	67·1				
		30	0	68·1	213·0			
		35	0	68·2				
		40	0	68·0	213·6			
		45	0	67·2				
		50	0	66·2	213·1			
		55	0	65·8				
	14	0	0	65·8	212·4	60		
		5	0	65·6				
		10	0	65·5	212·0			
		15	0	65·6				
		20	0	66·9	218·2			
		25	0	66·9				
		30	0	66·9	220·2			
		35	0	66·9				
		40	0	67·9	223·8			
		45	0	67·9				
		50	0	67·5	222·7			
		55	0	67·9				
	15	0	0	68·1	222·0			
		5	0	67·4				
		10	0	67·3	221·3			
		15	0	66·3				
		20	0	66·2	219·7			
		25	0	66·1				
		30	0	66·1	218·9			
		35	0	66·2				
		40	0	65·6	216·2			
		45	0	65·0				
		50	0	65·3	212·5			
		55	0	65·2				
	16	0	0	65·5	211·2	59		
		5	0	65·0				
		10	0	65·3	212·9			
		15	0	65·2				
		20	0	64·0	209·5			
		25	0	64·0				
		30	0	64·2	209·0			
		35	0	65·1				
		40	0	64·8	207·5			
		45	0	65·0				
		50	0	65·2	206·9			
		55	0	65·4				
	17	0	0	65·2	206·5			
		5	0	65·8				
		10	0	65·9	204·0			
		15	0	66·8				
		20	0	67·0	201·9			

August 28 and 29, 1840.

M. Gött. Time d. h. m. s.	Decl. Sc.-Div.	Hor. Force. Sc.-Div.	Ther. °	Vert. Force. Sc.-Div.	Ther. °
28 17 25 0	67·0				
30 0	67·9	204·1			
35 0	66·0				
40 0	69·1	207·8			
45 0	69·5				
50 0	69·4	207·9			
55 0	69·0				
18 0 0	68·0	209·0	59		
5 0	68·0				
10 0	68·0	208·9			
15 0	68·1				
20 0	68·2	208·8			
25 0	68·5				
30 0	68·5	209·2			
35 0	68·9				
40 0	68·4	208·9			
45 0	69·0				
50 0	69·1	208·7			
55 0	70·0				
19 0 0	71·2	208·8			
5 0	72·5				
10 0	73·3	207·0			
15 0	73·8				
20 0	74·1	205·0			
25 0	74·4				
30 0	75·4	201·3			
45 0	75·7				
50 0	76·0	201·5			
55 0	75·1				
20 0 0	76·9	199·0	59		
5 0	77·1				
10 0	67·5	218·7			
15 0	76·9				
20 0	76·5	200·0			
25 0	75·5				
30 0	74·2	201·3			
35 0	73·0				
40 0	72·5	202·9			
45 0	72·2				
50 0	71·9	202·8			
55 0	71·9				
21 0 0	71·4	201·8			
5 0	70·9				
10 0	70·6	200·9			
15 0	70·2				
20 0	69·9	201·0			
25 0	69·2				
30 0	69·0	201·1			
35 0	69·1				
40 0	67·8	201·6			
45 0	68 4				
50 0	68·2	202·1			
55 0	67·0				
22 0 0	67·2	204·0	58		
5 0	67·2				
10 0	67·8	205·4			
15 0	67·9				
20 0	67·2	206·0			
25 0	66·3				
30 0	66·5	206·5			
35 0	66·1				
40 0	66·1	206·8			
45 0	65·6				
50 0	65·2	207·0			
55 0	65·1				
23 0 0	65·4	207·9			
5 0	65·0				

August 28 and 29, 1840.

M. Gött. Time d. h. m. s.	Decl. Sc.-Div.	Hor. Force. Sc.-Div.	Ther. °	Vert. Force. Sc.-Div.	Ther. °
28 23 10 0	65·0	209·0			
15 0	64·8				
20 0	64·8	211·0			
25 0	64·2				
30 0	63·0	208·5			
35 0	62·2				
40 0	62·2	209·2			
45 0	62·1				
50 0	62·0	209·0			
55 0	62·0				
29 0 0 0	62·0	209·0	60		
5 0	62·0				
10 0	61·9	210·0			
15 0	61·5				
20 0	61·2	210·9			
25 0	61·1				
30 0	61·0	210·1			
35 0	60·9				
40 0	60·9	210·9			
45 0	60·9				
50 0	60·5	210·9			
55 0	60·8				
1 5 0	60·9				
10 0	60·9	212·0			
15 0	60·9				
20 0	61·0	212·0			
25 0	60·5				
30 0	60·4	210·8			
35 0	60·2				
40 0	60·0	211·0			
45 0	60·0				
50 0	60·0	211·0			
55 0	60·1				
2 0 0	60·1	210·8	62		
5 0	60·1				
10 0	60·2	211·2			
15 0	60·3				
20 0	60·3	211·0			
25 0	60·3				
30 0	60·3	210·0			
35 0	60·3				
40 0	60·4	209·5			
45 0	60·5				
50 0	60·4	208·0			
55 0	60·3				
3 0 0	60·5	207·8			
5 0	60·5				
10 0	60·3	207·0			
15 0	60·6				
20 0	60·5	207·0			
25 0	60·9				
30 0	60·9	207·8			
35 0	60·9				
40 0	61·0	207·5			
45 0	61·0				
50 0	61·0	207·0			
55 0	61·1				
4 0 0	61·5	207·7	63		
5 0	61·8				
10 0	61·9	206·2			
15 0	62·0				
20 0	62·2	206·2			
25 0	62·6				
30 0	62·9	206·3			
35 0	62·9				
40 0	63·0	206·8			
45 0	63·0				

August 28 and 29, 1840.

M. Gött. Time d. h. m. s.	Decl. Sc.-Div.	Hor. Force. Sc.-Div.	Ther. °	Vert. Force. Sc.-Div.	Ther. °
29 4 50 0	62·9	206·0			
55 0	62·9				
5 0 0	62·9	206·2			
5 0	62·9				
10 0	62·9	206·5			
15 0	62·9				
20 0	62·9	206·2			
25 0	62·9				
30 0	62·9	206·5			
35 0	63·0				
40 0	63·0	206·3			
45 0	63·0				
50 0	63·0	206·0			
55 0	63·0				
6 0 0	63·0	206·0	62		
5 0	63·1				
10 0	63·1	206·2			
15 0	63·2				
20 0	63·8	206·2			
25 0	63·8				
30 0	64·0	206·2			
35 0	64·4				
40 0	64·8	206·7			
45 0	65·0				
50 0	65·2	207·0			
55 0	65·3				
7 0 0	65·6	206·9			
5 0	65·7				
10 0	65·8	206·9			
15 0	65·8				
20 0	65·8	207·0			
25 0	65·7				
30 0	65·7	207·0			
35 0	65·7				
40 0	65·9	207·0			
45 0	65·9				
50 0	65·9	206·9			
55 0	65·8				
8 0 0	65·9	207·4	61		
5 0	65·8				
10 0	63·9	206·9			
15 0	65·7				
20 0	65·5	207·1			
25 0	65·5				
30 0	65·6	209·1			
35 0	66·4				
40 0	67·4	211·0			
45 0	66·5				
50 0	66·5	211·6			
55 0	66·6				
9 0 0	66·5	211·0			
5 0	66·5				
10 0	66·4	210·4			
15 0	66·0				
20 0	63·9	209·6			
25 0	65·8				
30 0	66·0	210·0			
35 0	66·0				
40 0	65·9	210·0			
45 0	65·9				
50 0	65·9	210·0			
55 0	66·1				
10 0 0	66·0	209·8	60		
12 0 0	66·1	207·4	60		

The magnetometers were not in adjustment during a sufficient number of consecutive days in this month to give mean positions.

AUGUST 28 and 29, 1840.

ANTARCTIC EXPEDITION AT HOBARTON.

Decl. 1 Scale Division = 0'·73

H. F. k = ·000176; q =

V. F. k = ; q =

The three Instruments were observed simultaneously.

M. Gött. Time.	Decl.	Hor. Force.		Vert. Force.	
d. h. m. s.	Sc.-Div^s.	Sc.-Div^s.	Ther.	Sc.-Div^s.	Ther.
28 0 0 0	45·5	67·3	49	46·7	50
1 0 0	45·9	66·4		46·0	
2 0 0	46·4	71·7	49	46·5	51
3 0 0	44·1	68·1		47·1	
4 0 0	46·2	67·5	48	50·9	48
5 0 0	45·2	68·3		53·7	
6 0 0	45·2	69·4	47	51·8	48
7 0 0	45·0	69·8		51·6	
8 0 0	49·7	73·8	47	51·8	48
9 0 0	47·1	72·0		51·8	
10 0 0	47·4	67·6	47	52·5	48
2 30	47·5	67·5		50·9	
5 0	41·9	67·7		50·2	
7 30	47·5	67·7		54·3	
10 0	47·5	66·7		54·0	
12 30	46·7	66·4		48·1	
15 0	46·3	67·5		46·9	
17 30	47·1	67·8		46·0	
20 0	47·3	67·8		47·5	
22 30	46·7	66·7		44·4	
25 0	45·4	65·1		43·3	
27 30	46·8	66·6		42·6	
30 0	46·6	65·6		42·1	
32 30	45·5	64·7		42·6	
35 0	44·5	65·4		41·4	
37 30	44·9	65·8		40·6	
40 0	43·1	64·6		39·4	
42 30	45·7	64·1		39·2	
45 0	43·9	65·1		38·5	
47 30	44·1	65·0		38·3	
50 0	44·5	64·8		38·6	
52 30	44·5	64·4		37·9	
55 0	44·6	65·9		37·9	
57 30	44·5	65·9		37·4	
11 0 0	44·6	66·2		38·0	
2 30	44·6	66·4		37·0	
5 0	44·5	66·4		36·7	
7 30	44·9	66·7		35·1	
10 0	44·6	66·4		34·7	
12 30	45·2	65·4	51	34·5	57
15 0	42·9	64·4		34·6	
17 30	45·8	67·6		34·6	
20 0	44·8	65·6		34·3	
22 30	46·3	67·4		34·0	
25 0	46·3	68·7		33·7	
27 30	46·5	67·9		33·4	
30 0	47·0	66·9		33·4	
32 30	47·3	66·5		32·7	
35 0	47·6	66·1		33·1	
37 30	44·8	65·9		32·8	
40 0	45·9	65·8		32·2	
42 30	43·7	65·6		32·1	
45 0	45·7	66·1		31·8	
47 30	44·7	63·1		31·8	
50 0	45·1	65·1		31·7	
52 30	46·9	66·1		32·5	
55 0	47·5	67·1		31·9	
57 30	48·2	67·4		31·8	
12 0 0	48·5	68·4	50	31·3	50
2 30	48·4	67·8		31·5	

AUGUST 28 and 29, 1840.

M. Gött. Time.	Decl.	Hor. Force.		Vert. Force.	
d. h. m. s.	Sc.-Div^s.	Sc.-Div^s.	Ther.	Sc.-Div^s.	Ther.
28 12 5 0	47·7	67·4		31·4	
7 30	47·6	67·0		30·7	
10 0	46·9	67·9		31·2	
12 30	47·2	67·3		30·9	
15 0	46·9	68·6		30·5	
17 30	46·5	67·6		30·3	
20 0	45·8	66·7		29·6	
22 30	45·3	66·8		29·3	
25 0	44·0	66·6		29·2	
27 30	44·3	66·8		29·4	
30 0	43·6	66·8		29·8	
32 30	43·2	66·4		30·7	
35 0	42·9	66·6		29·9	
37 30	43·3	67·9		28·8	
40 0	43·2	66·5		29·5	
42 30	42·9	66·5		29·0	
45 0	42·0	66·5		29·1	
47 30	42·6	66·6		28·9	
50 0	42·1	66·9		29·1	
52 30	41·9	65·9		28·4	
55 0	43·3	67·2		28·7	
57 30	42·8	66·6		28·7	
13 0 0	42·3	65·9	50	27·5	60
2 30	42·2	66·8		27·7	
5 0	41·9	66·6		27·6	
7 30	42·3	66·9		27·2	
10 0	41·4	67·0		26·5	
12 30	40·9	66·1		25·8	
15 0	47·2	69·1		29·7	
17 30	43·4	66·7		28·5	
20 0	43·8	66·7		29·5	
22 30	45·8	66·9		30·6	
25 0	45·7	67·8		31·4	
27 30	42·7	66·5		30·4	
30 0	40·2	64·8		27·4	
32 30	41·0	66·0		28·7	
35 0	41·6	67·4		29·8	
37 30	43·1	68·3		29·7	
40 0	45·3	67·1		29·0	
42 30	46·5	66·1		29·5	
45 0	47·0	64·9		29·8	
47 30	47·9	64·5		30·8	
50 0	47·8	64·1		31·2	
52 30	47·8	63·3		31·7	
55 0	47·1	62·8		31·5	
57 30	47·2	62·1		32·1	
14 0 0	46·5	60·7	51	32·7	50
2 30	46·8	59·6		33·0	
5 0	46·5	58·7		34·3	
7 30	45·8	57·6		34·5	
10 0	46·6	59·1		35·8	
12 30	45·9	56·8		34·0	
15 0	44·8	54·3		35·1	
17 30	43·7	52·1		35·1	
20 0	45·4	51·6		36·7	
22 30	46·2	51·1		37·3	
25 0	46·5	51·6		39·5	
27 30	47·2	50·3		41·8	
30 0	47·5	50·2		42·3	
32 30	47·0	50·1		42·9	
35 0	47·5	47·3		43·1	
37 30	47·1	45·5		43·8	
40 0	47·4	43·9		44·0	
42 30	47·3	43·2		45·1	
45 0	47·0	41·3		45·7	
47 30	47·0	39·9		46·8	
50 0	47·8	36·6		48·7	

AUGUST 28 and 29, 1840.

M. Gött. Time.	Decl.	Hor. Force.		Vert. Force.	
d. h. m. s.	Sc.-Div^s.	Sc.-Div^s.	Ther.	Sc.-Div^s.	Ther.
28 14 52 30	49·3	37·1		53·5	
55 0	48·7	38·7		51·5	
57 30	48·8	38·8		51·8	
15 0 0	49·4	38·0	51	52·0	61
2 30	49·9	37·3		52·7	
5 0	49·2	38·1		53·0	
7 30	49·1	37·3		53·2	
10 0	48·6	35·2		51·7	
12 30	49·4	31·5		51·4	
15 0	48·7	31·0		51·5	
17 30	48·9	31·9		52·3	
20 0	49·4	31·1		52·3	
22 30	49·7	31·5		52·4	
25 0	49·9	32·7		52·1	
27 30	49·2	33·1		51·1	
30 0	50·1	34·4		51·1	
32 30	50·5	33·3		52·5	
35 0	50·5	33·4		49·9	
37 30	50·7	32·3		50·0	
40 0	50·2	28·9		49·4	
42 30	50·2	29·9		50·0	
45 0	50·2	31·1		50·3	
47 30	50·3	34·6		50·1	
50 0	50·8	36·2		51·5	
52 30	49·6	34·1		48·0	
55 0	51·5	34·7		47·7	
57 30	50·2	34·1		46·2	
16 0 0	50·1	34·1	54	46·0	64
2 30	51·9	34·4		45·8	
5 0	52·2	36·1		45·6	
7 30	53·3	36·9		44·1	
10 0	54·1	39·7		43·8	
12 30	56·4	38·4		42·9	
15 0	57·7	35·4		42·9	
17 30	57·5	33·2		41·6	
20 0	56·2	33·7		41·6	
22 30	57·4	36·1		42·0	
25 0	59·3	39·2		42·3	
27 30	59·9	39·9		42·0	
30 0	62·3	40·8		42·0	
32 30	62·1	41·5		41·6	
35 0	61·8	39·8		40·0	
37 30	61·2	40·3		40·0	
40 0	61·4	40·3		39·8	
42 30	62·9	41·1		39·8	
45 0	63·5	40·7		39·6	
47 30	62·9	38·2		39·4	
50 0	63·3	38·2		39·5	
52 30	63·2	37·3		39·8	
55 0	62·9	37·3		40·4	
57 30	61·3	36·1		40·1	
17 0 0	59·6	37·4	57	39·9	65
2 30	60·5	38·9		40·4	
5 0	62·1	39·9		40·1	
7 30	62·5	40·4		40·1	
10 0	63·3	40·8		40·3	
12 30	64·2	42·5		40·3	
15 0	65·9	45·4		40·3	
17 30	65·9	44·4		38·9	
20 0	65·8	44·7		37·9	
22 30	67·1	45·4		38·4	
25 0	67·1	47·2		38·0	
27 30	68·8	45·4		37·7	
30 0	68·1	42·5		37·9	
32 30	66·7	38·2		38·5	
35 0	64·8	41·5		38·6	
37 30	72·3	43·1		43·3	

D

AUGUST 28 and 29, 1840.

d.	h.	m.	s.	Decl. Sc.-Div^ns	Hor. Force Sc.-Div^ns	Hor. Ther.	Vert. Force Sc.-Div^ns	Vert. Ther.
28	17	40	0	68·4	36·4		42·1	
		42	30	64·7	36·5		41·8	
		45	0	65·4	37·4		42·4	
		47	30	65·6	36·2		44·1	
		50	0	64·1	37·6		44·1	
		52	30	65·1	32·1		46·2	
		55	0	61·5	26·3		46·8	
		57	30	55·9	26·4		47·3	
18	0	0	0	54·7	27·8	58	48·3	66
		2	30	53·4	29·1		49·1	
		5	0	50·8	32·3		48·3	
		7	30	51·1	36·5		47·7	
		10	0	52·3	38·6		47·5	
		12	30	53·2	40·1		46·4	
		15	0	56·3	44·4		45·9	
		17	30	69·2	45·9		45·7	
		20	0	61·5	45·1		44·7	
		22	30	63·2	42·5		43·8	
		25	0	60·7	41·0		43·3	
		27	30	59·0	40·9		43·1	
		30	0	59·6	44·4		44·1	
		32	30	61·5	46·9		44·1	
		35	0	62·9	46·1		44·8	
		37	30	60·9	44·8		44·0	
		40	0	60·3	46·6		43·8	
		42	30	63·2	48·7		44·9	
		45	0	66·3	49·1		45·1	
		47	30	68·1	47·2		45·7	
		50	0	68·9	44·1		46·7	
		52	30	64·1	41·2		46·4	
		55	0	61·2	42·0		46·4	
		57	30	60·9	43·3		47·0	
19	0	0	0	67·8	45·4	57	48·0	63
		2	30	63·6	46·2		48·3	
		5	0	62·9	46·2		47·9	
		7	30	63·1	47·6		47·9	
		10	0	65·4	47·3		48·5	
		12	30	66·5	47·3		49·2	
		15	0	67·9	45·5		54·1	
		17	30	66·5	43·2		52·5	
		20	0	65·1	42·0		52·9	
		22	30	62·8	41·1		53·1	
		25	0	61·5	39·6		52·9	
		27	30	56·7	38·9		54·2	
		30	0	54·6	41·2		53·0	
		32	30	55·6	43·7		52·9	
		35	0	56·8	45·6		53·2	
		37	30	57·5	44·7		52·5	
		40	0	57·6	45·2		53·0	
		42	30	55·9	44·8		52·5	
		45	0	59·5	45·6		52·3	
		47	30	55·3	50·3		52·3	
		50	0	66·0	53·6		53·0	
		52	30	61·2	52·2		52·9	
		55	0	66·0	53·6		53·3	
		57	30	71·5	42·0		52·6	
20	0	0	0	67·3	37·7	56	59·4	61
		2	30	65·8	35·2		57·0	
		5	0	60·1	32·1		58·0	
		7	30	54·8	34·2		58·7	
		10	0	54·3	36·9		57·8	
		12	30	56·1	38·0		58·6	
		15	0	55·0	35·5		58·9	
		17	30	50·3	35·5		58·1	
		20	0	50·0	37·3		58·0	
		22	30	51·7	39·2		58·4	
		25	0	54·0	40·2		58·7	

AUGUST 28 and 29, 1840.

d.	h.	m.	s.	Decl. Sc.-Div^ns	Hor. Force Sc.-Div^ns	Hor. Ther.	Vert. Force Sc.-Div^ns	Vert. Ther.
28	20	27	30	53·7	40·1		58·0	
		30	0	50·3	41·0		58·5	
		32	30	51·4	43·5		58·3	
		35	0	55·4	44·8		58·5	
		37	30	56·3	44·5		58·4	
		40	0	57·0	44·3		58·4	
		42	30	58·1	44·4		57·5	
		45	0	58·7	43·8		55·2	
		47	30	58·1	43·4		59·2	
		50	0	58·2	44·2		59·2	
		52	30	59·1	44·0		59·0	
		55	0	56·8	43·7		58·1	
		57	30	57·8	45·3		53·0	
21	0	0	0	57·6	44·7	53	51·9	60
		2	30	56·7	44·7		53·2	
		5	0	57·4	44·8		52·9	
		7	30	55·6	44·4		54·2	
		10	0	54·4	45·2		49·0	
		12	30	54·1	46·6		46·6	
		15	0	54·6	46·4		48·2	
		17	30	55·0	47·7		47·7	
		20	0	55·0	48·1		47·0	
		22	30	55·1	48·2		46·7	
		25	0	55·5	47·6		46·5	
		27	30	54·9	46·9		46·3	
		30	0	54·1	46·7		45·7	
		32	30	53·3	46·8		45·5	
		35	0	53·6	47·3		45·8	
		37	30	53·2	47·6		45·1	
		40	0	52·4	47·2		44·4	
		42	30	52·0	47·6		43·8	
		45	0	52·1	47·4		43·4	
		47	30	51·9	47·8		43·1	
		50	0	51·9	48·3		42·5	
		52	30	51·8	48·7		42·3	
		55	0	51·4	48·5		40·9	
		57	30	51·3	48·4		40·6	
22	0	0	0	51·1	48·4	52	40·8	59
		2	30	50·3	48·1		41·3	
		5	0	49·6	47·7		41·0	
		7	30	50·4	47·7		40·4	
		10	0	50·8	47·1		40·8	
		12	30	50·6	46·4		41·5	
		15	0	50·3	45·4		42·7	
		17	30	49·4	45·4		43·1	
		20	0	49·2	45·6		43·4	
		22	30	48·8	44·1		43·7	
		25	0	47·8	43·3		43·4	
		27	30	46·6	42·3		43·2	
		30	0	44·5	41·8		44·2	
		32	30	42·8	42·7		44·3	
		35	0	42·5	43·6		44·0	
		37	30	41·3	43·9		44·1	
		40	0	38·8	44·8		42·6	
		42	30	38·7	46·4		42·3	
		45	0	39·3	47·7		42·1	
		47	30	39·9	49·8		41·0	
		50	0	40·7	50·0		40·9	
		52	30	40·7	51·4		39·8	
		55	0	40·7	52·7		39·1	
		57	30	44·3	52·8		39·9	
23	0	0	0	45·4	52·4	51	40·0	59
		2	30	46·6	52·5		40·0	
		5	0	48·5	51·8		40·7	
		7	30	48·8	50·4		41·3	
		10	0	47·8	49·6		41·3	
		12	30	46·8	49·4		41·2	

AUGUST 28 and 29, 1840.

d.	h.	m.	s.	Decl. Sc.-Div^ns	Hor. Force Sc.-Div^ns	Hor. Ther.	Vert. Force Sc.-Div^ns	Vert. Ther.
28	23	15	0	46·2	50·6		41·3	
		17	30	46·0	52·1		41·5	
		20	0	46·5	52·2		41·7	
		22	30	46·6	52·3		41·7	
		25	0	48·4	50·9		41·7	
		27	30	46·8	48·4		41·4	
		30	0	45·4	47·8		41·5	
		32	30	43·6	48·3		41·7	
		35	0	42·8	49·3		42·0	
		37	30	42·6	50·3		42·3	
		40	0	43·6	51·1		42·4	
		42	30	44·2	51·2		42·3	
		45	0	45·1	51·3		42·4	
		47	30	45·3	50·7		41·9	
		50	0	45·2	50·1		41·9	
		52 30	0	45·4	49·8		42·0	
		55	0	45·2	49·0		41·8	
		57	30	44·9	48·5		41·7	
29	0	0	0	43·9	48·5	51	41·9	57
		2	30	43·6	48·6		42·0	
		5	0	46·3	49·1		42·5	
		7	30	47·4	49·3		43·9	
		10	0	48·0	49·1		44·5	
		12	30	48·2	48·9		45·5	
		15	0	48·5	49·3		45·5	
		17	30	48·8	49·9		45·5	
		20	0	49·1	50·8		45·7	
		22	30	48·8	51·4		45·9	
		25	0	48·8	51·8		45·4	
		27	30	48·1	51·8		44·2	
		30	0	48·1	52·3		44·0	
		32	30	47·8	52·4		43·9	
		35	0	47·8	51·4		44·2	
		37	30	47·1	50·5		44·1	
		40	0	45·9	49·8		43·8	
		42	30	45·9	49·9		43·8	
		45	0	46·2	49·7		44·5	
		47	30	45·9	50·1		44·5	
		50	0	45·5	50·2		44·4	
		52	30	46·5	51·1		44·4	
		55	0	47·4	51·3		44·3	
		57	30	47·3	52·3		44·4	
	1	0	0	46·8	52·8	50	44·2	55
		2	30	46·9	53·9		44·2	
		5	0	47·6	54·5		44·0	
		7	30	48·5	54·4		43·7	
		10	0	48·5	54·7		43·3	
		12	30	48·5	54·8		43·2	
		15	0	48·5	54·8		43·8	
		17	30	48·7	55·2		42·1	
		20	0	48·6	54·7		43·0	
		22	30	48·7	54·3		42·6	
		25	0	48·6	54·5		43·1	
		27	30	48·9	54·2		42·1	
		30	0	49·1	54·5		42·1	
		32	30	49·6	55·1		42·5	
		35	0	49·7	55·3		42·4	
		37	30	50·1	55·5		42·6	
		40	0	50·0	55·7		41·4	
		42	30	50·1	56·1		41·1	
		45	0	50·0	55·5		41·2	
		47	30	50·1	55·8		40·7	
		50	0	50·0	56·4		40·4	
		52	30	50·1	56·3		40·4	
		55	0	50·1	56·4		40·4	
		57	30	50·1	55·4		40·5	
	2	0	0	50·1	55·4	50	40·7	55

August 28 and 29, 1840.

M. Gött. Time. d.	h.	m.	s.	Decl. Sc.-Divns.	Hor. Force. Sc.-Divns.	Hor. Force. Ther. °	Vert. Force. Sc.-Divns.	Vert. Force. Ther. °
29	2	2	30	50·1	55·8		40·4	
		5	0	49·6	56·5		40·3	
		7	30	49·6	57·3		40·6	
		10	0	49·9	57·1		40·8	
		12	30	50·1	56·7		41·4	
		15	0	50·1	57·1		41·1	
		17	30	50·2	56·7		41·6	
		20	0	50·0	56·6		41·4	
		22	30	49·6	56·6		42·2	
		25	0	49·8	56·7		42·2	
		27	30	50·1	56·9		42·2	
		30	0	50·2	56·4		42·2	
		32	30	50·1	56·8		42·5	
		35	0	50·1	57·0		42·3	
		37	30	50·5	57·2		42·3	
		40	0	50·5	56·8		42·1	
		42	30	50·6	56·7		42·3	
		45	0	50·6	56·1		42·4	
		47	30	50·6	56·2		42·4	
		50	0	50·5	56·8		42·3	
		52	30	50·6	56·9		42·2	
		55	0	50·4	56·8		42·2	
		57	30	50·6	57·3		43·0	
	3	0	0	50·6	57·5	50	42·7	54
		2	30	50·7	57·4		43·0	
		5	0	50·6	57·3		43·7	
		7	30	50·6	57·1		43·7	
		10	0	50·6	57·3		43·3	
		12	30	50·6	57·6		43·3	
		15	0	50·7	57·5		43·1	
		17	30	50·8	57·7		43·7	
		20	0	51·3	58·1		43·8	
		22	30	51·7	58·1		43·7	
		25	0	51·5	58·0		43·5	
		27	30	51·5	58·2		43·5	
		30	0	51·6	58·8		43·4	
		32	30	51·7	59·1		43·6	
		35	0	51·7	59·1		43·6	
		37	30	51·6	59·0		44·1	
		40	0	51·7	58·8		43·9	
		42	30	51·6	58·5		43·8	
		45	0	51·5	58·6		43·7	
		47	30	51·4	58·7		43·6	
		50	0	51·3	58·7		43·2	
		52	30	51·4	58·8		43·4	
		55	0	51·3	59·1		43·4	
		57	30	51·6	59·3		43·1	
	4	0	0	51·0	59·3	50	43·1	54
		2	30	51·6	59·2		43·2	
		5	0	51·9	59·1		43·6	
		7	30	52·3	59·1		43·9	
		10	0	52·4	59·1		43·9	
		12	30	52·4	58·8		43·9	
		15	0	52·1	59·2		44·1	
		17	30	52·1	59·0		44·4	
		20	0	52·1	59·2		43·5	
		22	30	52·1	59·4		43·3	
		25	0	52·1	59·5		43·2	
		27	30	52·1	59·4		43·7	
		30	0	52·4	59·5		44·1	
		32	30	52·5	59·6		44·2	
		35	0	52·6	59·5		43·9	
		37	30	52·6	59·6		43·9	
		40	0	52·6	59·6		44·0	
		42	30	52·6	59·3		43·9	
		45	0	52·5	59·3		43·4	
		47	30	52·5	59·5		43·4	

August 28 and 29, 1840.

M. Gött. Time. d.	h.	m.	s.	Decl. Sc.-Divns.	Hor. Force. Sc.-Divns.	Hor. Force. Ther. °	Vert. Force. Sc.-Divns.	Vert. Force. Ther. °
29	4	50	0	52·6	59·5		42·9	
		52	30	52·6	59·5		43·0	
		55	0	52·6	59·5		43·3	
		57	30	52·5	59·3		44·4	
	5	0	0	52·6	59·4	50	44·2	54
		2	30	52·5	59·5		44·0	
		5	0	52·5	59·6		44·3	
		7	30	52·9	59·7		44·3	
		10	0	52·8	59·6		44·2	
		12	30	52·8	59·8		44·2	
		15	0	52·9	59·8		44·3	
		17	30	53·0	59·9		44·1	
		20	0	53·1	60·0		44·2	
		22	30	53·0	59·6		44·0	
		25	0	53·0	59·5		43·7	
		27	30	53·0	59·6		43·7	
		30	0	53·0	59·5		43·8	
		32	30	53·0	59·5		43·9	
		35	0	53·1	59·6		43·9	
		37	30	53·0	59·6		44·2	
		40	0	53·1	59·6		44·0	
		42	30	53·0	59·6		43·8	
		45	0	53·0	59·6		43·8	
		47	30	53·1	59·5		44·0	
		50	0	53·1	59·6		44·1	
		52	30	53·1	59·6		44·4	
		55	0	53·1	59·6		44·4	
		57	30	53·1	59·6		44·8	
	6	0	0	53·1	59·7	50	44·9	54
		2	30	53·1	59·8		44·9	
		5	0	53·1	59·8		45·1	
		7	30	53·1	59·9		45·1	
		10	0	53·2	60·0		45·1	
		12	30	53·7	60·2		45·1	
		15	0	53·5	60·1		45·2	
		17	30	53·5	60·2		45·2	
		20	0	53·5	60·3		45·2	
		22	30	53·5	60·3		44·9	
		25	0	53·9	60·4		44·9	
		27	30	53·5	60·3		44·6	
		30	0	53·1	60·2		44·6	
		32	30	52·9	60·3		44·6	
		35	0	52·5	60·3		44·6	
		37	30	52·6	60·7		44·7	
		40	0	52·9	60·9		44·9	
		42	30	52·8	60·5		44·6	
		45	0	52·6	60·9		44·6	
		47	30	52·6	60·9		44·8	
		50	0	52·5	60·9		44·8	
		52	30	52·5	60·9		44·9	
		55	0	52·5	61·1		44·8	
		57	30	52·5	60·1		45·2	
	7	0	0	52·4	60·8	50	45·2	53
		2	30	52·1	60·6		45·1	
		5	0	52·1	60·5		45·3	
		7	20	52·3	60·3		45·1	
		10	0	52·4	60·3		44·8	
		12	30	52·2	60·5		44·9	
		15	0	52·3	60·8		45·1	
		17	30	52·5	60·8		45·1	
		20	0	52·6	60·9		45·1	
		22	30	52·6	60·9		45·4	
		25	0	52·5	61·1		45·7	
		27	30	52·4	61·2		45·7	
		30	0	52·4	61·3		45·7	
		32	30	52·3	60·9		45·7	
		35	0	52·4	61·1		45·8	

August 28 and 29, 1840.

M. Gött. Time. d.	h.	m.	s.	Decl. Sc.-Divns.	Hor. Force. Sc.-Divns.	Hor. Force. Ther. °	Vert. Force. Sc.-Divns.	Vert. Force. Ther. °
29	7	37	30	52·5	61·1		45·9	
		40	0	52·2	61·1		45·9	
		42	30	52·1	60·9		45·9	
		45	0	52·1	61·1		46·0	
		47	30	52·2	60·8		45·8	
		50	0	52·1	60·8		45·9	
		52	30	52·1	61·1		46·3	
		55	0	52·1	61·1		46·3	
		57	30	52·1	61·3		46·3	
	8	0	0	52·1	61·3	49	46·2	53
		2	30	51·9	61·3		46·1	
		5	0	52·0	61·7		45·9	
		7	30	52·1	61·8		45·8	
		10	0	52·0	61·4		45·5	
		12	30	52·0	61·5		45·4	
		15	0	52·2	61·6		45·6	
		17	30	52·1	61·5		45·4	
		20	0	52·1	61·6		45·5	
		22	30	52·2	61·8		45·8	
		25	0	52·2	61·7		45·9	
		27	30	52·0	62·0		46·0	
		30	0	51·1	63·7		45·9	
		32	30	53·2	64·9		45·8	
		35	0	54·8	64·9		45·8	
		37	30	53·5	65·3		45·6	
		40	0	53·5	65·8		44·0	
		42	30	52·5	64·7		43·1	
		45	0	51·3	65·3		42·8	
		47	30	52·5	66·0		42·4	
		50	0	52·1	66·2		42·3	
		52	30	51·4	66·2		41·8	
		55	0	52·2	66·2		41·4	
		57	30	51·9	66·1		41·4	
	9	0	0	51·6	65·4	49	40·8	54
		2	30	51·6	65·8		40·8	
		5	0	51·8	65·9		41·0	
		7	30	51·6	65·2		41·1	
		10	0	51·5	65·0		41·1	
		12	30	51·7	64·5		41·2	
		15	0	51·5	65·5		41·5	
		17	30	51·9	65·4		41·5	
		20	0	51·9	65·2		41·5	
		22	30	51·8	65·1		41·5	
		25	0	52·7	64·8		42·0	
		37	30	52·3	64·7		42·0	
		30	0	51·3	64·5		42·1	
		32	30	52·9	64·8		42·6	
		35	0	50·9	64·9		43·0	
		37	30	57·3	64·9		43·0	
		40	0	51·7	65·3		43·0	
		42	30	51·6	64·5		43·0	
		45	0	51·4	64·9		43·0	
		47	30	50·7	64·9		43·1	
		50	0	50·1	65·2		43·1	
		52	30	51·4	65·1		43·2	
		55	0	50·6	64·8		43·2	
		57	30	51·0	65·3		43·2	
	10	0	0	50·5	64·8	48	43·2	53
	11	0	0	48·5	64·8		50·3	

Mean Positions at the usual hours of observation, during the month of September.

h.	m.	s.	Decl. Sc.-Divns.	Hor. Force. Sc.-Divns.	Hor. Force. Ther. °	Vert. Force. Sc.-Divns.	Vert. Force. Ther. °
0	0	0	45·4	61·3	52	52·5	53
1	0	0	45·0	60·6		53·7	
2	0	0	46·7	62·4	51	54·3	52
3	0	0	47·2	63·4		54·2	

AUGUST 28 and 29, 1840.

M. Gött. Time. d.	h.	m.	s.	Decl. Sc.-Div.	Hor. Force. Sc.-Div.	Ther.	Vert. Force. Sc.-Div.	Ther.
4	0	0		46·0	64·3	49	54·4	50
5	0	0		48·2	64·3		55·9	
6	0	0		50·0	65·3	49	55·2	49
7	0	0		48·9	65·9		56·4	
8	0	0		50·0	66·0	48	56·8	49
9	0	0		48·6	66·4		56·5	
10	0	0		46·6	66·1	48	56·5	48
11	0	0		44·6	64·7		56·4	
12	0	0		42·1	64·4	48	55·4	49
13	0	0		44·7	61·0		56·0	
14	0	0		47·7	57·8	51	54·8	52
15	0	0		50·6	55·5		52·3	
16	0	0		54·1	53·9	56	49·8	56
17	0	0		56·3	54·2		46·8	
18	0	0		56·7	53·2	59	47·3	58
19	0	0		55·7	53·0		47·4	
20	0	0		53·5	55·4	58	48·3	57
21	0	0		50·0	57·3		48·7	
22	0	0		49·0	58·5	55	49·9	55
23	0	0		47·9	58·5		52·0	

SEPTEMBER 21 and 22, 1840.

TORONTO { Decl. 1 Scale Division = 0'·72
H. F. { The instruments were removing
V. F. { from the temporary to the permanent Observatory.

Regular and extra observations.[a]

d.	h.	m.	s.	Decl. Sc.-Div.
20	18	0	0	168·4
	20	0	0	171·8
	22	0	0	192·6
21	0	7	0	165·3
	2	0	0	161·4
	4	0	0	162·1
	6	0	0	153·4
	8	0	0	166·4
	10	0	0	169·6
	12	0	0	166·0
	14	0	0	204·1
	16	0	0	163·4
	18	0	0	151·7
	20	0	0	168·2
	21	16	26	168·9
		26	26	167·6
		31	26	168·3
		41	26	168·3
		46	26	164·5
		56	26	153·0
	22	1	26	139·3

SEPTEMBER 21 and 22, 1840.

M. Gött. Time. d.	h.	m.	s.	Decl. Sc.-Div.
21	22	11	26	117·7
		16	26	94·9
		26	26	102·2
		36	26	123·3
		41	26	144·5
		46	26	150·7
23	1	26		163·8
22	0	0	0	180·0
	2	0	0	161·0
	4	0	0	161·9
	6	0	0	150·2
	8	0	0	173·5
	10	0	0	172·7
	12	0	0	163·7
	14	0	0	172·5
	16	0	0	178·5
	18	0	0	165·4
	20	0	0	168·7
	22	0	0	155·9

Mean positions at the usual hours of observation, from the 18th to the 25th September inclusive.

d.	h.	m.	s.	Sc.-Div.
0	0	0		167·8
2	0	0		167·1
4	0	0		162·1
6	0	0		157·1
8	0	0		164·6
10	0	0		170·4
12	0	0		166·8
14	0	0		182·1
16	0	0		168·4
18	0	0		166·1
20	0	0		171·2
22	0	0		164·9

ST. HELENA { Decl. 1 Scale Division = 0'·71
H. F. k = ·00019; q = ·0003 }

Regular and extra observations.
The H. F. was observed 1m. 0s. after the times specified.

d.	h.	m.	s.	Decl.	Hor. Force.	Ther.
20	14	0	0	66·2	91·3	59
	15	0	0	67·8	91·9	59
	16	0	0	66·9	92·4	59
	18	0	0	67·1	93·1	59
	20	0	0	68·0	95·4	59
	22	0	0	64·8	94·3	60
	23	0	0	65·9	92·1	60
21	0	0	0	64·8		
	2	0	0	69·0	81·1	61

SEPTEMBER 21 and 22, 1840.

M. Gött. Time. d.	h.	m.	s.	Decl. Sc.-Div.	Hor. Force. Sc.-Div.	Ther.	Vert. Force. Sc.-Div.	Ther.
21	2	15	47	69·9	79·9			
		25	47	69·8	79·0			
		35	47	69·7	76·4			
		45	47	70·0	74·0			
		55	47	69·2	74·0			
3	0	0		69·2	74·0	62		
		5	47	67·2	74·0			
		10	47		75·0			
		15	47	69·9	73·9			
		20	47		73·2			
		25	47	69·8	72·2			
		30	47		71·0			
		35	47	67·9	71·0			
		40	47		71·0			
		45	47	68·0	69·9			
		50	47		68·9			
		55	47	68·9	67·8			
4	0	0		69·0	66·9	63		
		5	47	68·0	65·0			
		10	47		64·3			
		15	47	66·9	64·0			
		20	47		62·5			
		25	47		61·5			
		30	47		62·2			
		35	47		63·9			
		40	47		64·0			
		45	47		64·5			
		50	47		63·8			
		55	47		63·2			
5	0	47		66·1	63·3	67		
		5	47		63·0			
		10	47		63·0			
		15	47		62·1			
		20	47		62 9			
		25	47		61·2			
		30	47		62·1			
		35	47		62·0			
		40	47		61·0			
		45	47	68·3	61·0			
		50	47		61·0			
		55	47	67·2	62·0			
6	0	0		66·0	62·9	64		
		5	47	66·3	63·1			
		10	47		63·2			
		15	47	65·1	63·2			
		20	47		63·5			
		25	47	65·0	64·1			
		30	47		65·0			
		35	47	65·3	66·1			
		40	47		66·9			
		45	47	65·3	67·1			

d.	h.	m.	Wind.	
*21	20	0	Calm	Clouded round the horizon, clear in zenith. Auroral light in the North.
21	0		Calm	Clear; auroral light in the North.
	16		Calm	Aurora appearing in quick pulsations, covering the whole Northern part of the sky from N.W. to N.E., extending nearly to the zenith.
	26			Pulsations still continuing, mixed with streamers, very bright to Eastward of North; a bank of orange-coloured haze running along the North horizon. Pulsations shorter and quicker than when last observed.
	31			After a short pause of about three or four seconds, a succession of six or seven very bright flashes, and then another pause, &c.; brightest towards the West of North; a very bright bank due North.
	41			Nearly as at 21h. 31m., the pulsations rather quicker and not extending so high.

d.	h.	m.	Wind.	
21	21	46	Calm	Brilliancy greatly diminished; three large streamers West of North, alt. about 40°, connected by a bright bank with several smaller ones nearly N.E.; pulsations very slow, and a considerable interval between them.
		56	Calm	Pulsations greatly diminished; a very bright bank of an orange tinge extending from N.W. to N.E., in the form of an arch.
22	11		Calm	Arch remarkably bright, showing as much light as the moon at the full; pulsations very weak, brightest in N.W.
	16		Calm	Arch very bright; several very bright streamers of a reddish colour at the Western end of the arch; pulsations hardly visible.
	26		Calm	Arch still very bright, contracted in length, but more elevated; several very bright streamers at both extremities; pulsations very faint.
	35		Calm	Arch nearly disappeared; a few streamers rather faint in N.W., and also in N.E.; pulsations only just visible.
	41		Calm	No appearance of Aurora.

September 21 and 22, 1840.

M. Gött. Time. d. h. m. s.	Decl. Sc.-Div^s	Hor. Force. Sc.-Div^s	Hor. Force. Ther.	Vert. Force. Sc.-Div^s	Vert. Force. Ther.
21 6 50 47		68·0			
55 47	65·2	68·1			
7 0 0		68·4	62		
5 47	65·1	68·0			
10 47		67·9			
15 47	64·8	67·2			
20 47		67·2			
25 47	64·9	68·0			
30 47		67·8			
35 47	65·0	68·0			
40 47		67·9			
45 47	65·1	67·8			
50 47		67·5			
55 47	65·6	67·5			
8 0 0	65·8	67·4	61		
5 47	66·4	71·5			
10 47		74·4			
15 47	68·5	74·8			
20 47		73·0			
25 47	66·7	72·7			
30 47		73·3			
35 47	65·4	72·5			
40 47		72·6			
45 47	64·9	73·0			
50 47		73·1			
55 47	64·7	73·7			
9 0 0		73·5	60		
5 47	64·6	73·9			
10 47		73·8			
15 47	63·0	74·0			
20 47		72·3			
25 47	62·9	71·9			
30 47		71·3			
35 47	63·0	70·0			
40 47		69·6			
45 47	62·5	68·9			
50 47		68·0			
55 47	62·7	67·3			
10 0 0	63·0	68·0	60		
5 47	63·1	68·1			
10 47		68·2			
15 47	64·0	69·0			
20 47		70·1			
25 47	64·3	71·2			
30 47		72·1			
35 47	64·8	73·0			
40 47		73·5			
45 47	64·9	74·5			
50 47		75·0			
55 47	65·9	76·5			
11 0 0	65·9	73·0	59		
5 47	66·2	81·5			
10 47		82·0			
15 47	66·0	82·5			
20 47		82·7			
25 47	65·9	82·1			
30 47		81·2			
35 47	65·1	80·9			
40 47		80·5			
45 47	64·5	80·4			
50 47		80·3			
55 47	65·1	80·0			
12 0 0	64·0	79·2	59		
5 47	63·9	78·9			
10 47		78·9			
15 47	63·9	79·0			
20 47		79·1			

September 21 and 22, 1840.

M. Gött. Time. d. h. m. s.	Decl. Sc.-Div^s	Hor. Force. Sc.-Div^s	Hor. Force. Ther.	Vert. Force. Sc.-Div^s	Vert. Force. Ther.
21 12 25 47	64·0	80·0			
30 47		80·0			
35 47	63·9	80·0			
40 47		80·2			
45 47	63·0	80·5			
50 47		80·0			
55 47	62·9	79·0			
13 0 0		77·6	59		
5 47	63·0	76·9			
10 47		77·5			
15 47	64·0	80·8			
35 47	65·1	90·1			
40 47		91·0			
45 47	65·7	91·3			
50 47		91·0			
55 47	65·4	90·0			
14 0 0	65·1		59		
5 47	64·7				
10 47		88·7			
15 47	64·4	88·0			
20 47		87·0			
25 47	64·9	87·3			
30 47		85·2			
35 47	64·0	85·0			
40 47		84·9			
45 47	64·0	84·3			
50 47		84·0			
55 47	64·7	83·3			
15 0 0	64·6	82·9	59		
5 47	64·4	82·5			
10 47		82·9			
15 47	64·2	82·1			
20 47		82·7			
25 47	64·2	82·0			
30 47		81·9			
45 47	64·2	82·0			
55 47	64·8	82·2			
16 0 0	64·0	82·3	59		
5 47	64·5	82·5			
15 47	64·5	82·9			
25 47	64·4	83·3			
35 47	64·6	84·0			
45 47	64·6	84·1			
55 47	64·7	84·6			
17 5 47	64·5	85·5			
15 47	64·1	85·9			
25 47	64·1	86·1			
35 47	64·9	87·2			
45 47	64·9	89·4			
55 47		90·8			
18 5 47	67·1	91·2	59		
15 47	66·3	91·0			
25 47	66·7	90·4			
35 47	67·7	90·0			
45 47	68·6	89·5			
55 47	70·0	89·0			
19 5 47	71·0	88·7			
15 47	71·1	88·3			
25 47	71·0	87·0			
35 47	70·9	85·7			
45 47	70·6	85·2			
55 47	71·0	84·8			
20 5 47	69·9	82·9	59		
15 47	70·0	82·0			
25 47	69·2	81·5			
35 47	68·9	81·4			
45 47	68·2	82·6			

September 21 and 22, 1840.

M. Gött. Time. d. h. m. s.	Decl. Sc.-Div^s	Hor. Force. Sc.-Div^s	Hor. Force. Ther.	Vert. Force. Sc.-Div^s	Vert. Force. Ther.
21 20 55 47	68·2	83·6			
21 5 47	68·8	83·8			
15 47	68·1	84·1			
25 47	67·9	84·7			
35 47	67·0	84·0			
45 47	66·4	82·2			
55 47	66·5	80·7			
22 5 47	65·9	79·2	59		
15 47	65·3	78·2			
25 47	64·5	79·0			
35 47	64·5	79·0			
45 47	64·5	80·2			
55 47	64·8	82·9			
23 5 47	64·4	84·0			
15 47	65·0	85·8			
25 47	64·8	86·2			
35 47	63·0	86·0			
45 47	63·1	87·0			
55 47	63·1	88·1			
22 5 47	63·0	88·6	60		
15 47	63·0	90·0			
25 47	64·0	90·1			
35 47	63·1	90·5			
45 47	63·9	90·7			
55 47	63·1	90·0			
1 5 47	63·1	90·2			
15 47	63·5	90·3			
25 47	63·4	90·4			
35 47	64·6	90·1			
45 47	63·9	89·8			
55 47	63·0	88·6			
2 5 47	64·2	86·9	60		
15 47	64·1	87·3			
25 47	64·0	87·0			
35 47	64·4	88·2			
45 47	64·8	88·0			
55 47	65·0	86·5			
3 5 47	64·1	86·0			
15 47	63·0	85·7			
25 47	63·0	86·5			
35 47	62·5	86·1			
45 47	62·9	85·8			
55 47	63·0	86·0			
4 5 47	62·9	85·0	61		
15 47	62·9	83·6			
25 47	62·4	82·9			
35 47	62·0	81·0			
45 47	61·5	77·5			
55 47	60·9	73·9			
5 5 47	61·3	73·9			
15 47	61·0	75·0			
25 47	61·0	75·8			
35 47	61·1	76·0			
45 47	61·8	75·0			
55 47	62·8	75·0			
6 5 47	62·8	76·0	60		
15 47	62·8	76·5			
25 47	63·1	78·3			
35 47	63·5	79·8			
45 47	63·9	80·2			
55 47	63·3	79·1			
7 5 47	63·4	78·0			
15 47	63·0	77·3			
25 47	63·5	81·5			
35 47	65·0	84·1			
45 47	65·0	84·5			
55 47	65·0	83·1			

SEPTEMBER 21 and 22, 1840.

M. Gött. Time (d. h. m. s.)	Decl. (Sc.-Div^{ns})	Hor. Force (Sc.-Div^{ns})	Ther.	Vert. Force (Sc.-Div^{ns})	Ther.
22 8 5 47	64·8	83·7	59		
15 47	65·4	83·5			
25 47	64·9	82·9			
35 47	64·5	82·1			
45 47	64·7	81·7			
55 47	64·2	81·1			
10 0	66·1	88·2	59		
11	65·9	87·1	59		
12	65·0	85·5	59		
14	65·1	88·9	59		
15	64·8	88·7	58		
16 0 0	64·5	88·9	58		
18	66·2	92·0	58		
20	67·1	88·1	57		
22	65·0	88·7	58		
23 0 0	65·1	95·4	59		

Mean Positions at the usual hours of observation during the Month.*

M. Gött. Time (d. h. m. s.)	Decl.	Hor. Force	Ther.
0 0 0	65·8	89·4	61
2 0 0	67·0	87·7	62
3 0 0	66·2	84·9	63
4 0 0	66·1	82·3	63
5 0 0	65·9	79·2	64
6 0 0	65·7	77·7	63
8 0 0	66·7	78·0	61
10 0 0	66·7	79·0	60
11 0 0	67·0	79·8	60
12 0 0	66·8	80·7	60
14 0 0	66·9	81·7	59
16 0 0	66·8	82·8	59
18 0 0	67·6	83·6	59
20 0 0	68·6	83·9	59
22 0 0	65·3	87·1	59
23 0 0	65·6	89·7	60

ANTARCTIC EXPEDITION AT HOBARTON, VAN DIEMEN ISLAND.

Decl. 1 Scale Division = 0'·73
H. F. k = ·000271; q =
V. F. k =   ; q =

Positions at the usual hours of observation from September 20th, 12h. to September 22nd, 23b.

M. Gött. Time (d. h. m. s.)	Decl. (Sc.-Div)	Hor. Force (Sc.-Div)	Ther.	Vert. Force (Sc.-Div)	Ther.
20 12 0 0	39·5	65·6	47	60·4	47
13 0 0	41·4	65·4	48	57·6	48
14 0 0	43·8	64·4	51	52·7	52
15 0 0	48·4	58·9	57	45·7	57
16 0 0	51·0	53·8	61	44·9	60
17 0 0	53·0	53·8	63	38·6	62
18 0 0	53·6	52·7	64	41·0	62
19 0 0	52·3	53·0	63	42·1	61
20 0 0	51·9	54·2	61	47·1	59
[21 0 0	50·4	56·5	59	49·5	57
22 0 0	51·7	48·7	57	56·8	56
23 0 0	49·8	36·3	56	62·3	56
21 0 0 0	08·2	50·3	55	50·7	55
1 0 0	30·3	43·6	54	57·0	54
2 0 0	40·7	47·3	53	60·5	54
3 0 0	43·7	56·0	53	39·6	54

SEPTEMBER 21 and 22, 1840.

M. Gött. Time (d. h. m. s.)	Decl. (Sc.-Div^{ns})	Hor. Force (Sc.-Div^{ns})	Ther.	Vert. Force (Sc.-Div^{ns})	Ther.
21 4 0 0	18·4	47·4	53	38·7	53
5 0 0	26·2	41·6	52	45·3	53
6 0 0	49·2	44·6	52	36·3	53
7 0 0	43·5	51·6	52	42·4	52
8 0 0	40·5	49·2	51	47·4	52
9 0 0	46·1	49·7	51	45·7	52
10 0 0	48·7	43·4	51	48·7	51
11 0 0	41·9	45·4	52	49·6	52
12 0 0	43·2	50·3	52	49·6	53
13 0 0	46·5	40·2	54	59·7	55
14 0 0	51·5	45·9	55	56·6	56
15 0 0	51·8	45·3	62	41·8	61
16 0 0	54·8	47·3	65	37·6	64
17 0 0	54·4	42·7	67	41·5	65
18 0 0	52·6	37·8	68	47·0	66
19 0 0	56·5	34·9	66	55·1	64
20 0 0	54·4	35·8	63	58·8	61
21 0 0	44·5	37·6	60	44·2	59
22 0 0	40·3	49·4	58	44·2	58
23 0 0	43·8	43·1	56	59·6	56
22 0 0 0	45·9	52·6	54	54·3	55
1 0 0	37·7	54·4	53	53·5	54
2 0 0	53·0	60·5	52	51·2	53
3 0 0	47·7	55·4	51	51·3	52
4 0 0	45·2	58·9	51	53·4	52
5 0 0	48·7	69·4	50	49·4	50
6 0 0	37·4	64·0	48	45·5	50
7 0 0	43·1	53·4	49	55·4	49
8 0 0	51·0	53·7	49	60·4	49
9 0 0	49·7	56·2	49	55·8	49
10 0 0	45·2	54·8	49	56·9	49
11 0 0	44·4	52·8	51	58·5	50
12 0 0	44·2	55·9	52	53·9	51
13 0 0	43·3	54·7	54	50·3	54
14 0 0	47·0	50·5	58	45·4	54
15 0 0	49·3	47·3	63	44·3	62
16 0 0	53·4	42·1	67	36·9	66
17 0 0	57·6	37·3	71	31·8	72
18 0 0	58·0	34·6	72	26·1	75
19 0 0	58·4	33·2	71	39·5	71
20 0 0	54·2	39·3	68	39·8	69
21 0 0	40·9	39·4	65	36·1	65
22 0 0	41·3	41·2	62	44·5	63
23 0 0	41·5	36·0	60	52·5	61

Mean Positions at the same hours during the Month.

M. Gött. Time (d. h. m. s.)	Decl.	Hor. Force	Ther.	Vert. Force	Ther.
0 0 0	45·4	61·3	52	52·5	53
1 0 0	45·0	60·6		53·7	
2 0 0	46·7	62·4	50	54·3	52
3 0 0	47·2	63·4		54·2	
4 0 0	46·0	64·0	49	54·4	50
5 0 0	48·2	64·3		55·9	
6 0 0	50·0	65·3	49	55·2	49
7 0 0	48·9	65·9		56·4	
8 0 0	50·0	66·0	48	56·4	49
9 0 0	48·6	66·4		56·5	
10 0 0	46·6	66·1	48	56·5	48
11 0 0	44·6	64·7		56·4	
12 0 0	42·1	64·4	48	55·4	49
13 0 0	44·7	61·0		56·0	
14 0 0	47·7	57·8	51	54·8	52
15 0 0	50·6	55·5		52·3	

SEPTEMBER 21 and 22, 1840.

M. Gött. Time (d. h. m. s.)	Decl. (Sc.-Div^{ns})	Hor. Force (Sc.-Div^{ns})	Ther.	Vert. Force (Sc.-Div^{ns})	Ther.
16 0 0	54·1	53·9	56	49·8	56
17 0 0	56·3	54·2		46·8	
18 0 0	56·7	53·2	59	47·3	58
19 0 0	55·7	53·9		47·4	
20 0 0	53·5	55·4	58	48·3	57
21 0 0	50·0	57·3		48·7	
22 0 0	49·0	58·5	55	49·9	55
23 0 0	47·9	58·5		52·0	

OCTOBER 19, 20, 21, and 22, 1840.

TORONTO { Decl. 1 Scale Division = 0'·72
H. F. k = ·000074; q = ·0002
V. F. k = ·00009; q = ·00018ᵇ

Regular and extra observations.

The V. F. was observed at 1m. 3s. before, and the H. F. 2m. 30s. after the times specified.

M. Gött. Time (d. h. m. s.)	Decl. (Sc.-Div^{ns})	Hor. Force (Sc.-Div^{ns})	Ther.	Vert. Force (Sc.-Div^{ns})	Ther.
18 18 0 0	26·3	394·7	52	49·4	52
20 0 0	61·0	388·5	53	66·2	53
22 0 0	30·1	409·0	53	57·6	53
19 0 0 0	15·5	406·5	53	56·2	53
2 0 0	31·7	399·0	54	57·8	54
4 0 0	15·8	366·3	55	56·1	55
45 30	20·1	364·7		56·7	
50 30	23·8	342·7		55·9	
55 30	26·6	336·0		56·2	
5 5 30	27·2	338·7	55	55·3	
10 30	23·9	322·7		54·2	
20 30	23·6	320·9		50·9	
30 30	20·0	316·4		50·3	
40 30	6·6	336·2		51·1	
50 30	10·9	335·3		48·2	
55 30	14·9	334·1		45·4	
6 0 30	15·7	336·4	56	42·7	56
5 30	19·7	350·4		45·1	
10 30	14·7	353·0		46·8	
15 30	16·4	357·3		48·1	
20 30	19·3	386·4		47·7	
25 30	14·8	363·8		42·5	
30 30	19·0	349·6		44·9	
35 30	16·6	350·0		43·6	
40 30	19·7	341·1		43·9	
45 30	19·9	359·0		45·0	
50 30	18·5	372·8		45·6	
55 30	18·4	354·7		44·1	
7 0 30	17·0	358·7	57	48·0	56
5 30	16·4	358·9		48·7	
10 30	18·3	366·0		46·7	
15 30	17·0	359·5		47·3	
20 30	16·3	358·5			
25 30	17·4	367·6		45·0	
30 30	17·8	378·4		43·6	
35 30	18·7	367·5		43·0	
40 30	17·8	346·5		44·2	
45 30	16·8	371·7		48·4	
50 30	13·3	380·4		41·4	
55 30	15·1	386·7		38·4	
8 0 30	17·4	375·5	57	37·2	56
5 30	20·2	364·3		40·3	
10 30	19·7	374·7		36·0	
15 30	20·1	375·2		38·9	

* The mean positions during the month, of the H. F. magnet, are not strictly comparable with those of the 21st and 22nd, in consequence of the stretching of the suspension wires.

ᵇ This value is approximate.

OCTOBER 19, 20, 21, and 22, 1840.

M. Gött. Time.	Decl.	Hor. Force.		Vert. Force.	
m. s.	Sc.-Div^ns.	Sc.-Div^ns.	Ther.	Sc.-Div^ns.	Ther.
19 8 20 30	20·5	377·6		40·8	
25 30	18·6	370·0		40·6	
30 30	19·7	369·7		43·2	
10 0 0	24·9	384·9	57	48·1	57
20 30	25·0	387·7		47·4	
25 30	25·1	387·6		47·2	
12 0 0	46·8	361·2	57	41·8	57
14 0 0*	38·4	350·6	57	49·1	57
16 0 0	38·8	387·7	56	56·4	56
14 0	33·8	360·3^b			
19 0	27·8	351·0		63·4	
24 0	19·0	342·4		62·3	
34 0	19·9	336·8		60·7	
39 0	26·1	347·0		59·6	
44 0	29·7	342·7		54·7	
49 0	24·1	321·7		55·7	
54 0	27·4	318·2		62·8	
59 0	30·1	339·2		69·3	
17 4 0	23·9	341·2	56	72·6	57
9 0	24·5	333·0		73·0	
14 0		353·5		65·8	
19 0	26·1	374·2		67·3	
24 0	25·5	376·0		66·7	
29 0	24·1	373·7		67·7	
34 0	23·4	371·0		68·0	
39 0	28·2	342·7		66·8	
44 0	27·2	379·7		63·8	
49 0	29·4	386·0		61·1	
18 0 0	30·6	391·9	56	58·2	57
20 0 0	19·6	371·4	55	62·8	55
22 0 0	26·4	397·2	54	55·3	54
20 0 0 0	27·4	407·2	53	50·1	55
2 0 0	13·7	380·4	54	58·7	56
20 0	19·3	367·4			
25 0	18·8	359·9		57·0	
30 0	15·6	362·7		56·2	
35 0	15·3	369·8		56·0	
40 0	16·9	381·0		54·4	
45 0	17·1	388·2		53·0	
50 0	17·6	390·8		52·5	
55 0	18·0	384·5		53·1	
3 0 0	20·8	377·4	55	54·1	56
5 0	24·4	379·2		54·6	
10 0	23·2	379·5		54·4	
4 0 0	23·5	383·6	54	58·7	
6 0 0	21·9	371·7	54	52·6	55
30 0	23·6	370·1	55	50·9	55
7 0 0	26·6	384·4	55	50·0	55
30 0	25·4	287·0	55	46·9	56
8 0 0	25·7	385·3	56	50·8	57
30 0	24·7	392·6	56	49·9	56

OCTOBER 19, 20, 21, and 22, 1840.

M. Gött. Time.	Decl.	Hor. Force.		Vert. Force.	
m. s.	Sc.-Div^ns.	Sc.-Div^ns.	Ther.	Sc.-Div^ns.	Ther.
20 9 0 0	24·7			50·3	56
10 0 0	22·5	341·1	59	49·8	59
12 0 0	34·2	367·2	58	49·6	58
14 0 0	28·9	384·0	57	49·5	57
16 0 0	47·0	408·8	56	59·1	56
18 0 0	35·7	398·2	55	58·5	55
20 0 0	21·9	388·6	54	55·4	54
22 0 0	22·2	402·4	54	53·3	54
21 0 0 0	15·1	363·7	53	57·7	53
2 0 0	32·4	403·6	53	51·2	54
2 30 0	34·3	399·2	53	51·0	54
3 0 0	30·2	396·2	53	50·6	53
3 30 0	27·9	377·3	53	51·1	53
4 0 0	22·7	363·0	53	51·1	54
4 30 0	18·4	342·7	53	50·7	54
5 0 0	16·7	354·5	53	47·7	54
5 30 0	12·0	364·3	53	46·9	54
6 0 0	13·0	378·5	53	43·8	54
6 30 0	15·9	375·2	53	43·8	54
7 0 0	16·9	374·1	53	43·0	54
7 30 0	20·4	392·4	53	38·6	54
8 0 0	21·7	403·1	53	42·1	54
8 30 0	18·0	382·2	53	44·5	54
9 0 0	18·3	384·9	53	43·1	54
9 30 0	17·2	383·5	53	42·8	54
10 0 0	19·6	383·1	53	42·7	53
12 0 0	24·0	381·2	53	43·7	53
14 0 0	29·9	383·8	53	46·5	53
16 0 0	59·6	405·0	52	51·5	52
18 0 0	34·0	388·7	53	53·3	53
20 0 0	36·9	383·2	51	57·4	51
22 0 0	31·1	410·3	51	54·5	51
22 0 0 0	31·6	413·4	51	47·5	51
2 0 0	38·4	402·4	51	47·5	51
4 0 0	24·2	346·3	52	50·8	52
6 0 0	20·8	372·7	53	45·8	53
8 0 0	22·3	384·9	53	45·4	53
10 0 0	26·6	391·0	53	45·4	53
12 0 0	27·7	418·6	53	46·0	53
14 0 0	39·7	418·6	53	46·0	54
15 34 30*	36·3	369·3		47·0	
39 30	37·8	369·8		47·0	
44 30	45·1	374·7		47·8	
49 30	52·2	386·2		48·4	
54 30	46·0	390·5		48·0	
16 9 0	41·9	380·7	52	55·0	54
14 30	43·3	379·2		55·1	
19 30	42·5	381·8		54·3	
24 30	39·4	382·2		54·3	
29 30	38·1	381·5		54·3	
34 30	36·0	378·5		54·0	

OCTOBER 19, 20, 21, and 22, 1840.

M. Gött. Time.		Decl.	Hor. Force.		Vert. Force.	
d. h. m. s.		c.-Div^ns.	Sc.-Div^ns.	Ther.	Sc.-Div^ns.	Ther.
22 16 39 30		35·4	375·0		54·6	
17 19 30		34·4	358·3	53	59·5	54
24 30		35·3	355·9		60·6	
29 30		36·0	353·7		61·6	
34 30		37·0	355·1		64·1	
39 30		39·0	363·7		64·9	
44 30		38·3	371·6		64·6	
18 0 0		23·3	377·5	53	69·4	53
20 0 0		26·9	374·3	53	58·8	54
22 0 0		33·0	399·3	53	61·5	53

Breaks having occurred in the monthly series, there are no mean positions with which the observations on October 19, 20, 21, and 22, are comparable.

ST. HELENA { Decl. 1 Scale Division = 0'·71 / H. F. k = ·00019; q = ·0003

Regular and extra observations.

d. h. m. s.	c.-Div^ns.	Sc.-Div^ns.	Ther.	Sc.-Div^ns.	Ther.
18 14 0 0	64·4	27·1	60		
15 0 0	62·1	26·2	60		
16 0 0	62·5	27·9	60		
18 0 0	62·1	28·9	60		
19 0 0	60·5	29·5	60		
20 0 0	57·9	31·0	60		
5 0	58·0	29·1			
10 0	59·0				
15 0	59·0	28·9			
20 0	58·1				
25 0	58·0	29·0			
30 0	57·9				
35 0	57·4				
40 0	57·2				
45 0	57·3				
50 0	57·9	29·0			
55 0					
21 0 0	57·8				
5 0	57·8				
10 0					
15 0					
20 0	58·9	30·1	60		
25 0	59·0				
30 0	59·4	30·2			
35 0	58·9				
40 0	59·0	30·1			
45 0	59·0				
50 0	59·0				
55 0	59·1				
22 0 0	59·4	32·0	60		
5 0	60·5	33·4			
10 0	61·5	34·4			

d. h. m.	Wind	
*19 14 0	W. light	Clear; auroral light to North.
16 0	W. light	A few scattered cumuli, remainder clear; auroral light to N.
25	W. light	Faint arch of light to North, alt. about 8°, two streamers issuing from it; a few cumuli in North horizon.
30	W. light	Two concentric arches of light in North, alt. of exterior arch 10°, of interior 8°; bright patch in N.E., small streamers issuing from it.
35	W. light	Arches breaking up; bright patches and small streamers from N.E. to N.W., average alt. about 15°.
45	W. light	General appearance changing rapidly from arches to streamers and patches; slight pulsations.
50	W. light	A bank of well-defined strati appearing in North, rendered very distinct by auroral light; cumuli rising in N.W.
55	W. light	Three bright patches proceeding from N.W. towards East; horizon clouded with strati.
17 20	W. light	Streamers from N.W. to N.E.; slight pulsations.

d. h. m.	Wind	
20 17 40	W. light	Three or 4 streamers in N.; bank of cumu-strati in N. horizon.
45	W. light	Faint auroral light in the North only.
55	W. light	Bank of strati in North horizon. Auroral light disappeared.
18 0	W. light	A dense bank of strati from N.W. to N.E.; remainder clear.

b The H. F. was observed at 1m. 30s. after the times specified, from 19d. 10h. 14m. to 19d. 17h. 49m., and 2m. after the times specified, from 20d. 2m. 20m. to 23d. 3h. 10m.

d. h. m.	Wind	
c 22 15 30	Calm	Dark rounded masses ranged in concentric circles, the centre in the line of the meridian, alt. 60°, much broken. A faint auroral light. A few stars dimly visible.
16 0	S.W. light	Clouded as before, circles broken up, heavier masses in N.
40	S. light	Clouded, with dark base, partially clear to S. and near zenith; faint streaks of light to N.; a fringe of faint light towards zenith, stretching from E. to W. Auroral light to S.W. very faint.
17 0	S.W.	Moderate gusts. Densely clouded. Very dark.
18 0	S.	Moderate. Densely clouded with cir-cumuli. Broken cumuli in North, edges of clouds tinged with faint auroral light.

October 19, 20, 21, and 22, 1840.

d.	h.	m.	s.	Decl. Sc.-Div	Hor. Force Sc.-Div	Ther.	Vert. Force Sc.-Div	Ther.
18	22	15	0	61·3	34·1			
		20	0	60·1	33·0			
		25	0	60·4	33·0			
		30	0	60·9	32·8			
		35	0	61·8	33·6			
		40	0	62·6	35·0			
		45	0	63·4	35·9			
		50	0	63·1	36·2			
		55	0	63·7	36·7			
	23	0	0	63·9		61		
		5	0	63·8	34·4			
		10	0	64·0	35·3			
		15	0	63·5	35·0			
		20	0	63·3	34·5			
		25	0	64·0	35·0			
		30	0	65·0	35·2			
		35	0	65·6	36·2			
		40	0	66·1	36·3			
		45	0	66·0	34·9			
		50	0	65·5	33·4			
		55	0	64·8	32·7			
19	0	0	0	66·6	33·2	60		
		5	0	67·1	33·3			
		10	0	67·5	33·4			
		15	0	67·4	33·2			
		20	0	67·8	33·1			
		25	0	67·9	32·5			
		30	0	67·6	31·7			
		35	0	67·2	31·6			
		40	0	66·6	31·0			
		45	0	66·1	30·1			
		50	0	65·4	29·4			
		55	0	65·2	29·1			
	1	0	0	66·0	29·0	61		
		5	0	66·3	28·8			
		10	0	66·8	28·2			
		15	0	67·1	28·1			
		20	0	67·6	28·5			
		25	0	68·0	29·1			
		30	0	66·9	28·7			
		35	0	66·2	28·4			
		40	0	65·9	28·4			
		45	0	65·4	28·0			
		50	0	65·5	28·0			
		55	0	65·6	28·0			
	2	0	0	66·8		62		
		5	0	66·9				
		10	0	67·0	28·0			
		15	0	66·0	26·2			
		20	0	65·3	26·0			
		25	0	65·8	26·0			
		30	0	66·1	25·8			
		35	0	66·0	25·0			
		40	0	66·0	24·6			
		45	0	65·9	24·5			
		50	0	65·8	24·9			
		55	0	65·8	24·8			
	3	0	0	65·2	24·8	63		
		5	0	65·0	24·9			
		10	0	65·0	25·0			
		15	0	65·5	25·1			
		20	0	65·9	25·1			
		25	0	66·5	25·1			
		30	0	66·6	25·0			
		35	0	66·4	24·2			
		40	0	66·6	23·7			
		45	0	67·1	23·1			

October 19, 20, 21, and 22, 1840.

m.	s.	Decl. Sc.-Div	Hor. Force Sc.-Div	Ther.	Vert. Force Sc.-Div	Ther.
19 3 50	0	67·2	22·4			
55	0	66·9	20·7			
4 0	0	66·5	20·2	63		
5	0	65·0	19·9			
10	0	64·4	19·9			
15	0	64·1	19·0			
20	0	63·9	20·0			
25	0	63·7	20·0			
30	0	63·8	21·0			
35	0	64·0	21·9			
40	0	64·9	21·9			
45	0	65·2	21·0			
50	0	65·8	20·9			
55	0	66·0	20·8			
5 0	0	66·0	20·0	64		
5	0	66·2	19·9			
10	0	66·1	19·9			
15	0	66·0	19·8			
20	0	65·9	18·8			
25	0	65·1	17·6			
30	0	65·1	16·5			
35	0	64·9	15·0			
40	0	63·9	13·9			
43	0	64·0	14·0			
50	0	63·6	15·1			
55	0	63·5	15·5			
6 0	0	63·4	15·0	63		
5	0	63·5	14·9			
10	0	63·9	14·0			
15	0	63·9	13·8			
20	0	63·9	13·0			
25	0	64·0	13·8			
30	0	65·7	12·9			
35	0	63·1	12·4			
40	0	63·2	12·2			
45	0	63·5	12·2			
50	0	63·8	12·8			
7 0	0	63·8	12·0	63		
5	0	63·8	12·1			
10	0	63·9	12·1			
15	0	64·0	12·2			
20	0	64·0	12·0			
25	0	64·2	11·6			
30	0	64·5	11·2			
35	0	64·7	11·0			
40	0	64·3	10·5			
45	0	64·1	10·8			
50	0	64·2	12·6			
55	0	64·1	14·1			
8 0	0	64·1	15·5	62		
5	0	64·2	16·1			
10	0	64·2	15·3			
15	0	64·9	15·0			
20	0	64·9	14·2			
25	0	64·9	14·0			
30	0	64·6	14·0			
35	0	64·2	14·6			
40	0	64·2	15·3			
45	0	64·4	15·3			
50	0	65·0	16·9			
55	0	64·9	16·9			
10 0	0	65·4	16·4	61		
11 0	0	63·8	18·2	61		
12 0	0	65·3	20·1	61		
13 0	0	63·9	21·7	61		
14 0	0	62·9	23·5	60		

October 19, 20, 21, and 22, 1840.

d.	h.	m.	s.	Decl. Sc.-Div	Hor. Force Sc.-Div	Ther.	Vert. Force Sc.-Div	Ther.
19	15	0	0	64·1	22·0	60		
	16	0	0	62·0	22·8	60		
	18	0	0	63·0	23·3	60		
	19	30	0	63·9	22·7	60		
	20	0	0	62·7	25·0	60		
	20	30	0	62·9				
	22	0	0	64·1	29·8	60		
	23	0	0	65·8	30·9	61		
20	0	0	0	65·9	32·6	61		
	2	0	0	64·0	37·0	62		
	3	0	0	63·2	26·0	63		
	4	0	0	59·9	27·0	62		
	5	0	0	59·2	27·5	62		
	6	0	0	62·1	23·0	62		
	8	0	0	64·1	24·9	62		
	10	0	0	64·2	25·5	60		
	11	0	0	65·0	26·9	60		
	12	0	0	64·9	26·1	60		
	13	0	0	64·9	25·9	60		
	15	0	0	63·9	25·6	60		
	16	0	0	63·4	27·0	60		
	18	0	0	63·4	27·2	60		
	19	30	0	63·4	27·0	60		
	20	0	0	63·5	27·0	60		
	20	30	0	62·6	29·0	60		
	22	0	0	63·0	26·9	60		
	23	0	0	64·9	35·0	60		
21	0	0	0	63·1	31·0	61		
	2	0	0	66·8	37·0	62		
	3	0	0	65·8	34·7	63		
	4	0	0	64·1	28·5	63		
	5	0	0	62·8	22·8	63		
	6	0	0	62·7	26·0	62		
	8	0	0	63·1	19·1	61		
	10	0	0	64·0	17·8	61		
	11	0	0	65·1	23·6	61		
	12	0	0	64·9	25·6	60		
	13	0	0	65·1	27·8	60		
	14	0	0	64·9	26·9	60		
	15	0	0	63·0	25·5	60		
	16	0	0	63·3	31·1	60		
	18	0	0	62·6	31·0	60		
	19	30	0	61·9	31·4	60		
	20	0	0	61·5	31·0	60		
	20	30	0	61·2	30·8	60		
	22	0	0	64·9	32·5	60		
	23	0	0	68·0	33·9	61		
22	0	0	0	70·1	34·0	62		
	2	0	0	68·8	32·9	64		
	3	0	0	66·6	30·2	62		
	4	0	0	64·2	22·6	63		
	5	0	0	62·0	16·2	63		
	6	0	0	62·9	17·1	62		
	8	0	0	64·0	23·9	61		
	10	0	0	65·0	25·0	60		
	11	0	0	65·1	26·0	60		
	12	0	0	64·0	26·0			
	13	0	0	64·0	28·1	60		
	14	0	0	63·1	24·9	60		
	15	0	0	62·8	26·1	60		
	16	0	0	62·5	27·0	60		
	18	0	0	60·8	26·7	60		
	19	30	0	58·1	26·5	59		
		35	0	58·0	26·5			
		40	0	57·9	26·5			
		45	0	57·9	26·6			
		50	0	57·8	26·3			

October 19, 20, 21, and 22, 1840.

M. Gött. Time (m. s.)	Decl. (Sc.-Div)	Hor. Force (Sc.-Div)	Ther.	Vert. Force (Sc.-Dis)	Ther.
22 19 55 0	57·2	26·3			
20 0 0	56·8	26·8	60		
5 0	56·1	26·9			
10 0	56·0	27·0			
15 0	56·1	27·0			
20 0	56·1	27·0			
25 0	56·5	27·0			
30 0	56·0	27·0	60		
35 0	56·2	27·0			
40 0	56·2	27·0			
45 0	56·6	27·0			
50 0	57·0	27·0			
55 0	57·0	27·2			
21 0 0	57·0	28·5			
5 0	57·0	29·0			
10 0	57·3	29·2			
15 0	57·8	30·0			
20 0	58·0	30·4			
25 0	58·2	30·8			
30 0	58·8	31·0			
35 0	59·0	31·1			
40 0	59·1	31·1			
45 0	59·5	31·1			
50 0	60·2	31·6			
55 0	60·9	31·6			
22 0 0	61·0	31·6	60		
10 0	62·0				
15 0					
20 0	63·0				
25 0	63·2				
30 0	63·3				
35 0	63·9				
40 0	63·9				
45 0	64·0				
50 0	64·4				
55 0	64·9				
23 0 0	65·5	31·8	61		

Mean Positions at the usual hours of observation during the Month.

M. Gött. Time (h. m. s.)	Decl.	Hor. Force	Ther.	Vert. Force	Ther.
0 0 0	66·9	34·3	63		
2 0 0	69·3	32·4	64		
3 0 0	68·2	29·2	65		
4 0 0	65·6	26·7	66		
5 0 0	63·8	24·4	66		
6 0 0	63·8	23·0	65		
8 0 0	64·9	22·7	63		
10 0 0	65·2	22·8	62		
11 0 0	65·2	23·6	62		
12 0 0	65·4	24·3	62		
13 0 0	65·1	25·1	62		
14 0 0	65·2	25·0	61		
15 0 0	65·0	25·2	61		
16 0 0	65·0	26·0	61		
18 0 0	64·9	26·4	61		
19 30 0	64·7	27·0	61		
20 0 0	63·3	28·1	61		
20 30 0	62·4	28·9	61		
22 0 0	62·6	31·8	61		
23 0 0	64·5	33·6	62		

October 19, 20, 21, and 22, 1840.

VAN DIEMEN ISLAND. { Decl. 1 Scale Div. = $0'\cdot71$ / H.F. $k = \cdot0003$; $q =$

Regular and extra observations.

The H.F. was observed at $2^m\cdot 20^s$ after the times specified.*

M. Gött. Time (d. h. m. s.)	Decl. (Sc.-Div)	Hor. Force (Sc.-Div)	Ther.	Vert. Force (Sc.-Dis)	Ther.
18 14 0 0	57·2	64·6	62		
15 0 0	63·9	65·9			
16 0 0	72·0	65·0	65		
17 0 0	72·4	68·5			
18 0 0	71·7	72·8	65		
19 0 0	72·0	71·3			
20 0 0	72·5	72·0	65		
21 0 0	65·0	67·3			
22 0 0	61·6	71·7	63		
23 0 0	60·3	70·7			
19 0 0 0	52·4	69·4	60		
1 0 0	49·0	66·1			
2 0 0	54·6	72·7	60		
3 0 0	48·9	72·5	59		
5 0	51·0	72·2			
10 0	49·2	72·0			
15 0	49·1				
20 0	49·4				
4 0 0	39·3	72·2	58		
5 0	38·1				
10 0	40·0	72·1			
15 0	40·7				
20 0	42·1	72·0			
25 0	50·0				
30 0	50·6	71·3			
35 0	51·2				
40 0	51·2				
45 0	51·1				
5 0 0	57·7	67·1	58		
10 0	41·3				
15 0	42·5				
25 0	43·7	78·0			
32 30	41·6	76·1			
37 30	43·2	75·6			
42 30	46·1	74·3			
50 0	46·7	72·1			
55 0	46·6				
6 0 0	46·6	71·1	57		
7 30	41·6	69·4			
12 30	39·2	68·8			
17 30	38·6	69·1			
22 30	39·9	69·1			
30 0	42·5	70·5			
37 30	45·5	71·7			
42 30	47·3	72·1			
47 30	48·9	71·6			
52 30	50·3	71·7	57		
7 0 0	49·1	73·4			
7 30	51·1	71·6			
12 30	51·3	72·3			
17 30	51·3	71·8			
8 0 0	48·7	69·5	57		
5 0	49·9	70·1			

October 19, 20, 21, and 22, 1840.

M. Gött. Time (d. h. m. s.)	Decl. (Sc.-Div)	Hor. Force (Sc.-Div)	Ther.	Vert. Force (Sc.-Dis)	Ther.
19 8 10 0	53·2	69·5			
15 0	53·6	69·0			
20 0	54·0	68·0			
25 0	52·9	67·2			
30 0	52·8	69·3			
35 0	55·4	68·9			
40 0	54·6	69·3			
45 0	53·8	67·5			
50 0	51·2	68·1			
55 0	50·9				
9 0 0	50·9	69·5	56		
10 0 0	47·1	66·6	56		
11 0 0	46·9	65·9			
12 0 0	52·0	64·6	57		
13 0 0	54·4	62·3	61		
14 0 0	63·7	59·3			
15 0 0	64·6	65·0			
16 0 0	68·0	65·4	65		
17 0 0	66·4	65·9			
18 0 0	66·3	67·5	68		
19 0 0	65·6	66·5			
20 0 0	65·1	69·0	71		
21 0 0	57·2	69·9			
22 0 0	61·3	68·5	68		
23 0 0	61·3	69·9			
20 0 0	60·6	70·2	66		
1 0 0	55·7	69·6			
2 0 0	49·0	65·7	62		
3 0 0	51·4	72·3			
4 0 0	54·9	68·4	62		
5 0 0	58·2	68·8			
6 0 0	58·2	69·1	60		
7 0 0	63·0	71·4			
8 0 0	57·6	72·5	58		
9 0 0	52·5	73·3			
10 0 0	54·7	71·2	58		
11 0 0	53·3	68·3			
12 0 0	52·6	65·6	57		
13 0 0	54·2	67·9			
14 0 0	57·5	69·1	57		
15 0 0	62·5	68·9			
16 0 0	68·5	70·6	59		
17 0 0	66·9	72·2			
18 0 0	66·7	73·8	60		
19 0 0	67·2	76·0			
20 0 0	61·3	75·1	59		
21 0 0	59·7	75·0			
22 0 0	61·9	74·2	57		
23 0 0	60·5	74·2			
21 0 0	46·8	77·9	56		
1 0 0	57·3	74·5			
2 0 0	58·2	74·5	55		
3 0 0	57·8	75·5			
4 0 0	62·2	79·9	52		
5 0 0	61·5	80·6			
6 0 0	49·6	77·9	49		
7 0 0	58·7	77·1			
8 0 0	59·2	79·3	50		
9 0 0	56·4	79·9			
10 0 0	53·1	77·8	46		

* "A clear star-light, with great visibility of distant objects, and but little wind. At 18d. 22h. 50m. a faint yellowish light visible in the S.S.E. which remained with little or no change in its intensity until 19d. 3h. 50m., when it increased considerably for a quarter of an hour, and threw out a few faint coruscations towards the zenith to an altitude of about 40°, these latter only lasting a short time. On going to the Observatory, I found that the Declination magnet was vibrating considerably, and rapidly changing its mean position; consecutive observations were then taken, until it was considered to have regained its proper position. The H. F. magnet was observed during the same period, as also the V. F., but the latter did not exhibit any symptom of being affected in the same manner as the two former.

(Signed) "J. C. Ross."

E

October 19, 20, 21, and 22, 1840.

M. Gött. Time. (d. h. m. s.)	Decl. (Sc.-Div^n.)	Hor. Force. (Sc.-Div^n.)	Hor. Force (Ther.)	Vert. Force. (Sc.-Div^n.)	Vert. Force (Ther.)
21 11 0 0	51·5	74·7			
12 0 0	50·5	71·7	51		
13 0 0	52·7	68·0			
14 0 0	58·5	69·2	56		
15 0 0	65·5	68·5			
16 0 0	71·0	68·8	61		
17 0 0	69·9	72·3			
18 0 0	71·1	72·2	63		
19 0 0	70·5	71·3			
20 0 0	66·6	70·9	65		
21 0 0	54·9	71·4			
22 0 0	60·3	70·8	63		
23 0 0	60·4	72·8			
22 0 0 0	59·5	73·4	63		
1 0 0	59·6	74·2			
2 0 0	59·8	74·1	61		
3 0 0	60·3	75·9			
4 0 0	51·3	76·0	60		
5 0 0	51·3	74·7			
6 0 0	59·9	72·4	59		
7 0 0	60·1	70·9			
8 0 0	57·8	72·6	56		
9 0 0	55·1	72·8			
10 0 0	52·6	72·7	58		
11 0 0	48·6	70·7			
12 0 0	47·8	68·2	60		
13 0 0	52·2	63·6			
14 0 0	58·7	62·8	66		
15 0 0	67·9	63·0			
16 0 0	74·3	64·8	69		
17 0 0	76·4	66·0			
18 0 0	74·0	69·8	74		
19 0 0	70·1	68·1			
20 0 0	69·3	66·0	72		
21 0 0	65·8	63·5			
22 0 0	61·0	65·0	70		
23 0 0	61·7	67·6			

Mean Positions at the usual hours of observation during the Month.[a]

M. Gött. Time. (d. h. m. s.)	Decl. (Sc.-Div^n.)	Hor. Force. (Sc.-Div^n.)	Hor. Force (Ther.)
0 0 0	56·4	68·7	62
1 0 0	55·3	69·1	
2 0 0	55·5	69·3	60
3 0 0	56·1	69·5	
4 0 0	54·9	69·6	59
5 0 0	57·2	69·8	
6 0 0	57·0	70·1	57
7 0 0	58·2	70·4	
8 0 0	57·7	71·3	56
9 0 0	55·8	71·6	
10 0 0	53·3	71·3	56

October 19, 20, 21, and 22, 1840.

M. Gött. Time. (d. h. m. s.)	Decl. (Sc.-Div^n.)	Hor. Force. (Sc.-Div^n.)	Hor. Force (Ther.)
11 0 0	50·7	69·9	
12 0 0	50·4	67·7	57
13 0 0	52·5	65·9	
14 0 0	57·5	64·6	61
15 0 0	62·8	65·0	
16 0 0	67·4	65·7	63
17 0 0	68·7	67·4	
18 0 0	68·8	68·3	65
19 0 0	66·8	68·3	
20 0 0	64·1	68·3	66
21 0 0	61·1	67·7	
22 0 0	59·9	67·6	64
23 0 0	59·1	68·5	

November 13 and 14, 1840.

TORONTO { Decl. 1 Scale Division = $0'\cdot72$; H. F. $k = \cdot00075$; $q = \cdot0002$; V. F. $k = \cdot00009$; $q = \cdot00018$

Extra observations.

The V. F. was observed at $1^m. 40^s.$ before, and the H. F. $2^m. 2^s.$ after the times specified.

M. Gött. Time. (d. h. m. s.)	Decl. (Sc.-Div^n.)	Hor. Force. (Sc.-Div^n.)	Hor. Force (Ther.)	Vert. Force. (Sc.-Div^n.)	Vert. Force (Ther.)
13 12 40 0^b	53·2	465·9	51	35·2	52
45 0	53·6	469·6		35·1	
50 0	52·3	469·7		35·1	
55 0	50·8	470·4		35·1	
13 0 0	49·0	469·5	51	35·4	52
5 0	47·3	465·4		35·4	
10 0	46·8	462·1		35·0	
15 0	46·9	461·4		35·5	
20 0	45·5	463·0		34·7	
25 0	43·6	463·3		32·7	
30 0	45·1	460·3		32·4	
35 0	45·5	460·1		32·3	
40 0	47·1	457·8		30·1	
45 0	46·3	451·3		30·6	
50 0	43·4	451·5		29·9	
55 0	55·5	455·7		29·0	
14 0 0	59·0	460·4	51	28·1	52
5 0	61·6	457·2		28·7	
10 0	62·9	460·9		29·6	
15 0	61·4	459·2		28·3	
20 0	62·9	463·1		30·3	
25 0	65·0	464·6		32·3	
30 0	68·5	470·0		32·3	
35 0	64·2	471·0		34·4	
40 0	57·7	460·8		37·7	
45 0	55·2	460·8		39·0	
50 0	54·0	462·8		37·1	

November 13 and 14, 1840.

M. Gött. Time. (d. h. m. s.)	Decl. (Sc.-Div^n.)	Hor. Force. (Sc.-Div^n.)	Hor. Force (Ther.)	Vert. Force. (Sc.-Div^n.)	Vert. Force (Ther.)
13 14 55 0	55·0	462·7		37·1	
15 0 0	54·1	460·1	51	37·1	52
5 0	54·8	458·4		37·1	
10 0	55·6	458·9		37·1	
15 0	55·8	459·4		37·1	
20 0	56·5	459·4		37·1	
25 0	57·9	458·2		37·1	
30 0	57·0	460·5		38·8	
35 0	56·7	460·3		38·8	
40 0	54·8	461·5		38·9	
45 0	55·4	464·7		38·9	
50 0	58·0	464·8		38·9	

Positions at the usual hours of observation, November 13 and 14.

M. Gött. Time. (d. h. m. s.)	Decl. (Sc.-Div^n.)	Hor. Force. (Sc.-Div^n.)	Hor. Force (Ther.)	Vert. Force. (Sc.-Div^n.)	Vert. Force (Ther.)
13 0 0 0	52·8	497·0	49	40·7	51
2 0 0	51·1	485·3	49	39·3	50
4 0 0	53·1	480·5	49	40·9	50
6 0 0	47·7	469·6	51	42·8	51
8 0 0	44·2	485·5	52	41·5	52
10 0 0	47·3	477·7	52	39·7	52
12 0 0	42·5	467·1	51	34·2	52
14 0 0	55·5	460·4	51	28·1	52
16 0 0	56·7	468·0	50	40·5	52
18 0 0	47·8	477·0	49	38·8	50
20 0 0	45·3	474·8	48	37·3	50
22 0 0	41·4	458·1	47	42·3	49
14 0 0 0	46·5	492·4	47	42·5	49
2 0 0	50·7	481·5	47	36·1	48
4 0 0	40·9	467·2	46	39·3	48
6 0 0	43·7	449·7	47	36·8	48
8 0 0	43·1	476·8	47	33·2	48
10 0 0	45·0	458·7	47	27·4	48
12 0 0	51·8	489·0	47	32·1	48
14 0 0	52·0	482·3	47	31·2	48
16 0 0^c	55·3	468·6	47	32·4	48

Mean Positions at the same hours during the Month.[d]

M. Gött. Time. (d. h. m. s.)	Decl. (Sc.-Div^n.)	Hor. Force. (Sc.-Div^n.)	Hor. Force (Ther.)	Vert. Force. (Sc.-Div^n.)	Vert. Force (Ther.)
0 0 0	54·3	493·3	46	42·6	46
2 0 0	56·2	487·6	46	41·0	46
4 0 0	50·4	474·5	46	41·1	47
6 0 0	46·7	467·4	47	41·0	47
8 0 0	46·3	479·2	48	39·7	48
10 0 0	50·1	477·3	48	39·9	48
12 0 0	52·4	482·8	48	39·8	48
14 0 0	56·4	483·8	48	39·8	48
16 0 0	55·4	481·7	47	40·9	47
18 0 0	52·9	481·4	46	40·5	47
20 0 0	51·8	479·7	46	42·8	47
22 0 0	52·6	483·5	46	43·7	46

[a] The mean positions of the H. F. magnet are possibly not strictly comparable with those of the 19th, 20th, 21st, and 22nd, in consequence of the stretching of the suspension wires at so early a date after the adjustment of the instrument.

d. h. m. s.	Wind.	
[b] 13 2 0	N.W. light	Clear ; faint auroral light.
10		Faint streamers in North, one of them visible for about 20^m. or 30^m. 5° West of North.
3 0	Calm	Faint auroral light in North.
30	Calm	Arch of auroral light, extending from N.W. to N.E. alt. of centre about 8°; four small streamers at West extremity visible for a few minutes.
45	Calm	Eight very bright streamers at East end of arch, remaining for a short time, alt. of streamers about 38°, no alteration in the arch.
55	Calm	The whole extent of the arch brightened very suddenly, throwing up streamers, none exceeding 15° in alt. Haze thickening in North horizon.

d. h. m.	Wind.	
13 14 2	Calm	Haze increasing, double arch of light, alt. of lower about 8°; upper arch very faint, alt. 10° to 15°, dying away from West and in the centre.
14 10 0		Arches nearly broken up, faint light remaining, bright patch steady in N.E. near the point of the moon's rising.
15 0		A small bright patch appearing in N.N.E. visible for a few seconds only.
20 0		A very bright patch in North, remainder nearly extinct.
25 0		Nothing remaining, but faint auroral light.
30 0	Calm	Aurora brightening again, faint streamers.
32 0		One very bright streamer moving slowly to Eastward.
35 0		Streamers growing very faint, bank of light totally gone, moon rising above horizon.
37 0	Calm	Aurora no longer visible.

Saturday midnight at Toronto.

[c] The mean positions are from the 7th to the 25th November inclusive.

November 13 and 14, 1840.

St. Helena { Decl. 1 Scale Division = 0'·71 ; H. F. k = ·00019; q = ·0003 }

Regular and extra observations
The H. F. was observed at 2m. 30'. after the times specified.

M. Gött. Time.				Decl.	Hor. Force.		Vert. Force.	
d.	h.	m.	s.	Sc.-Div.	Sc.-Div.	Ther.	Sc.-Div.	Ther.
13	0	0	0	77·9	34·0	65		
		30	0	78·1	34·0			
	1	0	0	79·8	32·0	66		
		30	0	80·2	29·8			
	2	0	0	79·3	27·8	67		
		30	0	79·1	27·9			
	3	0	0	80·2	28·0	68		
		30	0	80·8	27·5			
	4	0	0	80·5	27·7	68		
		30	0	80·0	27·0			
	5	0	0	79·1	26·3	68		
		30	0	78·0	24·0			
	6	0	0	76·9	21·0	68		
		30	0	76·1	19·3			
	7	0	0	75·5	16·8	67		
		30	0	76·4	16·4			
	8	0	0	77·2	18·0	65		
		30	0	76·9	18·0			
	9	0	0	76·2	17·1	65		
		30	0	76·2	15·2			
	10	0	0	76·0	14·6	65		
		30	0	76·0	15·4			
	11	0	0	76·2	09·5	65		
		30	0	76·1	11·0			
	12	0	0	75·6	11·0	64		
		30	0	75·8	11·4			
	13	0	0	75·5	11·4	64		
		30	0	75·0	09·3			
	14	0	0	73·8	10·8	64		
		30	0	74·0	12·5			
	15	0	0	73·1	12·5	63		
		30	0	73·2	13·8			
	16	0	0	73·2	14·2	63		
		30	0	73·2	15·0			
	17	0	0	74·0	15·5	63		
		30	0	74·0	15·6			
	18	0	0	73·9	15·2	63		
		30	0	74·0	15·8			
	19	0	0	73·8	16·0	63		
		30	0	73·0	17·0			
	20	0	0	71·0	17·5	63		
		30	0	69·5	20·5			
	21	0	0	69·1	20·5	63		
		30	0	69·8	23·1			
	22	0	0	69·7	23·8	63		
		30	0	70·5	21·2			
	23	0	0	71·1	22·8	64		
		30	0	73·1	25·9			
14	0	0	0	74·8	25·4	65		
		30	0	75·8	24·0			
	1	0	0	78·7	24·7	66		
		30	0	78·8	23·7			
	2	0	0	80·5	23·7	66		
		30	0	80·4	20·4			
	3	0	0	79·4	15·6	67		
		30	0	78·0	15·0			
	4	0	0	76·2	15·0	67		
		30	0	75·7	14·0			

November 13 and 14, 1840.

M. Gött. Time.				Decl.	Hor. Force.		Vert. Force.	
d.	h.	m.	s.	Sc.-Div.	Sc.-Div.	Ther.	Sc.-Div.	Ther.
14	5	0	0	74·3	14·2	67		
		30	0	74·9	12·7			
	6	0	0	74·3	08·3	66		
		30	0	74·1	06·6			
	7	0	0	74·0	10·6	65		
		30	0	75·1	16·5			
	8	0	0	74·0	11·0	65		
		30	0	75·0	09·0			
	9	0	0	74·9	04·0	65		
		30	0	74·4	07·1			
	10	0	0	74·6	05·6	64		
		30	0	75·2	07·5			
	11	0	0	75·2	10·8	64		
		30	0	75·2	11·5			

Mean Positions at the regular hours of observation, during the Month.[a]

	0	0	0	76·1	24·8	65		
	2	0	0	77·6	22·9	66		
	3	0	0	76·3	20·5	67		
	4	0	0	74·7	18·6	67		
	5	0	0	73·6	17·0	67		
	6	0	0	73·5	14·6	67		
	8	0	0	74·6	13·4	65		
	10	0	0	75·2	13·8	64		
	11	0	0	75·1	14·3	64		
	12	0	0	75·0	15·1	64		
	13	0	0	74·6	15·6	64		
	14	0	0	74·1	16·1	63		
	15	0	0	73·7	15·9	63		
	16	0	0	74·0	16·3	63		
	18	0	0	72·6	16·6	63		
	19	30	0	70·9	18·1	63		
	20	0	0	69·7	19·2	63		
	20	30	0	69·4	20·3	63		
	22	0	0	71·2	23·4	63		
	23	0	0	74·1	24·2	64		

Van Diemen Island. { Decl. 1 Scale Division = 0'·71 ; H. F. k = ·0003; q = ; V. F. k = }

Positions at the usual hours of observation, November 13th and 14.

				Decl.	Hor. Force.		Vert. Force.	
13	0	0	0	60·8	87·5	59	60·6	58
	1	0	0	60·5	88·0		63·8	
	2	0	0	50·3	85·2	58	72·8	57
	3	0	0	54·5	84·9		74·0	
	4	0	0	56·9	86·2	57	74·0	56
	5	0	0	57·3	87·1		72·7	
	6	0	0	57·3	85·5	57	72·7	56
	7	0	0	57·7	86·5		71·9	
	8	0	0	56·4	86·8	56	71·9	55
	9	0	0	53·8	86·5		71·9	
	10	0	0	54·8	84·7	56	71·9	54
	11	0	0	54·9	87·5		72·4	
	12	0	0	51·5	84·8	57	72·8	57
	13	0	0	52·5	79·4		73·5	
	14	0	0	59·3	79·6	62	73·5	62
	15	0	0	65·7	78·9		73·1	
	16	0	0	70·9	79·3	68	73·7	67
	17	0	0	71·4	78·7		73·6	

November 13 and 14, 1840.

M. Gött. Time.				Decl.	Hor. Force.		Vert. Force.	
d.	h.	m.	s.	Sc.-Div.	Sc.-Div.	Ther.	Sc.-Div.	Ther.
13	18	0	0	69·5	80·8	71	73·6	68
	19	0	0	66·7	78·5		85·8	
	20	0	0	64·3	78·3	72	85·2	69
	21	0	0	61·5	77·8		85·2	
	22	0	0	55·8	79·1	71	85·2	69
	23	0	0	51·0	77·8		85·2	
14	0	0	0	53·4	77·6	68	85·0	67
	1	0	0	55·0	77·3		85·0	
	2	0	0b	51·6	78·7	67	84·6	65

Mean Positions at the same hours during the Month.

	0	0	0	58·4	82·3	63	70·7	61
	1	0	0	57·2	82·8		70·8	
	2	0	0	56·7	83·3	61	71·2	60
	3	0	0	55·9	83·2		70·0	
	4	0	0	56·9	83·2	59	70·2	58
	5	0	0	57·7	84·0		69·6	
	6	0	0	57·9	84·4	58	70·1	57
	7	0	0	57·6	84·6		69·9	
	8	0	0	57·0	85·2	57	70·0	56
	9	0	0	55·4	85·3		69·7	
	10	0	0	54·1	84·4	56	69·8	56
	11	0	0	52·0	83·8		70·0	
	12	0	0	50·8	81·4	58	69·9	58
	13	0	0	53·8	79·7		70·6	
	14	0	0	58·5	79·0	61	71·5	61
	15	0	0	63·8	79·0		70·4	
	16	0	0	67·7	80·6	64	70·2	62
	17	0	0	69·1	81·7		69·9	
	18	0	0	68·0	82·3	66	69·8	63
	19	0	0	66·5	82·0		70·8	
	20	0	0	63·7	82·0	66	70·5	64
	21	0	0	61·2	81·8		70·7	
	22	0	0	59·8	81·8	65	70·6	63
	23	0	0	58·5	82·2		70·8	

November 20 and 21, 1840.

Toronto { Decl. 1 Scale Division = 0'·72 ; H. F. k = ·000075; q = ·0002 ; V. F. k = ·00009; q = ·00018 }

Extra observations.

The V. F. was observed at 1m. 20s. before, and the H. F. 2m. 40s. after the times specified.

				Decl.	Hor. Force.		Vert. Force.	
20	14	30	0	80·9	442·2	Reduced to 44·4°	35·5	Reduced to 45·8°
		35	0	82·6	452·9		37·1	
		40	0	84·7	459·2		33·6	
		45	0	80·9	457·0		31·7	
		50	0	80·5	455·2		29·7	
		55	0	81·5	468·2		28·9	
	15	0	0	78·3	471·4		28·8	
		5	0	72·7	459·2		29·7	
		10	0	69·4	456·3		32·1	
		15	0	66·8	458·3		30·9	
		20	0	65·1	460·8		30·8	
		25	0	64·8	463·3		31·1	
		40	0	62·1	477·0		30·0	
		45	0	60·7	480·0		29·8	
		50	0	58·6	483·3		29·9	

---

[a] The mean positions of the Declination magnet are from the 10th to 30th November inclusive. [b] Saturday midnight at Van Diemen Island. [c] The observations of this disturbance were sent to England reduced for temperature as in the text.

NOVEMBER 20 and 21, 1840.

In the following observations, the V. F. was observed at 1ᵐ. 30ˢ. before, and the H. F. at 2ᵐ. 20ˢ. after the times specified.

M. Gött. Time. d. h. m. s	Decl. Sc.-Div^ss	Hor. Force. Sc.-Div^ss	Ther.	Vert. Force. Sc.-Div^ss	Ther.
20 22 23 40	48·1	460·6		65·8	
28 40	51·0	438·6		65·2	
33 40	54·0	438·4		65·8	
38 40	50·0	422·7		65·9	
43 40	49·5	435·3		69·7	
48 40	42·2	420·8		63·4	
53 40	37·2	413·2		63·5	
58 40	34·2	422·8		60·3	
23 3 40	28·6	413·4		64·4	
8 40	25·6	421·2		62·1	
13 40	22·3	432·1		68·3	
18 40	23·9	446·9		68·3	
23 40	25·0	443·8		61·4	
28 40	25·2	443·1		66·2	
33 40	24·5	438·1		62·4	
38 40	26·7	440·6		62·4	
43 40	28·4	409·9		62·4	
48 40	33·0	411·8		67·5	
53 40				61·8	
21 0 8 40	49·3	472·5		50·1	
13 40	52·1	473·5		48·8	
18 40	52·4			50·8	
23 40	52·6	473·4		50·8	
28 40	50·6	480·2		50·0	
33 40	50·6	487·4		48·4	
38 40	51·4	491·7		46·2	
43 40	49·6			45·6	
48 40	49·8	478·3			
53 40	55·6	477·8		42·3	
58 40	57·3	486·1		41·6	
1 3 40	54·5			41·3	
8 40	48·6			40·2	
13 40	47·0	471·0			
18 40	45·1	468·1		41·7	
23 40	44·9	467·4		40·7	
28 40	45·5	466·1		39·1	
33 40	46·1	477·6		38·7	
38 40	46·1	474·8		38·3	
43 40	45·4			38·1	

(Hor. Force column Reduced to 43·8; Vert. Force column Reduced to 45·0.)

Positions at the usual hours of observation, November 20 and 21.

M. Gött. Time. d. h. m. s	Decl.	Hor. Force.	Ther.	Vert. Force.
20 0 0 0	55·4	513·9	43	33·5
2 0 0	56·5	511·9	43	34·2
4 0 0	42·8	510·7	42	34·3
6 0 0	39·3	471·1	44	31·9
8 0 0	45·1	477·5	44	31·4
10 0 0	51·6	489·5	44	31·4
12 0 0	47·2	481·4	44	28·2
14 0 0	87·6	490·5	44	30·2
16 0 0	54·3	487·0	44	30·4
18 0 0	50·1	477·7	44	31·0
20 0 0	50·5	476·4	43	50·9
22 0 0	20·4	465·4	44	31·2
21 0 0 0	43·6	474·5	44	64·0
2 0 0	54·2	479·7	44	39·8
4 0 0	38·8	428·4	44	27·3
6 0 0	47·2	455·4	44	27·6
8 0 0	40·6	463·2	44	12·3
10 0 0	52·3	458·6	44	23·1

NOVEMBER 20 and 21, 1840.

M. Gött. Time. d. h. m. s.	Decl. Sc.-Div^ss	Hor. Force. Sc.-Div^ss	Ther.	Vert. Force. Sc.-Div^ss	Ther.
12 0 0	53·3	486·6	43	30·3	
14 0 0	54·8	481·7	42	29·3	
16 0 0*	60·8	500·8	42	37·1	

The Mean Positions at the same hours during the Month, are given in page 26.

ST. HELENA { Decl. 1 Scale Division = 0'·71   H. F. k = ·00019 ; q = ·0003 }

Positions at the usual hours of observation, November 20 and 21.

M. Gött. Time. d. h. m. s	Decl.	Hor. Force.	Ther.	Vert. Force.	Ther.
20 0 0 0	82·0	32·0	64		
2 0 0	81·6	33·1	65		
3 0 0	78·0	21·1	66·		
4 0 0	73·8	20·7	66		
5 0 0	75·2	15·7	66		
6 0 0	75·2	07·2	65		
8 0 0	77·0	08·8	64		
10 0 0	77·5	12·4	63		
11 0 0	76·9	13·2	63		
12 0 0	75·9	09·9	63		
13 0 0	75·4	16·2	63		
14 0 0	75·1	18·9	63		
15 0 0	74·7	16·8	63		
16 0 0	73·9	14·9	63		
18 0 0	72·2	15·1	62		
19 30 0	71·5	17·1	62		
20 0 0	69·8	18·0	62		
20 30 0	69·0	20·2	62		
22 0 0	71·8	19·5	63		
21 0 0 0	77·0	17·0	64		
2 0 0	77·1	13·0	64		
3 0 0	71·9	14·6	66		
4 0 0	73·1	17·8	65		
5 0 0	72·2	03·6	67		
6 0 0	71·3	02·0	67		
6 0 0	72·3	00·5	66		
8 0 0	76·6	03·7	65		
10 0 0	77·2	05·8	64		
11 0 0	76·1	09·0	67		
12 0 0ᵇ	75·9	11·3	63		

The Mean Positions at the same hour during the Month, are given in page 27.

VAN DIEMEN ISLAND. { Decl. 1 Scale Division = 0'·71   H. F. k = ·0001 ; q =   V. F. k = ; q = }

Extra observations.
The V. F. was observed at 2ᵐ. 30ˢ. before, and the H. F. 2ᵐ. 30ˢ. after the times specified.

M. Gött. Time. d. h. m. s	Decl.	Hor. Force.	Ther.	Vert. Force.	Ther.
20 3 10 0	38·1	84·6	64		
17 30	38·7	83·7		85·8	63
25 0	40·2	82·2		86·1	
32 30	41·3	81·2		86·4	
40 0	46·0	80·7		86·6	
47 30	49·7	80·1		87·1	
55 0	51·7			87·3	
4 10 0	54·1	81·3	63		
17 30	54·5	82·2			
25 0	54·4			87·3	63
23 5 0	54·0	82·3	63		

NOVEMBER 20 and 21, 1840.

M. Gött. Time. d. h. m. s	Decl. Sc.-Div^r	Hor. Force. Sc.-Div^r	Ther.	Vert. Force. Sc.-Div^r	Ther.
20 23 10 0	53·6				
15 0	51·3	79·0		87·7	63
20 0	50·7				
25 0	48·7	79·5		87·8	
30 0	47·7				
35 0	47·3	80·6		87·8	
40 0	47·5	81·4			

Positions at the usual hours of observation, November 20 and 21.

M. Gött. Time. d. h. m. s	Decl.	Hor. Force.	Ther.	Vert. Force.	Ther.
20 0 0 0	62·4	85·3	65	86·8	63
1 0 0	61·0	86·0		86·8	
2 0 0	58·1	87·0	64	86·9	62
3 0 0	39·6	88·3		85·2	
4 0 0	52·5	80·0	63	87·7	63
5 0 0	52·7	83·8		86·8	
6 0 0	45·2	81·3	63	86·7	62
7 0 0	50·9	81·4		87·4	
8 0 0	55·2	81·8	61	87·7	60
9 0 0	53·2	81·6		87·7	
10 0 0	51·9	80·1	60	87·6	60
11 0 0	51·7	78·3		88·9	
12 0 0	50·4	73·6	60	89·2	60
13 0 0	54·5	72·9		89·8	
14 0 0	61·1	73·5	61	89·8	61
15 0 0	72·6	77·4		89·8	
16 0 0	75·3	82·8	62	88·2	61
17 0 0	74·9	88·4		88·2	
18 0 0	72·7	86·2	65	88·9	63
19 0 0	71·0	86·2		89·4	
20 0 0	63·6	84·8	67	89·1	63
21 0 0	58·6	81·7		89·5	
22 0 0	61·8	80·4	63	89·2	61
23 0 0	50·0	83·8		87·2	
21 0 0 0	54·3	79·3	61	87·8	62
1 0 0	47·2	75·9		88·2	
2 0 0ᶜ	50·4	80·3	59	87·6	58

The Mean Positions at the same hours during the Month are given in page 27.

DECEMBER 11 and 12, 1840.

TORONTO { Decl. 1 Scale Division = 0'·72   H. F. k = ·000075 ; q = ·00022   V. F. k = ·00009 ; q = ·00018 }

Extra observations.
The V. F. was observed at 1ᵐ. 40ˢ. before, and the H. F. 2ᵐ. 2ˢ. after the times specified.

M. Gött. Time. d. h. m. s	Decl.	Hor. Force.	Ther.	Vert. Force.	Ther.
12 0 20 0	26·9	515·5	43	64·9	44
25 0	26·2	516·5		64·7	
30 0	25·8	517·5		65·3	
35 0	24·0	516·3		65·2	
45 0	26·4	515·9		64·7	
50 0	28·9	516·7		62·2	
55 0	29·9	513·6		62·2	
1 0 0	31·7	510·8	43	63·2	44
35 0	31·4	508·4		63·2	
5 0	30·0	507·2		62·1	
10 0	30·6	503·1		61·8	
20 0	34·1	501·5		61·4	
25 0	38·0	502·2		60·1	
30 0	38·7	498·7		60·3	
35 0	41·1	497·3		60·3	

* Saturday midnight at Toronto.   ᵇ Saturday midnight at St. Helena.   ᶜ Saturday midnight at Van Diemen Island.

DECEMBER 11 and 12, 1840.

d.	h.	m.	s.	Decl. Sc.-Div	Hor. Force Sc.-Div	Ther.	Vert. Force Sc.-Div	Ther.
12	1	40	0	40·6	497·4		60·4	
		45	0	39·9	497·4		60·5	
	2	20	0	36·6	499·9	43	56·3	44
		25	0	35·6	496·8		55·9	
		30	0	36·0	492·9		55·5	
		35	0	41·3	499·9		56·9	
		40	0	42·2	503·8		53·7	
		45	0	45·8	508·0		53·8	
		50	0	47·1	508·0		53·3	
	3	0	0	52·4	512·0	43	53·3	44
		5	0	52·6	512·5		53·0	
		10	0	52·5	509·8		52·7	
		15	0	54·0	511·1		52·7	
		20	0	54·3	507·7		52·2	
		25	0	53·7	502·8		52·3	
		30	0	52·5	493·9		52·7	
		35	0	50·1	487·0		52·8	
		40	0	50·1	485·8		53·3	
		45	0	48·1	487·0		52·9	
		50	0	44·9	488·3		52·3	
		55	0	43·8	490·3		52·0	
	4	0	0	42·3	485·5	43	51·7	44
		5	0	42·3	487·5		51·6	
		10	0	42·9	486·0		51·6	
		15	0	45·5	485·3		51·6	
		20	0	44·7	487·7		51·5	
		25	0	45·2	488·6		51·1	
		30	0	44·6	491·1		50·4	
		35	0	42·7	490·1		50·4	
		40	0	43·3	491·8		49·5	
		45	0	41·7	491·8		49·4	
		50	0	41·1	492·5		49·3	
		55	0	40·2	488·8		49·1	
	5	0	0	41·6	487·8	43	50·3	44
		5	0	43·3	486·1		50·5	
		10	0	41·1	486·1		50·0	
		15	0	43·0	490·5		50·0	
		20	0	44·6	489·5		49·0	
		25	0	45·9	491·1		49·0	
		30	0	46·5	491·6		48·0	
		35	0	43·8	491·2		47·5	
		40	0	44·5	486·7		48·4	
		45	0	43·1	486·3		48·4	
		50	0	43·3	483·1		48·4	
		55	0	43·2	479·5		48·9	
	6	0	0	41·8	482·0	43	48·8	44
		5	0	40·6	481·5		48·5	
		10	0	39·1	475·0		48·4	
		15	0	41·7	485·8		49·6	
		30	0	42·3	486·7		48·3	
	7	30	0	45·0	507·3	43	45·4	44
	8	30	0	45·7	491·9	43		
	9	0	0	46·0	502·6	43	46·8	44
		30	0	46·2	508·5	43	47·0	44
	10	30	0	51·4	511·9	43	46·6	44
	11	0	0	50·7	507·6	43	46·8	44
		30	0	50·0	497·6	43	45·8	44

Positions at the usual hours of observation, December 11 and 12.

d.	h.	m.	s.	Decl. Sc.-Div	Hor. Force Sc.-Div	Ther.	Vert. Force Sc.-Div	Ther.
11	0	0	0ᵃ	155·0	523·2	43		
	2	0	0	59·4	520·0	42		
	4	0	0		492·1	43		

DECEMBER 11 and 12, 1840.

d.	h.	m.	s.	Decl. Sc.-Div	Hor. Force Sc.-Div	Ther.	Vert. Force Sc.-Div	Ther.
11	6	0	0	42·2	502·4	44		
	8	0	0	43·7	499·0	44		
	10	0	0		503·0	45		
	12	0	0	52·6	502·9	45	47·6	45
	14	0	0	55·0	512·4	44	49·4	45
	16	0	0	52·0	506·3	43	48·3	44
	18	0	0	51·6	513·2	42	56·4	44
	20	0	0	42·4	487·9	42	58·6	43
	22	0	0	57·3	518·4	42	58·9	43
12	0	0	0	27·6	502·0	43	60·8	44
	2	0	0	39·9	499·8	43	59·9	43
	4	0	0	41·8	484·7	43	53·0	44
	6	0	0	42·1	477·4	44	50·8	45
	8	0	0	44·9	500·0	44	47·8	44
	10	0	0	49·2	504·6	44	49·1	45
	12	0	0	52·4	472·5	44	47·8	45
	14	0	0	65·9	479·7	45	40·9	45
	16	0	0	59·3	493·3	45	48·2	46

Mean Positions at the same hours during the Month.[b]

d.	h.	m.	s.	Decl. Sc.-Div	Hor. Force Sc.-Div	Ther.	Vert. Force Sc.-Div	Ther.
	0	0	0	46·3	515·8	38	53·9	41
	2	0	0	52·0	512·4	38	50·7	41
	4	0	0	52·4	500·5	39	47·7	41
	6	0	0	47·9	495·3	39	44·6	41
	8	0	0	46·4	503·4	40	43·4	42
	10	0	0	47·5	512·2	40	41·3	42
	12	0	0	50·1	510·5	40	44·8	42
	14	0	0	53·4	510·0	40	44·1	42
	16	0	0	54·7	507·9	39	44·9	42
	18	0	0	54·4	507·9	38	45·9	41
	20	0	0	50·3	506·7	38	50·7	41
	22	0	0	53·3	509·0	38	56·5	41

ST. HELENA { The pedestals which support the Magnetometers were undergoing removal at this period.

VAN DIEMEN ISLAND. { Decl. 1 Scale Division = 0'·71; H.F. k = 0003; q = ; V.F. k = ; q =

Extra observations.

The V.F. was observed at 2ᵐ·30ˢ·before, and the H.F. 2ᵐ·30ˢ·after the times specified.

d.	h.	m.	s.	Decl. Sc.-Div	Hor. Force Sc.-Div	Ther.	Vert. Force Sc.-Div	Ther.
11	5	25	0	36·0	51·4	68		
		30	0	37·0				
		35	0	38·4	51·3		55·9	67
		40	0	38·5				
		45	0	38·8	51·6		56·1	
		50	0	39·3				
		55	0	39·8	50·7		56·1	
	6	5	0	40·6	51·6	67	56·3	67
		10	0	41·9				
		15	0	42·0	51·3		56·3	
		20	0	42·5				
		25	0	42·8	51·4		56·3	
		30	0	42·5				

Positions at the usual hours of observation, December 11 and 12.

d.	h.	m.	s.	Decl. Sc.-Div	Hor. Force Sc.-Div	Ther.	Vert. Force Sc.-Div	Ther.
11	0	0	0	46·3	51·3	72	56·3	71
	1	0	0	46·3	50·4		56·8	
	2	0	0	44·9	52·0	70	56·2	69

DECEMBER 11 and 12, 1840.

d.	h.	m.	s.	Decl. Sc.-Div	Hor. Force Sc.-Div	Ther.	Vert. Force Sc.-Div	Ther.
11	3	0	0	43·2	53·4		56·2	
	4	0	0	41·1	52·5	68	56·2	67
	5	0	0	34·0	54·3		54·6	
	6	0	0	39·6	50·7	67	56·3	67
	7	0	0	40·0	51·3		56·3	
	8	0	0	41·0	52·6	66	56·3	65
	9	0	0	40·0	51·8		56·3	
	10	0	0	37·8	51·9	67	56·3	66
	11	0	0	37·8	51·5		56·3	
	12	0	0	37·4	50·5	67	56·0	66
	13	0	0	42·4	49·0		56·0	
	14	0	0	47·7	47·5	70	56·5	69
	15	0	0	50·6	49·1		55·1	
	16	0	0	54·3	48·0	72	56·1	70
	17	0	0	60·1	53·0		56·5	
	18	0	0	61·4	49·8	73	57·3	69
	19	0	0	62·9	56·6		56·1	
	20	0	0	61·8	58·2	71	56·6	69
	21	0	0	54·6	57·7		56·5	
	22	0	0	55·1	50·5	68	57·5	66
	23	0	0	52·9	53·1		57·5	
12	0	0	0	41·1	51·0	67	57·5	66
	1	0	0	39·7	48·4		57·5	
	2	0	0ᶜ	37·9	49·9	67	56·1	66

Mean Positions at the same hours during the Month.

d.	h.	m.	s.	Decl. Sc.-Div	Hor. Force Sc.-Div	Ther.	Vert. Force Sc.-Div	Ther.
	0	0	0	45·8	51·5	71	57·3	69
	1	0	0	45·1	51·6		57·8	
	2	0	0	45·0	52·3	70	57·3	68
	3	0	0	44·7	52·2		58·6	
	4	0	0	44·7	53·0	68	57·9	67
	5	0	0	44·0	53·7		57·7	
	6	0	0	44·3	53·6	66	58·3	66
	7	0	0	43·0	53·8		58·2	
	8	0	0	43·3	54·1	65	58·5	64
	9	0	0	40·6	53·7		58·7	
	10	0	0	39·2	53·6	65	58·6	64
	11	0	0	38·3	52·6		58·7	
	12	0	0	38·9	51·0	66	58·9	66
	13	0	0	42·0	49·3		59·0	
	14	0	0	46·3	48·8	69	59·1	69
	15	0	0	51·1	49·4		58·9	
	16	0	0	54·7	50·6	73	58·6	74
	17	0	0	56·0	51·5		58·3	
	18	0	0	55·5	51·9	75	58·8	75
	19	0	0	54·3	51·9		57·8	
	20	0	0	52·1	51·3	76	56·9	73
	21	0	0	49·8	51·5		57·9	
	22	0	0	48·7	51·4	74	58·0	71
	23	0	0	46·2	51·2		56·5	

DECEMBER 19, 20, 21, and 22, 1840.

TORONTO { Decl. 1 Scale Division = 0'·72; H.F. k = ·000075; q = ·0002; V.F. k = ·00009; q = ·00018

Regular and extra observations.

The V.F. was observed at 1ᵐ·30ˢ·before, and the H.F. 2ᵐ·after the times specified.

d.	h.	m.	s.	Decl. Sc.-Div	Hor. Force Sc.-Div	Ther.	Vert. Force Sc.-Div	Ther.
19	0	0	0	139·4	548·2		60·0	
		30	0	144·5	551·5		59·2	
	1	0	0	150·8	543·3		58·5	

ª The V.F. magnet was employed during the early hours of the 11th in experiments on temperature corrections.　　ᵇ The mean positions of the V.F. magnet are from December 11th to 22nd inclusive.　　ᶜ Saturday midnight at Van Diemen Island.

DECEMBER 19, 20, 21, and 22, 1840.

M. Gött. Time	Decl.	Hor. Force		Vert. Force	
d. h. m. s.	Sc.-Div^n	Sc.-Div^n	Ther.	Sc.-Div^n	Ther.
19 13 0 0	149.0	542.7	56.4		
2 0 0	145.6	542.9	54.7		
30	145.7	535.7	54.8		
3 0 0	143.3	515.2	53.5		
30	141.5	510.6	52.2		
4 0 0	142.3	510.0	51.7		
5 0 0	141.8	503.4	51.3		
30	139.6	507.1	49.3		
6 0 0	137.8	509.8	48.9		
7 0 0	143.9	511.0	48.9		
30	144.9	514.4	47.4		
8 0 0	145.0	516.9	48.1		
30	146.1	519.7	48.6		
9 0 0	146.3	513.0	48.7		
30	145.0	508.7	46.4		
10 0 0	144.2	519.5	43.6		
30	143.4	524.7	44.7		
11 0 0	147.8	524.8	44.7		
30	147.0	523.6	45.3		
12 0 0	148.6	523.8	45.0		
30	150.0	524.1	45.2		
13 0 0	152.9	515.6	44.9		
30	155.6	503.2	45.9		
14 0 0	157.1	504.6	44.8		
30	153.0	516.5	44.9		
15 0 0	153.5	513.5	44.9		
30	152.2	512.2	46.4		
16 0 0	154.7	512.0	46.9		
20 18 0 0*	195.1	465.1	40.6		
15 0	180.8	473.0	35.3		
20 0	172.5	488.9	40.5		
25 0	175.6	404.8	34.9		
30 0	173.2	493.9	34.8		
35 0	173.0	499.8	46.0		
40 0	169.7	498.6	49.6		
45 0	165.6	494.9	51.5		
50 0	164.2	490.5	53.2		
55 0	165.2	477.2	49.3		
19 0 0	166.5	471.1	58.3		
5 0	165.9	456.8	58.0		
10 0	162.7	441.9	57.9		
15 0	157.3	413.5	63.0		

(Hor. Force Ther. Reduced to 35°.b; Vert. Force Ther. Reduced to 36.5.b; later portion Reduced to 34.2.b and Reduced to 35.5.b)

DECEMBER 19, 20, 21, and 22, 1840.

M. Gött. Time	Decl.	Hor. Force		Vert. Force	
d. h. m. s.	Sc.-Div^n	Sc.-Div^n	Ther.	Sc.-Div^n	Ther.
20 19 20 0	154.4	418.3		68.8	
25 0	154.6	414.3		67.3	
30 0	152.5	400.7		74.6	
35 0	151.9	407.1		75.3	
40 0	145.5	422.8		68.6	
45 0	141.0	419.5		66.3	
50 0	136.7	419.4		68.8	
55 0	135.5	427.4		75.2	
20 0 0	137.3	417.7		77.9	
20 0	145.8	423.7		82.7	
25 0	144.0	450.0		78.4	
30 0	144.5	492.8		74.0	
35 0	148.7	473.1		71.8	
40 0	155.0	490.6		81.9	
45 0	162.5	491.9		77.5	
50 0	162.6	491.0		78.6	
55 0	161.7	500.8		81.0	
21 0 0	153.0	495.6		80.8	
5 0	148.3	501.1		81.3	
10 0	146.0	482.6		79.6	
15 0	145.4	489.9		82.2	
20 0	142.9	502.3		80.7	
25 0	134.9	515.2		78.6	
30 0	142.7	515.4		80.6	
35 0	138.5	504.9		85.0	
40 0	133.2	474.0		86.0	
45 0	122.5	412.0		92.2	
50 0	135.8	376.8		108.0	
55 0	133.2	368.2		117.6	
22 0 0	153.7	407.1		125.4	
5 0	166.1	422.4		109.6	
10 0	172.5	432.5		106.5	
15 0	170.5	462.6		105.2	
20 0	153.7	452.1		106.8	
25 0	131.0	430.8		107.5	
30 0	119.1	419.1		107.5	
35 0	111.0	360.7		104.6	
40 0	101.9	364.4		104.6	
50 0	100.5	390.4		104.3	
55 0	115.5	365.6		109.5	
23 0 0	123.0	418.5		109.6	

(Reduced to 34.2.b and Reduced to 33.5.b)

DECEMBER 19, 20, 21, and 22, 1840.

M. Gött. Time	Decl.	Hor. Force		Vert. Force	
d. h. m. s.	Sc.-Div^n	Sc.-Div^n	Ther.	Sc.-Div^n	Ther.
20 23 5 0	117.9	420.6		104.6	
10 0	106.8	450.2		104.6	
15 0	102.2	456.0		106.4	
20 0	91.6	470.4		108.4	
25 0	96.8	485.5		106.0	
30 0	94.9	468.3		102.5	
35 0	98.2	450.7		105.0	
50 0	108.1	459.8		100.4	
55 0	105.6	435.0		91.1	
21 0 0 0	88.3	452.0		88.5	
5 0	81.4	476.7		89.9	
10 0	93.8	499.3		88.4	
15 0	104.5	493.9		89.1	
20 0	109.6	480.4		88.5	
25 0	108.5	476.7		88.4	
30 0	106.9	482.5		87.1	
35 0	107.6	479.3		83.8	
40 0	105.7	463.8		82.1	
45 0	106.0	452.9		85.1	
50 0	117.2	452.2		85.9	
55 0	122.3	459.9		86.2	
1 5 0	120.6	432.1		73.0	
10 0	111.6	415.0		71.3	
15 0	111.0	414.7		73.9	
20 0	113.0	411.1		71.4	
25 0	121.0	429.5		81.6	
30 0	132.7	426.2		79.3	
35 0	137.3	426.9		74.7	
40 0	134.9	423.8		71.0	
45 0	130.6	410.9		69.4	
50 0	122.8			69.5	
2 0 0	127.5	397.4		86.0	
20 0	95.8	360.5		86.9	
25 0	99.6	364.4		85.8	
30 0	107.3	372.1		86.3	
35 0	99.4	359.8		85.9	
40 0	106.5	363.6		81.3	
45 0	103.3	387.4		71.0	
50 0	107.1	408.9		70.4	
55 0	114.3	412.4		70.4	
3 0 0	125.4	410.3		72.7	
5 0	123.1	403.7		68.8	

(Reduced to 34.2.b and Reduced to 35.5.b)

---

d. h. m.	Wind.	
*20 18 0	Calm	Unclouded. Aurora very bright, consisting of a succession of arches stretching from N.W. to N.E. reaching an alt. of 46°, intermixed with several very bright patches. Pulsations, like waves, meeting and forming a beautiful ridge of light across the zenith from East to West, continually changing in appearance and position; a few flashes in the East also.
25		Decreasing in brightness, the pulsations the same as before, none from the Eastward, a very bright bank in North.
45	Calm	Aurora considerably diminished, arches disappearing, highest ones first; a meteor of the first magnitude fell from an alt. of 30°, taking a due Northerly course and disappearing in the horizon; a few clouds rising in N.W. and stretching to East, a heavy bank of haze resting in North horizon, the flashes of auroral light appearing to rise from behind it.
19 0	Calm	Clouds rising rapidly from North and N.W. and stretching to zenith; the pulsations occasionally appearing through the clouds.
15	Calm	Clouded. Aurora scarcely visible, no pulsations.
35	Calm	Clouded in North. Aurora brightening up, pulsations appearing through the clouds and extending to the zenith, remainder of the sky clear.
19 55		Clouded in the North, light very faint and pulsations quicker; bright banks occasionally rising in East and progressing to West, and throwing out a few streamers.
20 40		Totally clouded so as to prevent the possibility of distinguishing any remarkable features of Aurora.
21 0	West, freshening	Clouded over; light still perceptible in North through the clouds.

d. h. m.	Wind.	
20 21 20	West, dying away	Clouded; light just perceptible near the Northern horizon.
45		A few breaks in the clouds, Aurora appearing to be in full activity.
22 25	S.W. light	Densely clouded, Aurora not visible, a light tinge on the edge of the clouds.
50		Densely clouded.
23 35	S.W. light	Densely clouded, pulsations visible for a few minutes in the zenith.
21 0 0	Calm	Densely clouded.
0 0 to 6 0	Generally calm	Heavily clouded with cirro-cumuli and cumuli.
7 0 10 0	Calm	Overcast with dense haze.
11 0 to 1 ,50	S.E., brisk	Snowing heavily.
18 0	S.E. brisk and squally	Densely clouded, ceased snowing.
19 0	S.W. very light	Densely clouded, very dark.
22 21 to 23 0 0	Calm	Continued densely clouded. Observations terminated.

b The observations of this disturbance were sent to England reduced for temperature as in the text.

December 19, 20, 21, and 22, 1840.

M. Gott. Time (d. h. m.)	Decl. (Sc.-Div)	Hor. Force (Sc.-Div)	Ther.	Vert. Force (Sc.-Div)	Ther.
21 3 10 0	125·3	398·4		67·1	
20 0	135·0	436·2		64·7	
25 0	139·6	425·7		65·1	
30 0	138·6	428·9		62·8	
35 0	137·0	429·2		69·0	
40 0	138·5	424·7		56·4	
45 0	134·8	401·9		55·2	
50 0	131·7	397·4		53·9	
55 0	134·2	406·9		54·8	
4 0 0	132·2	397·1		53·8	
5 0	131·2	392·7		54·0	
10 0	133·2	404·0		52·8	
15 0	131·4	412·2		49·2	
20 0	134·8	403·2		46·7	
25 0	133·1	405·2		46·0	
30 0	133·8	407·9		44·8	
5 0 0	131·4	421·5		41·6	
30 0	145·2	456·4		35·7	
6 0 0	152·0	490·5		31·5	
30 0	150·5	485·8		31·0	
7 0 0	144·2	495·4		36·1	
30 0	141·9	509·6		27·2	
8 0 0	146·7	492·1		27·7	
30 0	145·0	512·5		30·0	
9 5 0	145·3	514·4		27·7	
10 0	145·3	512·5		27·6	
15 0	145·4	511·8		27·7	
20 0	144·9	514·2		27·8	
25 0	143·5	525·1	Reduced to 34·2		
30 0	140·0	525·5			
35 0	139·3	531·0			
40 0	139·3	518·5			
50 0	147·7	519·9			
55 0	149·6	530·5			
10 0 0	148·3	558·1		39·8	Reduced to 35·5
10 0	142·2	562·7		8·7	
15 0	140·5	533·6		13·6	
20 0	151·2	553·6		12·8	
25 0	145·4	538·2		20·2	
30 0	148·9	543·1		18·9	
35 0	142·6	545·0		24·9	
40 0	143·4	522·9		26·6	
45 0	143·3	518·5		27·4	
50 0	142·9	514·6		29·0	
55 0	145·6	517·5		29·3	
11 0 0	145·0	515·8		30·4	
5 0	144·8	512·8		31·6	
10 0	145·0	512·2		31·5	
15 0	145·7	509·1		28·6	
20 0	146·2	511·4		34·4	
25 0	146·4	509·0		34·2	
12 0 0	146·9	518·7		35·0	
30 0	144·6	519·2		35·1	
13 0 0	148·0	516·3		34·6	
30 0	165·2	497·7		31·6	
14 0 0	166·4	515·5		37·1	
30 0	149·9	519·7		37·3	
15 0 0	151·0	513·8		33·3	
30 0	155·4	503·8		33·8	
16 0 0	157·9	499·7		35·5	
30 0	162·0	513·6		35·8	
17 0 0	162·5	506·0		36·5	
30 0	159·3	499·5		40·3	

December 19, 20, 21, and 22, 1840.

M. Gott. Time (d. h. m. s)	Decl. (Sc.-Div)	Hor. Force (Sc.-Div)	Ther.	Vert. Force (Sc.-Div)	Ther.
21 18 0 0	157·7	488·8		41·2	
30 0	159·1	503·5		42·6	
19 0 0	157·5	513·2	Reduced to 34·2	41·5	Reduced to 33·5
30 0	154·0	509·5		40·8	
20 0 0	153·6	512·6		40·9	
30 0	154·0	511·3		40·7	
21 0 0	153·2	512·5		41·1	
30 0	151·5	514·3		40·7	
22 0 0	151·2	518·6		40·7	
30 0	151·1	517·4		40·9	
23 0 0	153·9	520·8		41·5	
30 0	150·9	521·2		42·1	
22 0 0 0	150·3	525·0	41	42·6	41
2 0 0	155·0	511·9	40	52·1	41
4 0 0	152·1	503·9	40	53·6	41
6 0 0	147·5	488·6	41	50·7	42
8 0 0	147·0	490·5	41	52·6	42
10 0 0	146·6	503·6	41	51·4	42
12 0 0	150·1	513·2	40	51·4	40
14 0 0	153·0	514·4	40	50·4	40
16 0 0	155·5	516·9	38	48·8	39
18 0 0	151·8	518·6	38	48·6	39
20 0 0	149·6	522·8	37	49·1	38
22 0 0	150·1	525·8	37	48·6	37

The Mean Positions at the regular hours of observation during the Month, are given in page 29.

ST. HELENA. The pedestals which support the Magnetometers were undergoing removal at this period.

VAN DIEMEN ISLAND. Decl. 1 Scale Division = 0′·71; H. F. k = ·0003; q = ; V. F. k = ; q =

Regular and extra observations.

The V. F. was observed at 2h. 30m. before, and the H. F. 2h. 30m. after the times specified.[a]

M. Gott. Time	Decl.	Hor. Force	Ther.	Vert. Force	Ther.
19 0 0	38·7	48·7	79	50·8	77
1 0 0	44·5	45·3		53·1	
2 0 0[b]	45·6	47·0	77	51·9	75
20 3 0 0	35·1	39·5		54·3	
15 0	34·2	45·6			
20 0	31·5				
25 0	29·2	40·5		53·1	
30 0	27·7				
35 0	32·9	43·6		52·8	
40 0	35·5				
45 0	37·7	44·1		52·9	
50 0	37·7				
55 0	37·7			52·9	
4 0 0	36·2	47·5	72	51·0	73
10 0	31·5				
15 0	28·6	47·0			
20 0	23·0	45·2		49·3	
25 0	17·2				
30 0	17·6	49·4		48·7	
35 0	22·5				
40 0	25·0	49·4		48·7	
45 0	24·4			48·3	
50 0	26·0			48·5	
5 0 0	26·6	48·4		48·5	

December 19, 20, 21, and 22, 1840.

M. Gott. Time (d. s)	Decl.	Hor. Force	Ther.	Vert. Force	Ther.
20 5 10 0	26·0	48·6			
15 0	25·1				
20 0	24·2	48·4		48·8	
25 0	24·1				
30 0	24·5	46·5		48·8	
35 0	29·1				
40 0	31·4	47·5		50·1	
45 0	35·6				
50 0	33·1	48·3		59·4	
55 0	30·3				
6 0 0	30·5	49·1	73	48·9	73
10 0	25·6	46·4			
15 0	24·4				
20 0	24·0	46·5		48·8	
25 0	24·0				
30 0	24·0	46·5		48·5	
35 0	24·0				
40 0	24·0	46·7		48·5	
45 0	24·7				
50 0	26·0	46·8		48·5	
55 0	27·2				
7 0 0	31·4	48·7		48·9	
8 0 0	39·0	49·0	72	51·1	72
9 0 0	30·2	45·9		51·1	
10 0 0	30·3	46·6	70	51·1	70
11 0 0	35·7	46·1		51·1	
12 0 0	41·4	44·7	68	56·1	66
13 0 0	49·9	44·6		57·5	
14 0 0	56·0	43·9	67	58·1	65
15 0 0	59·7	46·1		57·2	
16 0 0	57·6	48·8	67	59·1	65
17 0 0	58·4	48·7		58·4	
18 0 0	57·0	50·4	68	67·4	66
19 0 0	63·1	57·1		57·0	
20 0 0	56·7	59·8	68	54·8	66
21 0 0	56·1	55·5		58·0	
22 0 0	52·6	57·0	67	56·2	65
23 0 0	52·4	50·9		56·9	
21 0 0 0	43·1	46·1	66	56·1	65
1 0 0	41·6	44·0		55·3	
2 0 0	37·5	43·2	67	50·3	67
3 0 0	42·3	47·2		42·5	
4 0 0	50·3	52·3	65	36·6	64
5 0 0	22·2	55·0		35·6	
10 0	22·8	50·6			
15 0	19·4				
20 0	23·1	49·7		35·6	
25 0	23·6				
30 0	25·3	49·9		37·6	
35 0	25·5				
40 0	29·3	49·6		39·1	
45 0	32·4				
50 0	33·6	48·8		40·1	
57 30	32·2				
6 0 0	32·2	46·4	64	41·4	64
15 0	40·3				
20 0	42·6	43·7		41·5	
25 0	40·7				
30 0	36·6	43·7		40·2	
35 0	33·6				
40 0	32·1	43·1		38·5	
45 0	30·3				
50 0	27·2	43·2		37·0	

[a] "December 20 and 21, (Göttingen): On conversing the day after this disturbance with Sir John Pedder (Chief Justice), he told me that the Aurora had been plainly seen the night before, and for a short time was very brilliant in some of its corruscations; it was not visible from the observatory.   (Signed)   "J. H. KAY."

[b] 19d. 3h. to 20d. 2h. fell on Sunday at Van Diemen Island.

DECEMBER 19, 20, 21, and 22, 1840.

M. Gött. Time.				Decl.	Hor. Force.		Vert. Force.	
d.	h.	m.	s.	Sc.-Div.	Sc.-Div.	Ther.	Sc.-Div.	Ther.
21	6	57	30	23·6				
	7	0	0	23·6	44·5		35·4	
		10	0	24·9				
		15	0	29·5				
		20	0	32·4	51·5		35·8	
		25	0	40·2				
		30	0	44·6	53·1		36·6	
		35	0	47·9				
		40	0	50·0	54·8		36·9	
		45	0	54·4				
		50	0	51·7	50·2		37·1	
		57	30	51·1				
	8	0	0	51·1	47·3	64	39·3	63
		10	0	46·9	42·2			
		15	0	53·0				
		20	0	57·9	47·1		43·5	
		25	0	57·1				
		30	0	54·4	48·5		42·3	
		35	0	50·7				
		40	0	46·2	46·0		40·0	
		45	0	45·3				
		50	0	49·1	45·3		41·1	
		57	30	50·6				
	9	0	0	50·2	44·5		42·5	
		15	0	50·2	45·5			
		30	0	49·6	46·1			
	10	0	0	52·5	46·2	64	44·2	63
	11	0	0	45·8	45·3		45·3	
	12	0	0	39·7	45·5	63	47·6	63
	13	0	0	40·7	42·9		48·4	
	14	0	0	44·2	41·6	66	50·8	66
	15	0	0	48·0	46·2		50·8	
	16	0	0	53·2	47·2	70	52·9	69
	17	0	0	56·7	47·6		52·9	
	18	0	0	56·7	49·4	73	52·7	69
	19	0	0	53·6	48·4		51·8	
	20	0	0	49·5	48·5	74	50·8	71
	21	0	0	47·1	48·9		50·8	
	22	0	0	47·0	48·8	74	51·5	69
	23	0	0	46·6	49·0		51·5	
22	0	0	0	43·3	50·4	69	50·1	67
	1	0	0	47·6	51·2		53·0	
	2	0	0	48·4	52·4	68	52·2	66
	3	0	0	49·4	52·5		52·6	
	4	0	0	48·6	52·8	67	52·6	64
	5	0	0	49·5	53·8		49·8	
	6	0	0	49·2	54·2	65	52·6	63
	7	0	0	47·3	53·2		52·6	
	8	0	0	41·9	54·1	63	52·8	62
	9	0	0	39·3	54·5		53·7	
	10	0	0	38·2	54·0	64	52·0	63
	11	0	0	38·9	53·1		51·8	
	12	0	0	43·1	51·6	67	52·1	67
	13	0	0	48·6	49·1		52·1	
	14	0	0	53·6	48·9	70	50·4	71
	15	0	0	57·9	48·8		49·8	
	16	0	0	57·6	50·5	76	48·7	74
	17	0	0	56·8	53·0		49·8	
	18	0	0	55·1	52·9	79	49·8	76
	19	0	0	53·3	52·5		49·8	
	20	0	0	53·3	52·8	79	49·8	77
	21	0	0	50·1	51·4		48·2	
	22	0	0	49·8	52·0	77	48·2	74
	23	0	0	48·9	52·3		49·8	

The Mean Positions at the regular hours of observation during the Month, are given in page 29.

JANUARY 13 and 14, 1841.

TORONTO { Decl. 1 Scale Division = 0'·72 ; H. F. k = ·000076; q = ·0002 ; V. F. k = ·00009; q = ·00018

Extra observations.

The V. F. was observed at 1m. 30s. before, and the H. F. 2m. after the times specified.

M. Gött. Time.				Decl.	Hor. Force.		Vert. Force.	
d.	h.	m.	s.	Sc.-Div.	Sc.-Div.	Ther.	Sc.-Div.	Ther.
13	2	30	0	50·6	459·3	40	91·4	41
		35	0	54·6	459·1		91·5	
		40	0	56·1	462·9		91·2	
		45	0	57·4	457·5		91·1	
		50	0	58·0	457·4		91·6	
		55	0	58·6	456·7		91·4	
	3	0	0	59·0	456·7	40	91·3	41
		5	0	59·8	455·9		91·5	
		10	0	59·0	453·6		91·7	
		15	0	59·5	455·5		91·8	
		20	0	59·1	452·8		91·9	
		25	0	59·2	452·0		92·0	
		30	0	59·5	450·4		92·5	
		35	0	59·1	453·1		91·6	
		40	0	55·9	445·3		92·6	
		45	0	58·1	443·3		92·7	
		50	0	59·3	447·3		92·8	
	4	0	0	55·6	440·7	40	92·5	41
		10	0	55·7	437·0		93·0	
		15	0	55·9	433·1		92·5	
		20	0	55·9	433·2		93·0	
		25	0	55·8	433·4		93·4	
		30	0	56·7	434·4		92·2	
		35	0	55·7	432·6		94·4	
		40	0	56·3	433·7		94·2	
		45	0	55·5	431·8		94·0	
		50	0	54·8	432·0		93·4	
		55	0	54·6	431·8		94·9	
	5	5	0	52·3	428·6	40	95·7	41
		10	0	53·0	428·4		95·7	

Positions at the usual hours of observation, January 13 and 14.

M. Gött. Time.				Decl.	Hor. Force.		Vert. Force.	
d.	h.	m.	s.	Sc.-Div.	Sc.-Div.	Ther.	Sc.-Div.	Ther.
13	0	0	0	53·1	443·1	41	88·7	42
	2	0	0	38·1	443·4	41	90·2	41
	4	0	0	55·6	440·7	41	92·5	41
	6	0	0	52·9	430·4	41	95·8	41
	8	0	0	48·8	431·6	41	96·8	41
	10	0	0	50·5	444·0	41	96·7	41
	12	0	0	53·5	443·0	41	97·2	41
	14	0	0	59·0	433·2	41	97·0	41
	16	0	0	56·0	432·7	41	97·1	41
	18	0	0	56·5	444·1	41	93·6	41
	20	0	0	54·8	434·8	41	94·9	41
	22	0	0	52·2	424·9	41	95·5	41
14	0	0	0	45·9	433·7	41	93·4	41
	2	0	0	56·9	455·1	40	95·9	41
	4	0	0	53·1	440·2	41	95·5	41
	6	0	0	48·7	423·3	41	97·5	41
	8	0	0	43·4	425·1	41	98·5	41
	10	0	6	44·9	439·0	41	99·6	41
	12	0	0	52·0	431·7	41	99·9	41
	14	0	0	48·6	427·1	41	97·9	41
	16	0	0	60·8	425·7	41	102·2	41
	18	0	0	62·1	434·5	41	99·0	42
	20	0	0	53·0	434·0	42	98·0	42
	22	0	0	53·3	442·4	42	88·3	42

JANUARY 13 and 14, 1841.

Mean Positions at the same hours during the Month.

M. Gött. Time.				Decl.	Hor. Force.		Vert. Force.	
h.	d.	m.	s.	Sc.-Div.	Sc.-Div.	Ther.	Sc.-Div.	Ther.
0	0	0		52·6	451·2	38	97·6	39
2	0	0		55·3	454·0	38	99·6	39
4	0	0		56·6	445·7	38	98·1	39
6	0	0		51·0	430·2	38	98·3	39
8	0	0		48·1	434·7	38	99·3	40
10	0	0		40·0	443·9	38	99·5	40
12	0	0		52·1	444·3	38	99·3	40
14	0	0		55·9	442·2	38	99·3	40
16	0	0		57·1	439·2	38	99·0	40
18	0	0		55·9	443·6	38	99·3	39
20	0	0		55·1	445·6	38	98·1	39
22	0	0		54·9	445·6	38	97·7	39

ST. HELENA { Decl. 1 Scale Division = 0'·71 ; H. F. k = ·00022; q = ·0003

Positions at the usual hours of observation, January 13 and 14.

M. Gött. Time.				Decl.	Hor. Force.		Vert. Force.	
d.	h.	m.	s.	Sc.-Div.	Sc.-Div.	Ther.	Sc.-Div.	Ther.
13	0	0	0	54·0	21·9	66		
	2	0	0	55·0	18·9	68		
	3	0	0	55·0	21·0	67		
	4	0	0	53·0	19·6	68		
	5	0	0	53·0	19·2	68		
	6	0	0	53·2	17·8	68		
	8	0	0	55·0	19·3	68		
	10	0	0	56·1	14·9	66		
	11	0	0	56·1	14·2	66		
	12	0	0	54·9	15·0	66		
	13	0	0	55·0	15·0	66		
	14	0	0	54·9	16·4	66		
	15	0	0	55·7	16·4	66		
	16	0	0	53·1	16·1	66		
	18	0	0	53·2	17·2	65		
	19	30	0	50·2	17·5	65		
	20	0	0	51·8	17·8	65		
	20	30	0	50·8	18·0	65		
	22	0	0	52·1	19·9	65		
	23	0	0	52·2	30·0	65		
14	0	0	0	55·1	22·0	66		
	2	0	0	53·8	20·5	67		
	3	0	0	54·8	18·3	68		
	4	0	0	53·5	14·2	68		
	5	0	0	52 5	11·2	68		
	6	0	0	53·4	10·2	68		
	8	0	0	55·7	9·0	68		
	10	0	0	56·7	11·0	66		
	11	0	0	56·1	10·7	66		
	12	0	0	56·2	12·2	66		
	13	0	0	55·3	18·0	66		
	14	0	0	55·7	14·3	66		
	15	0	0	55·1	16·4	65		
	16	0	0	54·0	16·0	65		
	18	0	0	54·0	16·2	65		
	19	30	0	54·0	17·0	65		
	20	0	0	53·9	17·6	65		
	20	30	0	53·9	19·6	65		
	22	0	0	53·8	20·5	64		
	23	0	0	54·0	20·5	65		

Mean Positions at the same hours during the Month.

M. Gött. Time.				Decl.	Hor. Force.		Vert. Force.	
0	0	0		55·4	21·0	66		
2	0	0		55·1	20·1	67		
3	0	0		54·0	19·6	68		

JANUARY 13 and 14, 1841.

M. Gött. Time. d. h. m. s.	Decl. Sc.-Div^ns	Hor. Force. Sc.-Div^ns.	Ther.	Vert. Force. Sc.-Div^ns.	Ther.
4 0 0	53·3	18·3	68		
5 0 0	53·6	17·2	68		
6 0 0	53·4	15·9	68		
8 0 0	54·5	15·1	67		
10 0 0	55·2	14·5	67		
11 0 0	55·4	14·0	67		
12 0 0	55·2	14·4	67		
13 0 0	54·8	15·1	66		
14 0 0	54·4	15·0	66		
15 0 0	54·3	16·1	66		
16 0 0	54·1	15·9	66		
18 0 0	53·4	16·3	66		
19 30 0	52·0	16·9	66		
20 0 0	51·4	17·7	65		
20 30 0	50·6	18·6	65		
22 0 0	52·2	19·9	65		
23 0 0	53·9	21·0	66		

VAN DIEMEN ISLAND { Decl. 1 Scale Division = 0'·71
H. F. k = ·0003; q =
V. F. k = ; q =

Extra and regular observations.
Positions at the usual hours of observation, January 13.

d. h. m. s.	Decl. Sc.-Div^ns	Hor. Force. Sc.-Div^ns.	Ther.	Vert. Force. Sc.-Div^ns.	Ther.
13 0 0 0	41·0	11·0	73	7·2	71
1 0 0	38·6	12·9		6·1	
2 0 0	28·2	06·0	71	6·9	69
3 0 0	33·7	8·7		8·1	
4 0 0	36·2	10·0	68	7·5	66
5 0 0	35·8	11·5		7·0	
6 0 0	35·8	12·9	66	6·5	66
7 0 0	33·2	·13·3		6·5	
8 0 0	32·6	13·0	64	6·5	64
9 0 0	30·6	14·1		6·3	
10 0 0	26·5	14·5	63	5·9	63
11 0 0	28·6	14·6		4·9	
12 0 0	29·8	12·9	67	5·2	66
13 0 0	35·4	11·4		5·1	
14 0 0	37·9	8·9	73	5·1	73
15 0 0	43·7	8·1		5·1	
16 0 0	48·2	9·4	78	4·8	75
17 0 0	49·2	10·0		4·8	
18 0 0	47·9	10·8	79	4·3	75
19 0 0	47·2	9·9		4·2	
20 0 0	47·6	9·6	79	4·1	75
21 0 0	45·6	9·8		4·1	
22 0 0	43·6	8·9	77	5·0	74
23 0 0	35·5	7·3		3·9	
14 0 0 0	32·8	9·4	74	3·2	73
1 0 0	33·0	8·7		4·3	
2 0 0	32·6	8·5	73	5·0	73
3 0 0	29·8	10·8		3·4	
4 0 0	25·8	9·7	71	3·4	70
5 0 0	32·8	7·7		5·3	
6 0 0	32·1	8·5	69	4·7	68
7 0 0	32·5	12·0		3·4	
8 0 0	29·3	11·1	66	5·0	66
9 0 0	31·3	13·3		4·3	
10 0 0	25·8	11·8	66	4·4	65
11 0 0	26·7	10·1		4·4	
12 0 0	28·4	9·3	66	4·9	65
13 0 0	30·8	7·5		5·1	
14 0 0	35·5	6·7	67	4·9	67
15 0 0	41·9	6·3		6·5	
16 0 0	47·1	9·5	70	5·2	68
17 0 0	49·1	13·0		4·4	

JANUARY 13 and 14, 1841.

M. Gött. Time. d. h. m. s.	Decl. Sc.-Div^ns	Hor. Force. Sc.-Div^ns.	Ther.	Vert. Force. Sc.-Div^ns.	Ther.
14 18 0 0	49·4	11·2	72	5·4	69
19 0 0	47·8	13·3		4·0	
20 0 0	43·3	11·6	74	4·0	72
21 0 0	43·3	11·6		4·0	
22 0 0	40·4	10·1	73	5·0	71
23 0 0	41·1	12·9		5·0	

Mean Positions at the same hours during the Month.

d. h. m. s.	Decl. Sc.-Div^ns	Hor. Force. Sc.-Div^ns.	Ther.	Vert. Force. Sc.-Div^ns.	Ther.
0 0 0	37·8	11·6	71	4·7	70
1 0 0	36·5	12·5		4·4	
2 0 0	36·5	12·4	69	4·7	68
3 0 0	36·8	13·0		4·5	
4 0 0	35·7	12·7	68	5·0	67
5 0 0	36·0	13·2		5·1	
6 0 0	36·5	13·6	66	5·3	65
7 0 0	36·0	14·0		5·3	
8 0 0	34·8	14·4	64	5·3	64
9 0 0	33·2	14·6		5·3	
10 0 0	30·9	13·4	64	5·6	63
11 0 0	30·4	12·2		5·4	
12 0 0	31·1	11·8	65	5·4	65
13 0 0	32·9	10·8		5·4	
14 0 0	38·3	9·4	69	5·5	69
15 0 0	43·1	9·3		5·2	
16 0 0	46·8	10·3	73	4·8	71
17 0 0	48·4	11·4		4·4	
18 0 0	47·3	11·9	75	4·3	73
19 0 0	45·0	11·9		3·9	
20 0 0	43·2	11·0	76	4·1	73
21 0 0	41·2	10·0		4·1	
22 0 0	39·9	10·9	74	4·2	72
23 0 0	39·2	11·5		4·3	

JANUARY 18 and 19, 1841.

TORONTO { Decl. 1 Scale Division = 0'·72
H. F. k = ·000076; q = ·0002
V. F. k = ·00009; q = ·00018

Extra observations.
The V. F. was observed at 1m. 30s. before, and the H. F. 2m. after the times specified.

d. h. m. s.	Decl. Sc.-Div^ns	Hor. Force. Sc.-Div^ns.	Ther.	Vert. Force. Sc.-Div^ns.	Ther.
18 0 15 0	50·7	463·5	32	103·3	34
20 0	51·7	463·4		103·7	
25 0	51·4	462·6		104·3	
30 0	52·3	463·8		105·0	
35 0	51·9	462·6		104·8	
40 0	52·8	464·4		104·1	
45 0	51·8	463·3		103·9	
50 0	52·4	464·6		103·9	
55 0	53·0	465·8		103·8	
1 0 0	52·6	465·7	31	103·8	34
5 0	54·2	463·6		103·8	
10 0	54·0	465·1		104·0	
15 0	53·7	464·6		104·0	
20 0	53·2	463·3		104·1	
25 0	53·6	464·3		104·1	
30 0	54·2	464·3		104·8	
35 0	54·3	463·2		104·7	
40 0	53·6	463·7		104·7	
45 0	53·5	463·0		104·8	
2 45 0	57·4	466·1	31	105·1	33

JANUARY 18 and 19, 1841.

M. Gött. Time. d. h. m. s.	Decl. Sc.-Div^ns	Hor. Force. Sc.-Div^ns.	Ther.	Vert. Force. Sc.-Div^ns.	Ther.
18 2 50 0	57·4	465·0		104·8	
55 0	57·2	465·0			
3 0 0	57·4	464·2	31	104·5	33
5 0	57·5	463·6		104·7	
10 0	58·7	462·5		104·7	
15 0	58·4	461·2		106·8	
20 0	59·4	460·7		106·8	
25 0	61·6	462·1		104·8	
30 0	59·3	459·7		105·3	
35 0	61·1	459·9		106·2	
40 0	53·6	459·3		106·5	
45 0	54·6	457·1		105·4	
50 0	56·0	456·2		105·3	
55 0	56·6	455·7		105·1	
4 0 0	53·8	455·2	31	104·6	33
5 0	54·1	452·0		104·6	
10 0	55·2	454·3		104·7	
15 0	55·2	454·5		104·4	
20 0	55·2	454·5		104·4	
25 0	54·3	454·1		105·3	
30 0	54·6	451·7		105·2	
35 0	54·2	450·5		104·6	
40 0	54·3	451·8		104·6	
45 0	54·8	449·7		104·5	
50 0	53·6	448·7		104·6	
55 0	54·0	449·7		105·0	
5 0 0	53·9	450·5	31	105·0	33
5 0	52·4	448·9		105·0	
10 0	53·5	449·3		104·9	
15 0	53·7	448·9		104·4	
20 0	52·6	448·7		104·6	
25 0	53·1	449·4		104·6	
30 0	52·3	448·7		103·8	
35 0	52·1	448·6		104·4	
40 0	51·1	448·0		104·2	
45 0	52·4	449·1		105·0	
50 0	52·7	451·5		105·0	
55 0	52·1	449·4		105·0	
6 0 0	51·5	450·1	30	105·5	32
5 0	51·3	450·0		105·0	
10 0	51·6	451·2		105·0	
15 0	52·4	454·3		105·6	
20 0	51·2	452·8		105·7	
25 0	51·2	451·3		105·7	
30 0	51·7	452·9		105·6	
35 0	51·1	453·6		106·3	
40 0	50·7	453·9		106·3	
45 0	51·1	453·6		106·4	
50 0	50·6	454·5		106·8	
55 0	51·3	453·9		106·8	
7 0 0	51·1	455·0	30	106·6	32
5 0	51·2	454·3		106·6	
10 0	51·3	457·6		107·0	
15 0	51·3	455·8		107·0	
20 0	51·1	457·5		107·1	
25 0	51·1	456·8		107·0	
30 0	51·1	458·2		107·0	
35 0	50·8	458·7		107·0	
40 0	50·2	458·7		106·8	
45 0	51·2	459·7		106·8	
50 0	51·5	459·1			
55 0	51·1	459·0		107·0	
8 0 0	50·9	459·5	30	108·5	32
5 0	50·4	459·4		107·5	
10 0	51·2	460·5		107·0	
15 0	51·1	461·7		106·9	
20 0	51·4	461·4		106·2	

F

JANUARY 18 and 19, 1841.

M. Gött. Time.	Decl.	Hor. Force.		Vert. Force.	
d. h. m. s.	Sc.-Div^ns.	Sc.-Div^ns.	Ther.	Sc.-Div^ns.	Ther.
18 8 25 0	51·3	462·0		106·9	
30 0	50·8	462·4		106·7	
35 0	51·0	462·5		106·5	
40 0	52·2	461·6		107·0	
45 0	51·3	463·7		106·2	
50 0	51·1	464·2		106·3	
55 0	51·5	464·2		106·3	
9 0 0	51·5	465·8	31	106·4	33
5 0	51·1	465·5		106·4	
10 0	51·2	466·3		106·2	
15 0	50·5	467·7		106·2	
20 0	51·9	467·7		106·2	
25 0	51·9	468·4		106·2	
30 0	51·0	469·7		105·8	
35 0	50·9	469·6		105·8	
40 0	51·3	468·6		105·8	
45 0	51·5	470·0		105·8	
50 0	51·3	470·0		105·8	
55 0	51·4	470·2		105·8	
10 0 0	51·4	471·0	31	105·8	33
5 0	51·9	471·2		106·1	
10 0	52·3	471·8		106·0	
15 0	51·8	469·7		100·0	
20 0	52·5	470·8		106·0	
25 0	52·7	470·5		106·0	
30 0	53·3	471·7		106·0	
35 0	53·6	470·0		106·2	
40 0	53·2	471·1		106·1	
45 0	53·3	471·9		105·9	
50 0	52·9	472·2		106·6	

Positions at the usual hours of observation, January 18 and 19.*

M. Gött. Time.	Decl.	Hor. Force.		Vert. Force.	
d. h. m. s.	Sc.-Div^ns.	Sc.-Div^ns.	Ther.	Sc.-Div^ns.	Ther.
18 0 0 0	52·5	463·2	32	103·8	34
2 0 0	54·8	463·1	31	106·5	33
4 0 0	56·4	454·5	31	104·7	33
6 0 0	51·8	450·2	30	105·5	32
8 0 0	51·0	459·7	30	107·0	32
10 0 0	52·5	470·6	31	105·8	33
12 0 0	53·7	472·8	31	106·0	32
14 0 0	53·5	473·4	31	106·0	32
16 0 0	52·9	475·3	29	107·0	31
18 0 0	53·2	474·5	28	107·2	30
20 0 0^b	58·1	475·0	29	106·3	30
22 0 0	56·6	482·0	28	108·3	29
19 0 0 0	55·4	485·4	27	108·7	30
2 0 0	52·3	497·4	27	108·6	29
4 0 0	55·3	481·7	28	106·1	30
6 0 0	41·7	448·5	29	107·0	30
8 0 0	47·0	460·8	30	108·1	31
10 0 0	48·5	465·6	32	107·8	32
12 0 0	55·4	469·2	32	114·8	33
14 0 0	60·0	443·1	33	114·7	33
16 0 0	55·6	447·7	33	113·4	33
18 0 0	54·3	444·2	33	117·8	33
20 0 0	53·8	448·1	32	117·6	33
22 0 0	53·2	448·1	32	115·1	33

The Mean Positions at the same hours during the Month are given in page 32.

JANUARY 18 and 19, 1841.

ST. HELENA {Decl. 1 Scale Division = 0'·71 / H. F. k = ·00021; q = ·0003}

Positions at the usual hours of observation, January 18 and 19.

M. Gött. Time.	Decl.	Hor. Force.		Vert. Force.	
d. h. m. s.	Sc.-Div^ns.	Sc.-Div^ns.	Ther.	Sc.-Div^ns.	Ther.
18 0 0 0	54·9	24·6	66		
2 0 0	55·0	24·3	68		
3 0 0	57·0	23·9	68		
4 0 0	53·0	22·6	68		
5 0 0	53·2	22·0	68		
6 0 0	54·2	21·2	69		
8 0 0	55·2	19·8	68		
10 0 0	56·0	19·9	67		
11 0 0	56·1	19·9	67		
12 0 0	55·9	19·6	67		
13 0 0	55·2	19·9	67		
14 0 0	53·2	20·0	66		
15 0 0	54·9	20·2	66		
16 0 0	55·1	20·8	66		
18 0 0	53·8	20·4	66		
19 0 0	51·8	23·8	66		
20 0 0	50·5	25·0	66		
21 0 0	50·2	25·0	66		
22 0 0	54·2	25·0	66		
23 0 0	55·9	26·8	66		
19 0 0 0	56·0	25·2	66		
2 0 0	53·5	25·3	67		
3 0 0	51·9	26·8	67		
4 0 0	51·8	24·0	68		
5 0 0	51·7	19·4	68		
6 0 0	50·0	15·8	68		
8 0 0	52·2	16·0	67		
10 0 0	53·0	18·0	67		
11 0 0	52·9	9·0	67		
12 0 0	51·8	9·6	66		
13 0 0	51·5	9·1	66		
14 0 0	51·2	14·8	66		
15 0 0	51·5	14·8	66		
16 0 0	50·8	14·5	66		
18 0 0	50·0	15·0	66		
19 0 0	50·0	16·5	66		
20 0 0	49·0	17·0	66		
21 0 0	48·6	18·0	66		
22 0 0	49·8	21·2	66		
23 0 0	48·9	23·2	66		

The Mean Positions at the same hours during the Month are given in pages 32 and 33.

VAN DIEMEN ISLAND {Decl. 1 Scale Division = 0'·71 / H. F. k = ·0003; q = / V. F. k = ; q = }

Extra observations.

The V. F. was observed at 2m. 30s. before, and the H. F. 2m. 30s. after the times specified.

M. Gött. Time.	Decl.	Hor. Force.		Vert. Force.	
d. h. m. s.	Sc.-Div^ns.	Sc.-Div^ns.	Ther.	Sc.-Div^ns.	Ther.
19 11 10 0	41·9	6·8	68		
15 0	43·3				
20 0	42·2	7·8		8·8	68
25 0	42·1				
30 0	39·9	6·8		7·8	

JANUARY 18 and 19, 1841.

M. Gött. Time.	Decl.	Hor. Force.		Vert. Force.	
d. h. m. s.	Sc.-Div^ns.	Sc.-Div^ns.	Ther.	Sc.-Div^ns.	Ther.
19 11 35 0	38·3				
40 0	38·3	5·1		7·0	
45 0	40·2				
50 0	44·6	6·7		7·8	
55 0	47·8				
12 10 0	45·3	9·7	69		
15 0	45·3				
20 0	43·9	8·9		6·0	68
25 0	42·5				
30 0	41·3	9·4		5·4	
35 0	42·1				
40 0	42·3	8·4		6·9	
45 0	41·0				
50 0	42·8	9·3		5·1	
13 10 0	40·0	8·5	69		
15 0	41·8				
20 0	42·6	8·6		4·3	69
25 0	40·4				
30 0	39·0			3·6	

Positions at the usual hours of observation, January 18 and 19.

M. Gött. Time.	Decl.	Hor. Force.		Vert. Force.	
d. h. m. s.	Sc.-Div^ns.	Sc.-Div^ns.	Ther.	Sc.-Div^ns.	Ther.
18 0 0 0	40·2	9·7	77	3·3	75
1 0 0	39·8	10·1		3·4	
2 0 0	38·0	11·1	75	3·3	73
3 0 0	38·6	10·6		3·5	
4 0 0	38·5	11·4	73	4·4	72
5 0 0	38·1	12·5		4·5	
6 0 0	37·8	13·0	70	4·5	70
7 0 0	37·1	14·0		5·2	
8 0 0	35·7	14·5	68	5·1	67
9 0 0	33·3	15·1		5·1	
10 0 0	30·6	14·4	67	5·0	66
11 0 0	29·9	13·8		4·8	
12 0 0	30·5	12·0	68	4·7	67
13 0 0	32·8	10·9		4·0	
14 0 0	36·7	9·7	74	4·5	73
15 0 0	41·6	9·0		3·0	
16 0 0	44·7	8·7	79	3·0	78
17 0 0	46·5	10·0		2·0	
18 0 0	46·3	11·9	82	2·3	77
19 0 0	45·3	10·6		2·3	
20 0 0	45·9	15·1	81	1·1	77
21 0 0	44·3	10·0		3·7	
22 0 0	43·9	9·8	78	3·9	76
23 0 0	43·5	13·0		2·3	
19 0 0 0	42·3	12·4	76	3·2	74
1 0 0	37·1	10·8		3·3	
2 0 0		9·8	74	5·5	74
3 0 0	39·0	11·2		4·5	
4 0 0	33·1	11·0	73	4·5	72
5 0 0	32·2	13·6		1·7	
7 0 0	35·5	12·5		6·1	
8 0 0	34·5	14·4	68	5·4	67
9 0 0	35·0	14·5		5·6	
10 0 0	32·5	10·8	66	5·4	67
11 0 0	45·1	8·2		9·0	
12 0 0	48·5	8·3	69	8·0	68
13 0 0	44·4	10·1		4·9	
14 0 0	44·5	8·5	70	4·6	69
15 0 0	48·8	9·1		3·8	
16 0 0	49·3	11·3	70	3·0	70

a "Although the abstract does not show any great change in the readings of the magnetometers, the magnets were, nevertheless, much disturbed; sometimes vibrating considerably, and at other times suddenly becoming perfectly quiet. The partial results were during the whole series very irregular."

b 18d. 20h. 0m. Faint auroral light in the North.

JANUARY 18 and 19, 1841.

M. Gött. Time. d. h. m. s.	Decl. Sc.-Div.	Hor. Force. Sc.-Div.	Ther.	Vert. Force. Sc.-Div.	Ther.
19 17 0 0	47·7	12·0		3·9	
18 0 0	46·8	12·4		4·8	71
19 0 0	43·3	12·6	72	4·5	
20 0 0	42·2	10·9		5·2	71
21 0 0	40·9	10·6	72	5·8	
22 0 0	40·8	10·8		4·3	69
23 0 0	39·2	12·8	70	5·2	

The Mean Positions at the same hours during the Month, are given in page 33.

JANUARY 25, 26, and 27, 1841.

TORONTO { Decl. 1 Scale Division = 0'·72
H. F. k = ·000076 ; q = ·0002
V. F. k = ·00009 ; q = ·00018 }

Extra observations.

The V. F. was observed at 1m. 30s. before, and the H. F. 2m. after the times specified.

M. Gött. Time. d. h. m. s.	Decl. Sc.-Div.	Hor. Force. Sc.-Div.	Ther.	Vert. Force. Sc.-Div.	Ther.
25 16 15 0	96·4	417·7	41	92·1	42
20 0	94·9	407·9		90·0	
25 0	94·0	404·6		89·5	
30 0	92·1	401·4		88·5	
35 0	87·8	392·1		88·2	
40 0	85·2	389·5		90·6	
45 0	80·0	389·2		90·6	
50 0	73·0	388·5		91·6	
55 0	65·6	390·7		95·5	
17 0 0	66·2	402·9	41	93·4	42
5 0	63·4	413·0		96·4	
10 0	63·7	419·2		94·9	
26 4 10 0	57·2	447·6	40	95·4	40
15 0	57·6	443·9		95·4	
20 0	59·1	447·6		95·6	
25 0	58·0	448·3		95·5	
30 0	57·3	446·0		95·3	
35 0	56·9	445·1		95·5	
40 0	58·8	448·0		95·2	
45 0	56·1	441·9		95·2	
50 0	58·3	446·5		94·9	
55 0	56·3	437·7		95·5	
5 0 0	56·3	434·5	40	95·1	40
5 0	56·6	435·5		94·6	
10 0	55·5	434·9		94·6	
15 0	53·4	438·7		95·3	
20 0	53·2	431·6		95·5	
25 0	51·3	428·4		94·5	
30 0	51·5	429·4		94·5	
35 0	52·0	424·9		94·9	
40 0	51·5	428·6		95·0	
45 0	50·9	425·6		95·1	
50 0	51·5	426·5		95·1	
27 3 0 0	58·5	437·9	45	92·8	45
5 0	55·7	423·8	45	89·9	45
7 0	48·0	416·7	45	91·1	45
9 0	45·3	425·8	46	92·2	46
13 0	52·3	422·5	47	96·6	46
15 0	63·0	432·0	45	93·2	45
17 0	55·8	430·6	45	93·6	45
19 0	57·1	428·9	45	83·9	45
21 0	57·2	430·8	45	87·7	45

JANUARY 25, 26, and 27, 1841.

Positions at the usual hours of observation, January 25, 26, and 27.

M. Gött. Time. d. h. m. s.	Decl. Sc.-Div.	Hor. Force. Sc.-Div.	Ther.	Vert. Force. Sc.-Div.	Ther.
25 0 0 0	50·7	450·4	40	94·9	41
2 0 0	55·0	444·0	40	95·5	40
4 0 0	53·1	449·4	40	95·2	40
6 0 0	49·1	429·5	41	92·9	41
8 0 0	48·5	436·9	42	93·2	42
10 0 0	50·1	439·7	42	93·3	42
12 0 0	53·0	442·8	41	93·8	42
14 0 0	54·2	440·9	41	94·3	42
16 0 0	80·3	413·2	41	94·4	42
18 0 0	67·5	420·3	42	91·9	43
20 0 0	52·6	433·0	42	94·9	42
22 0 0	53·2	422·2	41	93·8	42
26 0 0 0	56·1	451·7	40	93·0	41
2 0 0	52·1	448·9	38	97·5	40
4 0 0	58·3	445·7	40	95·4	40
6 0 0	51·4	418·4	40	95·0	41
8 0 0	44·5	420·8	41	99·5	41
10 0 0	44·1	441·5	41	98·6	41
12 0 0	50·2	440·0	41	97·5	41
14 0 0	57·9	431·9	41	97·0	41
16 0 0	59·9	449·7	41	91·4	42
18 0 0	57·9	431·7	42	93·2	42
20 0 0	52·9	432·3	42	92·6	43
22 0 0	58·7	420·4	43	85·8	43
27 0 0 0	48·2	435·6	43	88·8	44
2 0 0	58·6	438·7	43	94·1	44
4 0 0	57·6	434·6	45	90·1	45
6 0 0	51·3	416·2	45	88·9	45
8 0 0	47·3	419·0	46	89·1	45
10 0 0	47·2	408·2	46	94·6	46
12 0 0	46·7	411·5	46	91·8	46
14 0 0	60·5	449·9	46	88·7	47
16 0 0*	56·3	420·1	45	93·0	45
18 0 0	53·2	426·1	45	90·8	45
20 0 0	55·7	431·2	44	82·7	45
22 0 0	55·2	423·1	44	86·0	45

The Mean Positions at the same hours during the Month are given in page 32.

ST. HELENA { Decl. 1 Scale Division = 0'·71
H. F. k = ·00021 ; q = ·0003 }

Positions at the usual hours of observation, January 25, 26, and 27.

M. Gött. Time. d. h. m. s.	Decl. Sc.-Div.	Hor. Force. Sc.-Div.	Ther.
25 0 0 0	51·1	28·9	67
2 0 0	53·1	28·1	67
3 0 0	53·5	26·0	68
4 0 0	54·1	26·0	68
5 0 0	54·5	25·0	69
6 0 0	52·2	21·0	68
8 0 0	53·0	20·9	68
10 0 0	54·0	21·2	67
11 0 0	54·3	21·7	67
12 0 0	54·4	22·4	66
13 0 0	53·5	23·0	66
14 0 0	53·1	23·9	66
15 0 0	55·3	25·3	66
16 0 0	51·5	25·7	66
18 0 0	52·9	21·6	65
19 0 0	51·6	21·0	66
20 0 0	49·1	21·0	65

JANUARY 25, 26, and 27, 1841.

M. Gött. Time. d. h. m. s.	Decl. Sc.-Div.	Hor. Force. Sc.-Div.	Ther.	Vert. Force. Sc.-Div.	Ther.
25 21 0 0	48·9	22·0	66		
22 0 0	50·8	27·0	65		
23 0 0	52·1	27·0	66		
26 0 0 0	52·8	28·4	66		
2 0 0	54·1	25·6	67		
3 0 0	51·5	24·9	67		
4 0 0		23·9	67		
5 0 0		23·5	68		
6 0 0	52·4	22·7	68		
8 0 0	52·8	18·9	67		
10 0 0	52·6	16·7	67		
11 0 0	52·8	18·1	67		
12 0 0	52·7	18·2	67		
13 0 0	51·9	20·5	66		
14 0 0	51·1	21·1	66		
15 0 0	51·1	22·0	66		
16 0 0	51·1	22·9	66		
18 0 0	50·9	24·0	66		
19 0 0	50·0	22·9	66		
20 0 0	48·3	23·9	66		
21 0 0	47·0	24·1	66		
22 0 0	48·1	23·4	65		
23 0 0	50·1	23·0	66		
27 0 0 0	51·1	23·5	67		
2 0 0	52·7	25·0	68		
3 0 0	51·4	26·2	69		
4 0 0	51·6	27·5	69		
5 0 0	52·1	24·2	69		
6 0 0	52·6	23·6	69		
8 0 0	53·4	20·5	68		
10 0 0	52·4	18·9	68		
11 0 0	52·6	18·9	68		
12 0 0	53·4	16·8	67		
13 0 0	53·6	17·4	67		
14 0 0	54·0	19·6	67		
15 0 0	55·1	22·0	67		
16 0 0	53·1	21·0	67		
18 0 0	52·2	21·3	66		
19 0 0	51·9	21·3	67		
20 0 6	50·4	21·0	66		
21 0 0		22·2	67		
22 0 0	50·7	23·5	67		
23 0 0	51·2	22·5	67		

The Mean Positions at the same hours during the Month are given in pages 32 and 33.

VAN DIEMEN ISLAND { Decl. 1 Scale Division = 0'·71
H. F. k = ·0003 ; q =
V. F. k = ; q = }

Positions at the usual hours of observation, January 25, 26, and 27.

M. Gött. Time. d. h. m. s.	Decl. Sc.-Div.	Hor. Force. Sc.-Div.	Ther.	Vert. Force. Sc.-Div.	Ther.
25 0 0 0	40·5	11·4	76	3·7	73
1 0 0	39·9	12·6		2·5	
2 0 0	36·5	15·6	74	0·6	73
3 0 0	36·6	14·4		0·5	
4 0 0	35·2	12·8	73	1·6	72
5 0 0	36·0	13·5		2·3	
6 0 0	32·6	14·6	71	1·4	70
7 0 0	32·3	15·0		1·6	
8 0 0	33·1	15·8	71	1·9	70
9 0 0	34·3	14·5		2·5	
10 0 0	31·6	14·5	69	3·2	70

* From 27d. 16h. to 27d. 18h. the sky was partially clouded, and a faint auroral light was visible in the North; at 27d. 19h. clouds spread around the horizon, and at 27d. 21h. the sky became heavily overcast. Previous to 27d. 16h. the sky was densely clouded.

JANUARY 25, 26, and 27, 1841.

d.	h.	m.	s.	Decl. Sc.-Div	Hor. Force Sc.-Div	Ther.	Vert. Force Sc.-Div	Ther.
25	11	0	0	29·7	12·9		4·0	
	12	0	0	30·9	10·8	68	4·4	67
	13	0	0	34·5	10·7		4·5	
	14	0	0	41·8	12·2	68	4·0	67
	15	0	0	44·4	16·3		2·8	
	16	0	0	48·2	17·3	67	3·9	66
	17	0	0	48·4	13·4		6·1	
	18	0	0	48·0	16·7	68	4·5	66
	19	0	0	45·1	15·7		4·4	
	20	0	0	43·1	16·3	69	3·9	66
	21	0	0	41·3	16·1		5·2	
	22	0	0	34·6	19·4	67	3·4	65
	23	0	0	39·0	15·7		5·7	
26	0	0	0	38·0	15·6	64	6·0	62
	1	0	0	32·2	21·1		3·9	
	2	0	0	38·5	18·5	63	4·4	
	3	0	0	38·1	18·2		5·0	
	4	0	0	40·9	18·1	63	5·3	62
	5	0	0	40·1	18·8		5·3	
	6	0	0	40·1	17·6	60	5·8	60
	7	0	0	42·1	17·8		5·1	
	8	0	0	38·5	19·3	59	3·4	59
	9	0	0	32·9	20·5		2·5	
	10	0	0	29·9	15·2	59	2·7	59
	11	0	0	29·0	18·5		3·1	
	12	0	0	29·1	16·7	61	3·4	62
	13	0	0	33·4	15·8		3·7	
	14	0	0	38·4	13·7	65	3·9	66
	15	0	0	42·1	13·9		3·6	
	16	0	0	47·0	12·5	70	4·7	70
	17	0	0	48·2	12·6		4·5	
	18	0	0	49·9	14·1	72	4·0	70
	19	0	0	47·0	18·2		3·4	
	20	0	0	44·7	13·0	72	5·4	69
	21	0	0	42·7	17·2		3·2	
	22	0	0	36·7	15·0	72	3·2	69
	23	0	0	37·9	13·9		4·5	
27	0	0	0	34·8	13·7	69	4·6	67
	1	0	0	37·6	18·0		2·1	
	2	0	0	36·7	19·3	67	2·5	65
	3	0	0	39·0	16·2		4·6	
	4	0	0	38·3	15·8	66	5·3	64
	5	0	0	34·0	17·1		5·6	
	6	0	0	40·2	17·6	63	5·7	62
	7	0	0	40·3	18·5		4·5	
	8	0	0	37·4	18·0	62	5·2	61
	9	0	0	38·4	19·0		5·4	
	10	0	0	43·1	15·9	60	8·3	59
	11	0	0	36·4	15·7		5·7	
	12	0	0	39·5	13·5	63	5·3	63
	13	0	0	41·3	13·6		4·4	
	14	0	0	40·4	10·6	70	3·8	69
	15	0	0	44·9	8·7		4·7	
	16	0	0	46·5	9·8	75	3·8	72
	17	0	0	48·4	11·1		2·9	
	18	0	0	48·0	16·8	76	1·3	73
	19	0	0	44·3	10·5		4·5	
	20	0	0	40·7	13·6	76	2·9	74
	21	0	0	42·3	12·5		2·9	
	22	0	0	40·2	12·1	75	3·6	73
	23	0	0	35·1	11·3		3·5	

The Mean Positions at the same hours during the Month are given in page 33.

FEBRUARY 7, 1841.

VAN DIEMEN ISLAND { Decl. 1 Scale Division = 0'·71; H. F. $k = ·0003$; $q =$; V. F. $k =$; $q =$

Extra observations.[a]

The V. F. was observed at 2m. 30s. before, and the H. F. 2m. 30s. after the times specified.

d.	h.	m.	s.	Decl. Sc.-Div	Hor. Force Sc.-Div	Ther.	Vert. Force Sc.-Div	Ther.
7	3	5	0	20·6				
		10	0	19·9	60·2	64		
		15	0	21·3			94·2	
		20	0	21·6	58·6			
		25	0	20·1			94·5	
		30	0	15·5	56·4			
		35	0	13·8			93·9	
		40	0	16·0	60·4			
		45	0	20·9			92·9	
		50	0	23·0	60·9			
		55	0	23·3			92·6	
	4	5	0	22·7				64
		15	0	22·3				
		25	0b	21·4			93·1†	
		35	0	22·8				
		45	0	23·4			94·0†	
		55	0	26·0				
	5	10	0	37·4				
		25	0	40·2			94·7†	
		35	0	41·7				
		40	0	40·9	64·7		90·6	
		45	0	38·7				
		50	0	35·3			88·5	
	6	10	0	29·2	63·9	63		63
		15	0	27·5				
		20	0	29·2	64·2		89·6	
		25	0	28·7				
		30	0	29·6	64·9		91·1	
		35	0	30·4				
		40	0	30·4	65·4		92·0	
		45	0	30·9				
		50	0	32·2	64·4		92·9	
	7	5	0	34·6				
		10	0	34·5	66·2		94·9	
		15	0	35·1				
		20	0	35·3	65·8	63	94·7	62
	9	10	0	26·2	68·1	63		
		15	0	26·8				
		20	0	24·9	68·8		95·1	
		25	0	25·6				
		30	0	25·3	67·8		95·5	
		35	0	25·4				
		40	0	26·4	67·5		95·9	
		45	0	25·0				
		50	0	24·2				
	10	10	0	23·9	63·8	63	96·1	62
		15	0	26·7				
		20	0	32·6	64·9		98·0	
		25	0	34·6				
		30	0	35·4	64·2		98·5	
		35	0	36·7				
		40	0	35·3	62·8		98·0	
		45	0	40·3				
		50	0	40·3	64·1		98·9	
	11	10	0	43·4	66·1			

FEBRUARY 7, 1841.

d.	h.	m.	s.	Decl. Sc.-Div	Hor. Force Sc.-Div	Ther.	Vert. Force Sc.-Div	Ther.
7	11	15	0	40·5				
		20	0	37·3	59·4		95·9	
		25	0	27·0				
		30	0	32·0	63·0		95·4	
		35	0	32·9				
		40	0	32·0	64·5		94·3	
		45	0	33·4				
		50	0	29·3	58·2		93·2	64
	12	10	0	32·2	62·4	64		
		15	0	30·2				
		20	0	29·2	62·3		92·8	
		25	0	29·4				
		30	0	31·0	62·4		92·0	
		35	0	31·6				
		40	0	31·6	61·0		92·2	
		45	0	30·6				
		50	0	30·8	60·6		91·5	
	13	10	0	33·0	59·0			
		15	0	32·5				
		20	0	31·4	59·1		92·7	
		25	0	31·2				
		30	0	31·2	58·3		92·6	
		35	0	32·8				
		40	0	34·5	59·0		93·6	
		45	0	36·3				
		50	0	37·5		69	94·7	69

Positions at the usual hours of observation, February 7.

d.	h.	m.	s.	Decl. Sc.-Div	Hor. Force Sc.-Div	Ther.	Vert. Force Sc.-Div	Ther.
7	3	0	0b	24·6	60·5		95·0	
	4	0	0	23·4	61·0	64	92·8	64
	5	0	0	26·0	56·5		94·7	
	6	0	0	33·0	63·9	63	88·9	63
	7	0	0	34·8	65·0		94·8	
	8	0	0	32·0	66·6	62	95·5	61
	9	0	0	27·8	67·8		95·4	
	10	0	0	22·5	67·9	63	97·9	62
	11	0	0	39·0	63·5		98·2	
	12	0	0	27·7	60·2	64	94·9	64
	13	0	0	31·7	59·5		92·2	
	14	0	0	38·7	59·3	69	94·6	69
	15	0	0	45·3	59·9		95·2	
	16	0	0	48·3	60·2	72	95·7	71
	17	0	0	50·4	62·1		95·6	
	18	0	0	49·0	63·2	75	95·0	73
	19	0	0	48·8	62·4		94·0	
	20	0	0	45·8	61·8	77	93·7	76
	21	0	0	42·6	61·1		92·8	
	22	0	0	42·5	61·2	76	93·3	73
	23	0	0	41·9	62·8		94·2	

Mean Positions at the same hours during the Month.

d.	h.	m.	s.	Decl. Sc.-Div	Hor. Force Sc.-Div	Ther.	Vert. Force Sc.-Div	Ther.
	0	0	0	40·9	69·9	69	94·9	68
	1	0	0	39·6	70·1		94·7	
	2	0	0	38·7	70·3	68	94·7	67
	3	0	0	37·8	69·0		94·7	
	4	0	0	38·2	69·4	67	94·9	66
	5	0	0	38·2	69·6		94·8	
	6	0	0	38·4	70·1	66	95·8	66
	7	0	0	38·7	70·8		95·0	
	8	0	0	38·1	70·6	65	95·4	63
	9	0	0	36·0	70·8		95·5	

[a] The ten hours of the 7th of February, during which these extra observations were made at Van Diemen Island, formed part of Sunday at Toronto and St. Helena; there are consequently no comparative observations at those stations.

[b] The three observations marked thus † were taken 5m. before the times specified.

[c] The hours of 0, 1, and 2, on February 7th, fell on Sunday at Van Diemen Island.

FEBRUARY 7, 1841.

M. Gött. Time. d h m s	Decl. Sc.-Div	Hor. Force Sc.-Div	Ther.	Vert. Force Sc.-Div	Ther.
10 0 0	34·8	70·0	64	95·7	63
11 0 0	35·8	69·0		96·0	
12 0 0	35·2	67·5	65	95·7	64
13 0 0	38·1	67·0		95·9	
14 0 0	42·5	66·5	68	95·7	67
15 0 0	44·8	67·3		95·5	
16 0 0	46·9	67·9	70	95·3	69
17 0 0	47·7	68·7		94·9	
18 0 0	47·2	68·9	72	94·8	69
19 0 0	46·1	68·8		94·8	
20 0 0	44·3	68·7	73	94·5	70
21 0 0	43·3	68·3		94·8	
22 0 0	42·3	69·0	71	94·6	69
23 0 0	42·0	69·7		94·7	

FEBRUARY 8 and 9, 1841.

TORONTO { Decl. 1 Scale Division = 0'·72
H. F. k = ·000076; q = ·0002
V. F. k = ·00009; q = ·00018

Extra observations.

The V. F. was observed at 1m. 30s. before, and the H. F. 2m. after the times specified.

d h m s	Decl. Sc.-Div	Hor. Force Sc.-Div	Ther.	Vert. Force Sc.-Div	Ther.
9 2 30 0	47·0	453·0	40	107·6	41
35 0	46·3	449·5		105·9	
40 0	51·2	445·4		108·1	
45 0	53·9	441·9		108·5	
50 0	54·7	444·9		110·6	
55 0	53·8	447·1		112·9	
3 0 0	54·5	446·0	41	113·8	41
5 0	51·1	446·3		113·2	
10 0	46·6	441·8		114·1	
15 0	51·6	430·7		114·0	
20 0	54·0	430·6		114·3	
25 0	54·7	429·2		115·6	
30 0	55·1	427·1		116·1	
40 0	54·9	430·5		117·8	
45 0	54·8	429·1		118·1	
50 0	53·5	430·3		119·1	
55 0	55·2	427·2		119·4	

Positions at the usual hours of observation, from February 8th, 0h. to February 9th, 22h.

d h m s	Decl. Sc.-Div	Hor. Force Sc.-Div	Ther.	Vert. Force Sc.-Div	Ther.
8 0 0 0	50·5	433·0	37	124·5	39
2 0 0	53·5	444·9	37	130·2	38
4 0 0	53·9	422·3	38	124·4	39
6 0 0	42·8	412·0	40	124·7	40
8 0 0	45·5	425·0	41	122·9	41
10 0 0	48·3	430·6	42	122·1	42
12 0 0	52·2	426·2	42	123·4	42
14 0 0	50·0	429·8	42	123·2	42
16 0 0	50·3	431·1	41	122·5	41
18 0 0	48·3	434·1	40	122·0	41
20 0 0	47·7	444·2	40	123·6	40
22 0 0	52·4	449·4	40	109·5	41
9 0 0 0	72·9	462·7	40	108·3	42
2 0 0	25·0	423·2	40	108·0	41
4 0 0	54·9	422·1	41	118·1	41
6 0 0	47·7	400·5	42	120·9	42
8 0 0	40·5	395·2	42	123·8	42

FEBRUARY 8 and 9, 1841.

M. Gött. Time. d h m s	Decl. Sc.-Div	Hor. Force Sc.-Div	Ther.	Vert. Force Sc.-Div	Ther.
9 10 0 0	50·0	405·5	42	128·1	42
12 0 0	44·3	417·4	42	124·1	43
14 0 0	50·7	433·0	41	121·9	42
16 0 0	50·4	433·0	41	120·9	42
18 0 0	49·0	435·7	41	120·8	42
20 0 0	48·3	438·0	40	122·7	41
22 0 0	51·1	441·8	40	123·3	40

Mean Positions at the same hours during the Month.[a]

d h m s	Decl. Sc.-Div	Hor. Force Sc.-Div	Ther.	Vert. Force Sc.-Div	Ther.
0 0 0	55·0	453·4	38	124·2	39
2 0 0	56·3	452·6	38	125·7	38
4 0 0	55·6	442·6	39	125·0	38
6 0 0	49·8	427·8	40	125·3	39
8 0 0	46·7	430·6	41	124·9	39
10 0 0	48·0	441·0	44	125·7	40
12 0 0	50·6	448·9	41	125·1	41
14 0 0	53·6	449·0	41	125·6	40
16 0 0	56·5	449·2	41	125·1	40
18 0 0	54·3	439·6	40	124·5	40
20 0 0	55·5	443·1	39	123·7	39
22 0 0	53·5	446·5	39	123·3	39

ST. HELENA { Decl. 1 Scale Division = 0'·71
H.F. k = ·0002; q = ·0003

Positions at the usual hours of observation, from February 8th, 0h., to February 9th, 23h.

d h m s	Decl. Sc.-Div	Hor. Force Sc.-Div	Ther.
8 0 0	46·9	61·7	69
2 0 0	47·1	58·9	70
3 0 0	43·1	57·2	71
4 0 0	38·1	54·8	72
5 0 0	37·4	53·0	72
6 0 0	36·6	53·4	73
8 0 0	39·2	54·4	71
10 0 0	40·8	54·9	70
11 0 0	40·6	54·9	70
12 0 0	41·5	56·0	69
13 0 0	41·0	55·2	70
14 0 0	40·8	55·7	69
15 0 0	40·5	56·2	69
16 0 0	40·2	55·9	69
18 0 0	39·0	57·7	69
19 30 0	37·2	60·4	69
20 0 0	35·1	66·8	69
20 30 0	33·3	65·1	69
22 0 0	37·0	74·2	68
23 0 0	40·7	79·0	68
9 0 0 0	40·8	72·5	69
2 0 0	42·1	59·0	70
3 0 0	38·2	55·9	71
4 0 0	38·0	55·1	71
5 0 0	37·1	54·0	72
6 0 0	37·8	53·1	71
8 0 0	38·5	49·2	70
10 0 0	40·2	55·2	69
11 0 0	40·9	52·3	69
12 0 0	40·9	54·8	69
13 0 0	41·1	56·1	69
14 0 0	40·9	55·6	69
15 0 0	40·6	56·6	69
16 0 0	40·3	57·1	69

FEBRUARY 8 and 9, 1841.

M. Gött. Time. d h m s	Decl. Sc.-Div	Hor. Force Sc.-Div	Ther.	Vert. Force Sc.-Div	Ther.
9 19 30 0	38·8	59·0	68		
20 0 0	37·5	59·9	68		
20 30 0	36·9	60·1	69		
22 0 0	38·0	62·4	69		
23 0 0	38·6	64·2	69		

Mean Positions at the same hours during the Month.[b]

d h m s	Decl. Sc.-Div	Hor. Force Sc.-Div	Ther.
0 0 0	42·4	67·2	69
2 0 0	41·6	65·3	70
3 0 0	40·0	63·9	70
4 0 0	38·8	62·2	71
5 0 0	38·6	61·4	71
6 0 0	38·8	60·3	71
8 0 0	39·7	58·5	70
10 0 0	40·7	59·4	70
11 0 0	41·0	59·5	70
12 0 0	41·0	59·8	69
13 0 0	40·6	60·0	69
14 0 0	40·7	59·8	69
15 0 0	40·5	59·4	69
16 0 0	40·1	60·3	69
18 0 0	39·5	60·5	69
19 30 0	37·9	61·8	69
20 0 0	36·9	62·5	69
20 30 0	36·2	63·5	69
22 0 0	38·7	65·6	69
23 0 0	40·1	67·0	69

VAN DIEMEN ISLAND { Decl. 1 Scale Division = 0'·71
H. F. k = ·0003; q =
V. F. k = ; q =

Extra observations.

The V. F. was observed at 2m. 30s. before, and the H. F. 2m. 30s. after the times specified.

d h m s	Decl. Sc.-Div	Hor. Force Sc.-Div	Ther.	Vert. Force Sc.-Div	Ther.
9 1 10 0	27·9	65·5	69		
15 0	24·9			95·6	69
20 0	26·3	63·4		95·6	
25 0	26·4				
30 0	18·2	56·0		95·6	
35 0	7·6				
40 0	4·0			96·0	69
2 12 30 0	24·4			97·3	
17 30	27·2	59·5	69	98·4	
22 30	29·0				
27 30	29·1	58·7		98·1	
32 30	28·2				
37 30	28·0	59·2		99·0	
42 30	29·0				
47 30	29·5	60·0		97·1	68
3 10 0	30·4	60·6	68		
15 0	37·2				

Positions at the usual hours of observation, from February 8th, 8h. to February 9th, 23h.

d h m s	Decl. Sc.-Div	Hor. Force Sc.-Div	Ther.	Vert. Force Sc.-Div	Ther.
8 8 0 0	38·3	67·2	66	95·3	65
9 0 0	37·1	67·7		95·2	
10 0 0	35·6	67·4	66	95·0	65
11 0 0	33·3	65·3		95·7	
12 0 0	32·9	62·9	67	96·1	67
13 0 0	36·4	61·8		96·2	

---

[a] The mean positions of the H. F. magnet are between the 1st and 11th of February, inclusive; those of the V. F. between the 8th and 18th of February, inclusive.

[b] The connexion of the daily observations of the H. F. at St. Helena, with the mean positions during the month, cannot be considered as strictly determined prior to July 9th, 1841, owing, first, to the gradual stretching of the wires, and secondly, to the suspension-screw not having been tightly screwed down.

FEBRUARY 8 and 9, 1841.

M. Gött. Time.				Decl.	Hor. Force.		Vert. Force.	
d.	h.	m.	s.	Sc.-Div.	Sc.-Div.	Ther.	Sc.-Div.	Ther.
8	14	0	0	40·7	62·2	71	95·4	71
	15	0	0	44·0	62·5		95·1	
	16	0	0	47·2	62·8	75	95·3	74
	17	0	0	49·0	64·8		95·1	
	18	0	0	49·6	66·4	75	95·0	73
	19	0	0	49·8	68·3		94·7	
	20	0	0	50·6	72·7	73	93·3	72
	21	0	0	49·8	65·5		98·0	
	22	0	0	46·1	73·3	72	94·8	69
	23	0	0	43·2	70·7		97·3	
9	0	0	0	42·0	67·8	70	99·3	68
	1	0	0	29·2	64·9		95·7	
	2	0	0	21·9	65·8	69	95·3	69
	3	0	0	32·3	61·0		98·8	
	4	0	0	41·1	63·5	68	99·5	66
	5	0	0	40·2	64·0		99·6	
	6	0	0	40·0	65·4	66	99·8	65
	7	0	0	37·3	65·2		99·7	
	8	0	0	36·0	65·4	65	98·8	64
	9	0	0	33·4	64·5		97·6	
	10	0	0	30·1	63·7	63	96·8	63
	11	0	0	33·4	64·4		96·5	
	12	0	0	31·9	64·3	68	95·1	66
	13	0	0	32·0	64·0		94·6	
	14	0	0	37·0	62·3	72	95·5	71
	15	0	0	39·9	61·5		95·3	
	16	0	0	43·6	61·3	77	95·1	76
	17	0	0	46·3	62·3		94·8	
	18	0	0	46·5	64·0	79	94·1	76
	19	0	0	45·1	64·8		94·2	
	20	0	0	43·2	64·9	79	94·5	75
	21	0	0	42·5	64·6		94·4	
	22	0	0	42·5	64·6	77	95·4	75
	23	0	0	42·1	65·9		94·4	

The Mean Positions at the same hours during the Month are given in pages 36 and 37.

FEBRUARY 15 and 16, 1841.

TORONTO $\left\{\begin{array}{l}\text{Decl. 1 Scale Division} = 0'\cdot72 \\ \text{H. F. } k = \cdot000076\,; \ q = \cdot0002 \\ \text{V. F. } k = \cdot00009\,; \ q = \cdot00018\end{array}\right.$

Extra observations.

The V. F. was observed at 1ᵐ. 30ˢ. before, and the H. F. 2ᵐ. after the times specified.

M. Gött. Time				Decl.	Hor. Force.		Vert. Force.	
15	2	30	0	51·9	529·2	30	128·8	31
	35	0		48·3	518·1		128·4	
	40	0		49·9	519·3		128·5	
	45	0		47·4	513·6		126·9	
	50	0		45·0	506·7		126·9	
	55	0		43·0	504·1		126·9	
3	0	0		41·0	495·8	31	126·1	32
	5	0		41·1	493·1		128·5	
	10	0		42·7	492·7		128·6	
	15	0		46·3	491·2		128·2	
	20	0		48·4	489·3		128·3	
	25	0		48·6	485·1		128·4	
	30	0		50·8	484·8		128·8	
	35	0		50·6	479·1		128·8	

ᵃ Commencing after Sunday midnight at Toronto.

FEBRUARY 15 and 16, 1841.

M. Gött. Time.				Decl.	Hor. Force.		Vert. Force.	
d.	h.	m.	s.	Sc.-Div.	Sc.-Div.	Ther.	Sc.-Div.	Ther.
15	3	40	0	49·8	480·9		128·9	
	45	0		45·9	478·7		129·0	
	50	0		44·2	481·1		129·4	
	55	0		47·4	489·3		128·8	
4	5	0		47·2	504·5	31	128·6	32
	10	0		49·0	488·2		128·1	
	15	0		46·1	486·4		127·3	
	20	0		47·6	499·1		128·4	
	25	0		48·8	501·3		128·6	
	30	0		49·2	503·5		128·8	
	35	0		49·9	507·4		129·0	
	40	0		50·7	503·9		128·6	
	45	0		49·9	497·4		128·6	
	50	0		49·8	499·8		128·5	

Positions at the usual hours of observation, from February 14th, 18ᵇ. to February 16th, 22ᵇ.ᵇ

M. Gött. Time.				Decl.	Hor. Force.		Vert. Force.	
d.	h.	m.	s.	Sc.-Div.	Sc.-Div.	Ther.	Sc.-Div.	Ther.
14	18	0	0	53·3	503·4	30	134·9	31
	20	0	0	54·1	503·2	30	133·0	31
	22	0	0	55·8	515·6	29	133·7	31
15	0	0	0	56·0	511·0	29	132·5	31
	2	0	0	41·7	495·5	29	127·5	30
	4	0	0	48·6	492·9	30	127·4	31
	6	0	0	53·9	489·4	32	128·4	33
	8	0	0	44·9	460·7	35	131·5	34
	10	0	0	40·2	468·2	38	135·9	36
	12	0	0	48·7	474·5	38	132·2	37
	14	0	0	59·1	465·4	37	125·5	37
	16	0	0	57·1	472·6	37	131·5	37
	18	0	0	60·6	476·4	36	125·6	36
	20	0	0	59·2	471·0	35	126·0	36
	22	0	0	49·5	477·1	35	121·1	36
16	0	0	0	58·6	484·5	35	124·4	36
	2	0	0	59·3	485·5	34	129·5	35
	4	0	0	53·4	463·2	35	127·7	36
	6	0	0	46·7	452·0	38	125·3	37
	8	0	0	45·3	463·7	39	127·0	38
	10	0	0	41·9	456·8	39	127·3	38
	12	0	0	47·0	474·6	39	128·2	39
	14	0	0	50·8	473·0	40	128·2	39
	16	0	0	66·8	470·0	40	124·5	40
	18	0	0	63·6	459·5	41	119·2	41
	20	0	0	56·0	476·3	41	113·0	41
	22	0	0	56·9	467·7	41	120·2	41

Mean Positions at the same hours during the Month.ᵇ

				Decl.	Hor. Force.		Vert. Force.	
	0	0	0	55·0	473·0	38	124·2	39
	2	0	0	56·3	473·3	38	125·7	38
	4	0	0	55·6	457·7	39	125·0	38
	6	0	0	49·8	456·7	40	125·3	39
	8	0	0	46·7	457·7	41	124·9	39
	10	0	0	48·0	462·5	41	125·7	40
	12	0	0	50·6	463·2	41	125·1	41
	14	0	0	53·6	458·0	41	125·6	40
	16	0	0	56·5	463·1	41	125·1	40
	18	0	0	54·3	465·3	40	124·5	40
	20	0	0	52·5	468·2	39	123·7	39
	22	0	0	53·5	470·0	39	123·3	39

ᵇ The mean positions of the H. F. magnetometer are between the 12th and 27th February, both days inclusive.

FEBRUARY 15 and 16, 1841.

ST. HELENA $\left\{\begin{array}{l}\text{Decl. 1 Scale Division} = 0'\cdot71 \\ \text{H. F. } k = \cdot0002\,; \ q = \cdot0003\end{array}\right.$

Positions at the usual hours of observation, from February 14th, 18ᵇ., to February 16th, 23ᵇ.

M. Gött. Time.				Decl.	Hor. Force.		Vert. Force.	
d.	h.	m.	s.	Sc.-Div.	Sc.-Div.	Ther.	Sc.-Div.	Ther.
14	18	0	0	41·9	63·2	69		
	19	30	0	40·9	65·2	69		
	20	0	0	39·4	66·1	69		
	20	30	0	38·5	67·0	69		
	22	0	0	42·0	71·1	69		
	23	0	0	43·7	73·1	69		
15	0	0	0	46·7	69·9	70		
	2	0	0	40·0	62·4	71		
	3	0	0	38·2	60·1	72		
	4	0	0	36·6	54·6	73		
	5	0	0	38·1	53·3	73		
	6	0	0	39·0	55·6	74		
	8	0	0	39·3	50·8	73		
	10	0	0	40·2	58·0	72		
	11	0	0	41·3	59·4	72		
	12	0	0	40·0	56·1	71		
	13	0	0	40·0	56·9	71		
	14	0	0	41·0	59·2	71		
	15	0	0	40·0	58·6	70		
	16	0	0	39·1	59·1	71		
	18	0	0	39·0	59·8	70		
	19	30	0	38·9	61·1	70		
	20	0	0	38·1	61·3	70		
	20	30	0	37·2	62·9	70		
	22	0	0	40·7	61·2	70		
	23	0	0	41·9	62·3	70		
16	0	0	0	42·8	63·0	71		
	2	0	0	41·9	61·9	72		
	3	0	0	36·5	62·1	73		
	4	0	0	34·0	58·0	73		
	5	0	0	37·1	58·0	73		
	6	0	0	37·0	54·9	73		
	8	0	0	37·0	54·9	72		
	10	0	0	37·9	55·6	71		
	11	0	0	38·3	56·8	72		
	12	0	0	38·1	56·1	71		
	13	0	0	38·8	59·1	71		
	14	0	0	38·0	59·3	71		
	15	0	0	38·9	59·3	70		
	16	0	0	39·1	61·0	70		
	18	0	0	38·2	61·9	70		
	19	30	0	37·5	60·7	70		
	20	0	0	36·2	61·4	70		
	20	30	0	36·5	62·0	70		
	22	0	0	36·8	62·1	70		
	23	0	0	37·2	62·2	70		

The Mean Positions at the same hours during the Month are given in page 37.

VAN DIEMEN ISLAND $\left\{\begin{array}{l}\text{Decl. 1 Scale Division} = 0'\cdot71 \\ \text{H. F. } k = \cdot0003\,; \ q = \\ \text{V. F. } k = \quad\ ; \ q = \end{array}\right.$

Extra observations.

The V. F. was observed at 2ᵐ. 30ˢ. before, and the H. F. 2ᵐ. 30ˢ. after the times specified.

M. Gött. Time.				Decl.	Hor. Force.		Vert. Force.	
15	3	15	0	25·3				
	20	0		26·3	68·5	68	93·2	67
	25	0		27·0				

FEBRUARY 15 and 16, 1841.

M. Gött. Time. d. h. m. s.	Decl. Sc.-Div^ss	Hor. Force. Sc.-Div^ss	Ther.	Vert. Force. Sc.-Div^ss	Ther.
15 3 30 0	27·2	68·5		93·5	
35 0	26·1				
40 0	24·0	67·3		93·3	
45 0	22·8				
50 0				93·6	66
4 10 0	28·2	70·3	68		
15 0	27·0				
20 0	24·0	71·2		92·5	
25 0	22·5				
30 0	21·4	70·4		91·8	
35 0	21·1				
40 0	21·5	67·7		90·8	
45 0	21·7				
50 0	21·7			92·3	
5 15 0	29·2				
20 0	30·5	64·4		96·9	
25 0	33·4				
30 0	36·2	65·3		97·2	
35 0	36·0				
40 0	34·3	65·3		97·0	
45 0	33·5				
50 0	34·5	66·4	67	97·0	67
22 10 0	29·5	66·7	74		
15 0	29·9				
20 0	33·7	67·5		93·8	71
25 0	35·3				
30 0	36·3	67·0		93·8	
35 0	37·3				
40 0	37·6	66·4		94·4	
45 0	37·3				
50 0	37·3			94·5	
23 10 0	38·7	66·0			
15 0	38·5				
20 0	39·3	66·7		95·0	
25 0	39·7				
30 0	39·1	66·6	73	95·2	71

Positions at the usual hours of observation, from February 14th, 18h., to February 16th, 23h.

M. Gött. Time. d. h. m. s.	Decl. Sc.-Div^ss	Hor. Force. Sc.-Div^ss	Ther.	Vert. Force. Sc.-Div^ss	Ther.
14 18 0 0	46·9	70·9	73	94·9	70
19 0 0	46·0	71·5		94·7	
20 0 0	44·8	69·5	72	94·7	69
21 0 0	44·5	69·5		95·2	
22 0 0	45·9	72·8	71	93·6	68
23 0 0	47·3	74·1		94·0	
15 0 0 0	45·1	71·0	69	95·0	67
1 0 0	37·0	69·9		95·3	
2 0 0	34·6	74·2	68	90·5	66
3 0 0	25·9	68·0		93·8	
4 0 0	26·3	66·3	68	94·7	67
5 0 0	22·6	64·8		94·1	
6 0 0	35·6	66·8	67	96·7	66
7 0 0	38·4	69·3		94·7	
8 0 0	31·5	70·7	66	93·5	64
9 0 0	30·4	70·0		94·8	
10 0 0	29·8	64·9	65	96·6	64
11 0 0	35·0	63·0		98·3	
12 0 0	39·2	64·5	65	96·6	64
13 0 0	40·6	65·6		95·1	
14 0 0	45·7	65·6	67	95·4	66
15 0 0	47·4	66·1		95·3	
16 0 0	48·4	67·0	71	95·3	70
17 0 0	47·9	69·1		95·0	
18 0 0	46·1	65·4	75	96·0	72

FEBRUARY 15 and 16, 1841.

M. Gött. Time. d. h. m. s.	Decl. Sc.-Div^ss	Hor. Force. Sc.-Div^ss	Ther.	Vert. Force. Sc.-Div^ss	Ther.
15 19 0 0	44·4	66·9		94·4	
20 0 0	43·8	67·0	76	93·9	72
21 0 0	43·5	66·4		94·3	
22 0 0	28·7	65·0	74	94·1	71
23 0 0	38·4	65·9		95·3	
16 0 0 0	40·5	67·0	73	96·0	71
1 0 0	39·4	67·2		96·2	
2 0 0	42·2	69·3	71	95·6	70
3 0 0	41·0	69·5		95·8	
4 0 0	41·0	69·7	68	93·0	68
5 0 0	39·0	70·1		94·3	
6 0 0	34·6	68·8	67	94·9	66
7 0 0	39·7	71·2		94·7	
8 0 0	39·6	67·5	65	96·1	64
9 0 0	40·5	68·6		95·9	
10 0 0	42·5	69·0	65	96·1	63
11 0 0	38·6	69·7		94·7	
12 0 0	40·1	69·1	66	94·5	66
13 0 0	41·6	66·3		95·2	
14 0 0	44·0	66·5	70	94·0	70
15 0 0	43·2	65·4		93·3	
16 0 0	44·3	66·0	74	93·7	73
17 0 0	45·3	62·6		95·2	
18 0 0	45·7	63·5	76	95·0	73
19 0 0	45·0	65·7		93·7	
20 0 0	43·3	66·1	76	93·6	73
21 0 0	43·0	66·0		94·0	
22 0 0	41·8	66·4	75	94·2	73
23 0 0	39·2	66·9		93·6	

The Mean Positions at the same hours during the Month are given in pages 36 and 37.

FEBRUARY 22 and 23, 1841.

TORONTO { Decl. 1 Scale Division = 0'·72 ; H. F. k = ·000076; q = ·0002 ; V. F. k = ·00009; q = ·00018 }

EXTRA observations.
The V. F. was observed at 1m. 30s. before, and the H. F. 2m. after the times specified.

M. Gött. Time. d. h. m. s.	Decl. Sc.-Div^ss	Hor. Force. Sc.-Div^ss	Ther.	Vert. Force. Sc.-Div^ss	Ther.
23 14 25 0	72·5	485·9	40	112·5	41
30 0	68·2	474·9		114·1	
35 0	63·0	468·7		111·1	
40 0	63·1	462·6		112·5	
55 0	56·7	467·0		114·4	
15 0 0	54·0	460·5	40	114·2	42
5 0	53·1	455·0		114·4	
10 0	54·0	449·2		114·7	
15 0	56·9	448·7		114·6	
20 0	58·5	445·3		114·8	
25 0	59·0	445·1		115·7	
30 0	59·1	446·0		115·7	
35 0	58·6	444·5		115·8	

Positions at the usual hours of observation, February 22 and 23.

M. Gött. Time. d. h. m. s.	Decl. Sc.-Div^ss	Hor. Force. Sc.-Div^ss	Ther.	Vert. Force. Sc.-Div^ss	Ther.
22 0 0 0	59·2	470·7	42	109·3	44
2 0 0	61·5	467·7	43	111·8	43
4 0 0	55·2	455·7	44	110·4	44
6 0 0	48·2	449·0	45	110·4	44
8 0 0	47·0	460·5	45	110·6	44

FEBRUARY 22 and 23, 1841.

M. Gött. Time. d. h. m. s.	Decl. Sc.-Div^ss	Hor. Force. Sc.-Div^ss	Ther.	Vert. Force. Sc.-Div^ss	Ther.
22 10 0 0	45·1	479·2	46	109·8	45
12 0 0	40·2	467·0	46	119·5	46
14 0 0	57·2	443·2	45	112·4	45
16 0 0	51·6	436·0	45	115·6	46
18 0 0	52·0	437·2	46	111·8	46
20 0 0	52·8	440·8	45	110·9	46
22 0 0	49·8	433·6	46	110·9	46
23 0 0 0	65·1	454·6	45	107·4	45
2 0 0	55·0	432·3	43	109·6	44
4 0 0	53·0	421·2	42	109·0	43
6 0 0	46·4	426·6	41	118·8	42
8 0 0	35·9	434·1	40	122·4	41
10 0 0	45·2	451·5	41	125·7	41
12 0 0	47·1	437·0	42	119·8	42
14 0 0	82·4	417·0	40	128·6	41
16 0 0	54·7	439·0	40	118·4	42
18 0 0	50·0	449·5	39	111·6	39
20 0 0	47·2	458·5	38	115·9	39
22 0 0	52·7	466·5	37	119·8	38

Mean Positions at the same hours during the Month. *

M. Gött. Time. d. h. m. s.	Decl. Sc.-Div^ss	Hor. Force. Sc.-Div^ss	Ther.	Vert. Force. Sc.-Div^ss	Ther.
0 0 0	55·0	473·3	38	113·7	43
2 0 0	56·3	473·0	38	117·5	42
4 0 0	55·6	457·7	39	116·9	42
6 0 0	49·8	456·7	40	119·2	43
8 0 0	46·7	457·7	41	120·0	43
10 0 0	48·0	462·5	41	119·4	44
12 0 0	50·6	463·2	41	119·9	45
14 0 0	53·6	458·0	41	120·2	45
16 0 0	56·5	463·1	41	118·7	45
18 0 0	54·3	465·3	40	111·3	43
20 0 0	52·5	468·2	39	109·4	43
22 0 0	53·5	470·0	39	111·6	43

St. Helena { Decl. 1 Scale Division = 0'·71 ; H. F. k = ·0002; q = ·0003 }
Extra observations.
The H. F. was observed at 1m. after the times specified.

M. Gött. Time. d. h. m. s.	Decl. Sc.-Div^ss	Hor. Force. Sc.-Div^ss	Ther.	Vert. Force. Sc.-Div^ss	Ther.
23 8 1 47		46·1	69		
16 47		46·0			
31 47		48·2			
46 47		49·5			
9 1 47		49·2	69		
16 47	38·2	49·2			
31 47	38·0	49·8			
46 47	37·8	49·2			
10 1 47	37·2	49·3	69		
16 47	37·1	50·7			
31 47	37·6	51·3			
46 47	38·5	54·2			
11 1 47	38·0	55·8	68		
16 47	40·1	56·0			
31 47	40·1	56·2			
46 47	40·1	56·8			
12 1 47	40·0	56·9	68		
16 47	40·0	57·7			
31 47	39·2	57·7			
46 47	40·1	57·8			
13 1 47	40·2	57·3	68		
16 47	40·0	56·1			
46 47	40·0	56·9			
14 1 47	40·5	58·0	68		

* The mean positions of the V. F. magnet are from February 20th to 27th inclusive.

FEBRUARY 22 and 23, 1841.

M. Gött. Time (d. h. m. s.)	Decl. Sc.-Div	Hor. Force Sc.-Div	Ther.	Vert. Force Sc.-Div	Ther.
23 14 16 47	40·9	59·1			
31 47	41·0	60·0			
46 47	41·0	60·9			
15 1 47	41·0	61·0	68		
16 47	41·5	61·1			
31 47	41·6	60·9			
46 47	41·0	60·3			
16 1 47	40·8	60·1	68		
16 47	41·1	60·1			

Positions at the usual hours of observation, February 22 and 23.

M. Gött. Time (d. h. m. s.)	Decl. Sc.-Div	Hor. Force Sc.-Div	Ther.	Vert. Force Sc.-Div	Ther.
22 0 0	47·5	74·8	69		
2 0 0		71·5	69		
3 0 0		69·1			
4 0 0		68·9			
5 0 0		68·1	69		
6 0 0		67·3	69		
8 0 0	41·5	66·5	69		
10 0 0	42·6	68·0	69		
11 0 0	41·0	62·4	69		
12 0 0	39·4	59·2	69		
13 0 0	36·4	57·4	69		
14 0 0	38·9	57·2	68		
15 0 0	38·7	55·2	68		
16 0 0	39·1	56·3	68		
18 0 0	38·1	58·0	68		
19 30 0	35·1	60·9	68		
20 0 0	33·2	62·0	68		
20 30 0	32·0	65·0	68		
22 0 0	35·2	69·1	68		
23 0 0	39·0	74·8	68		
23 0 0 0	41·9	73·5	68		
2 0 0	38·9	65·3	68		
3 0 0	39·8	63·2	69		
4 0 0	40·4	56·2	69		
5 0 0	37·2	53·9	69		
6 0 0	37·0	53·0	69		
8 0 0	36·2	46·1	69		
10 0 0	37·1	49·3	69		
11 0 0	39·8	55·8	68		
12 0 0	40·2	57·0	69		
13 0 0	40·2	57·2	69		
14 0 0	40·8	58·0	68		
15 0 0	41·0	61·0	68		
16 0 0	40·9	60·6	68		
18 0 0	39·8	60·9	68		
19 30 0	37·7	61·1	68		
20 0 0	36·1	62·3	68		
20 30 0	36·1	63·0	68		
22 0 0	37·8	66·0	68		
23 0 0	39·8	69·0	68		

The Mean Positions at the same hours during the Month are given in page 37.

VAN DIEMEN ISLAND { Decl. 1 Scale Division = 0'·71 ; H. F. k = ·0003 ; = q ; V. F. k = ; = q }

The V. F. was observed at 2m. 30s. before, and the H. F. 2m. 30s. after the times specified.

M. Gött. Time (d. h. m. s.)	Decl. Sc.-Div	Hor. Force Sc.-Div	Ther.	Vert. Force Sc.-Div	Ther.
22 12 15 0	29·8				
20 0	29·2	61·4	62	100·0	61
25 0	31·1				

FEBRUARY 22 and 23, 1841.

M. Gött. Time (d. h. m. s.)	Decl. Sc.-Div	Hor. Force Sc.-Div	Ther.	Vert. Force Sc.-Div	Ther.
22 12 30 0	31·7	62·3		100·1	
35 0	32·1				
40 0	34·9	63·6		100·7	
45 0	35·1				
50 0	39·6			100·7	

Positions at the usual hours of observation, February 22 and 23.

M. Gött. Time (d. h. m. s.)	Decl. Sc.-Div	Hor. Force Sc.-Div	Ther.	Vert. Force Sc.-Div	Ther.
22 0 0	43·0	71·2	70	94·1	68
1 0 0	42·3	72·8		93·7	
2 0 0	40·4	72·9	69	92·9	67
3 0 0	40·2	72·5		93·4	
4 0 0	41·2	73·0	67	94·1	66
5 0 0	42·4	73·3		94·6	
6 0 0	40·3	74·1	65	94·0	64
7 0 0	39·8	74·4		94·1	
8 0 0	40·0	74·6	64	94·6	63
9 0 0	37·9	74·8		94·8	
10 0 0	34·6	74·7	63	94·0	62
11 0 0	34·2	70·0		97·7	
12 0 0	29·2	61·9	62	99·2	61
13 0 0	43·7	65·2		101·1	
14 0 0	52·1	67·2	62	99·4	60
15 0 0	46·7	72·5		97·0	
16 0 0	46·9	72·5	63	97·1	60
17 0 0	48·4	73·7		96·6	
18 0 0	48·0	72·6	64	97·8	62
19 0 0	45·5	70·8		97·8	
20 0 0	44·5	70·2	65	97·1	62
21 0 0	43·7	70·8		96·5	
22 0 0	37·0	70·2	64	97·5	62
23 0 0	39·4	73·2		94·7	
23 0 0	41·3	73·3	63	96·1	61

The Mean Positions at the same hours during the Month are given in pages 36 and 37.

FEBRUARY 26 and 27, 1841.

TORONTO { Decl. 1 Scale Division = 0'·72 ; H. F. k = ·000076 ; q = ·0002 ; V. F. k = ·00009 ; q = ·00018 }

Regular and term observations.
The V. F. was observed at 2m. 30s. before, and the H. F. 2m. 30s. after the times specified.

M. Gött. Time (d. h. m. s.)	Decl. Sc.-Div	Hor. Force Sc.-Div	Ther.	Vert. Force Sc.-Div	Ther.
26 0 0	63·0	428·8	42	102·0	43
2 0 0	55·6	464·6	41	103·6	42
4 0 0	62·5	446·3	43	103·1	45
6 0 0	49·9	425·8	44	108·1	45
8 0 0	43·6	445·7	45	110·2	45
10 0 0	50·1	457·7	46	109·2	46
5 0	50·2				
10 0	49·4	454·9	45	109·0	46
15 0	48·4				
20 0	48·1	448·9		108·9	
25 0	48·9				
30 0	48·1	448·1		109·9	
35 0	48·9				
40 0	49·2	448·3		110·5	
45 0	50·6				
50 0	50·9	453·5		111·8	
55 0	51·0				
11 0 0	50·2	452·3	45	112·2	46
5 0	48·4				

FEBRUARY 26 and 27, 1841.

M. Gött. Time (d. h. m. s.)	Decl. Sc.-Div	Hor. Force Sc.-Div	Ther.	Vert. Force Sc.-Div	Ther.
26 11 10 0	48·1	454·9		111·3	
15 0	48·0				
20 0	48·0	453·8		111·9	
25 0	45·9				
30 0	45·0	446·0		110·9	
35 0	44·7				
40 0	44·9	428·0		110·1	
45 0	46·6				
50 0	48·5	447·1		110·3	
55 0	49·0				
12 0 0	49·4	448·5	45	111·3	46
5 0	49·0				
10 0	47·8	450·7		111·5	
15 0	47·0				
20 0	47·1	447·1		111·6	
25 0	48·5				
30 0	48·5	448·4		112·1	
35 0	48·7				
40 0	46·5	443·0		112·2	
45 0	47·1				
50 0	53·9	426·7		118·6	
55 0	63·8				
13 0 0	77·2	438·3	45	118·7	46
5 0	75·3				
10 0	77·1	435·4		111·7	
15 0	78·2				
20 0	75·3	428·6		108·6	
25 0	73·7				
30 0	70·4	429·5		109·2	
35 0	64·1				
40 0	56·4	423·1		111·5	
45 0	54·7				
50 0	52·6	422·5		114·2	
55 0	54·2				
14 0 0	55·1	426·1	45	116·1	46
5 0	55·4				
10 0	56·1	432·0		115·4	
15 0	55·8				
20 0	56·9	427·8		115·3	
25 0	56·9				
30 0	55·4	427·1		115·0	
35 0	55·5				
40 0	54·6	430·1		116·7	
45 0	55·4				
50 0	55·9	434·2		116·6	
55 0	57·5				
15 0 0	59·9	432·5	45	115·9	46
5 0	63·5				
10 0	67·1	430·6		113·8	
15 0	67·8				
20 0	62·8	421·4		110·5	
25 0	51·3				
30 0	50·4	388·0		109·2	
35 0	46·0				
40 0	56·9	402·2		102·8	
45 0	64·1				
50 0	66·6	415·3		104·1	
55 0	76·4				
16 0 0	79·8	416·9	45	109·8	46
5 0	80·1				
10 0	73·1	427·2		107·7	
15 0	65·2				
20 0	62·9	428·7		111·1	
25 0	64·1				
30 0	60·7	442·6		115·2	
35 0	59·8				
40 0	56·2	441·4		114·4	

FEBRUARY 26 and 27, 1841.

d.	h.	m.	s.	Decl. Sc.-Div.	Hor. Force. Sc.-Div.	Ther.	Vert. Force. Sc.-Div.	Ther.
26	16	45	0	55·2				
		50	0	56·1	439·9		113·2	
		55	0	56·7				
	17	0	0	56·8	442·1	45	112·5	46
		5	0	56·0				
		10	0	57·0	439·6		111·7	
		15	0	57·6				
		20	0	57·0	435·6		110·0	
		25	0	57·2				
		30	0	58·1	438·8		110·2	
		35	0	57·7				
		40	0	57·2	438·2		110·3	
		45	0	57·6				
		50	0	56·7	438·2		110·7	
		55	0	55·2				
	18	0	0	54·1	437·9	45	111·7	46
		5	0	52·6				
		10	0	51·4	438·4		111·4	
		15	0	53·2				
		20	0	54·8	443·7		111·4	
		25	0	54·0				
		30	0	53·0	440·0		110·3	
		35	0	52·7				
		40	0	55·4	436·8		108·3	
		45	0	57·9				
		50	0	57·3	430·9		108·0	
		55	0	53·1				
	19	0	0	53·0	426·4	45	110·2	46
		5	0	54·6				
		10	0	53·9	428·2		108·9	
		15	0	48·9				
		20	0	45·1	421·4		108·6	
		25	0	45·6				
		30	0	47·8	433·8		108·2	
		35	0	50·0				
		40	0	51·0	438·9		109·0	
		45	0	49·6				
		50	0	50·4	437·1		106·5	
		55	0	50·9				
	20	0	0	53·8	441·0	45	107·9	46
		5	0	53·9				
		10	0	54·2	440·5		107·8	
		15	0	54·3				
		20	0	54·9	438·1		108·3	
		25	0	54·1				
		30	0	52·7	437·3		107·8	
		35	0	54·0				
		40	0	53·4	438·7		108·0	
		45	0	54·1				
		50	0	54·3	436·8		106·5	
		55	0	55·1				
	21	0	0	54·4	430·4	45	105·1	46
		5	0	49·8				
		10	0	50·2	435·4		105·5	
		15	0	49·5				
		20	0	48·9	445·6		107·0	
		25	0	47·2				
		30	0	47·0	444·6		105·7	
		35	0	48·4				
		40	0	49·8	444·6		105·6	
		45	0	51·0				
		50	0	52·1	447·4		105·6	
		55	0	52·2				
	22	0	0	53·6	443·8	45	105·0	46
		5	0	55·1				
		10	0	54·2	444·2		106·8	
		15	0	54·8				

FEBRUARY 26 and 27, 1841.

d.	h.	m.	s.	Decl. Sc.-Div.	Hor. Force. Sc.-Div.	Ther.	Vert. Force. Sc.-Div.	Ther.
26	22	20	0	55·4	444·0		106·7	
		25	0	56·2				
		30	0	57·0	445·4		107·9	
		35	0	56·7				
		40	0	56·4	444·3		108·6	
		45	0	56·1				
		50	0	56·5	443·7		109·9	
		55	0	57·6				
	23	0	0	57·4	445·7	45	111·4	46
		5	0	57·9				
		10	0	57·9	445·2		113·2	
		15	0	57·7				
		20	0	57·1	444·9		116·5	
		25	0	57·0				
		30	0	58·0	445·0		120·0	
		35	0	57·4				
		40	0	57·8	443·7		124·5	
		45	0	57·1				
		50	0	57·1	445·1		129·7	
		55	0	56·6				
27	0	0	0	56·1	445·9	45	133·7	46
		5	0	56·0				
		10	0	55·9	447·9		137·5	
		15	0	55·7				
		20	0	54·9	446·5		140·0	
		25	0	55·2				
		30	0	55·3	445·1		141·4	
		35	0	55·9				
		40	0	55·7	446·0		142·6	
		45	0	56·0				
		50	0	56·8	446·8		143·6	
		55	0	56·9				
	1	0	0	57·6	450·4	45	145·1	46
		5	0	57·2				
		10	0	58·4	448·8		145·7	
		15	0	59·1				
		20	0	59·0	449·6		147·1	
		25	0	59·5				
		30	0	59·7	449·3		148·1	
		35	0	60·9				
		40	0	61·5	449·0		148·0	
		45	0	62·0				
		50	0	60·7	447·2		148·7	
		55	0	61·2				
	2	0	0	61·4	445·6	45	148·5	46
		5	0	60·8				
		10	0	60·4	445·5		148·5	
		15	0	60·4				
		20	0	59·8	444·3		149·4	
		25	0	59·5				
		30	0	61·0	445·5		149·1	
		35	0	61·4				
		40	0	60·9	444·7		149·2	
		45	0	60·8				
		50	0	60·7	444·6			
		55	0	61·0				
	3	0	0	60·7	442·4	45	150·3	46
		5	0	61·4				
		10	0	61·5	447·2		150·7	
		15	0	61·0				
		20	0	60·7	446·7		149·7	
		25	0	60·7				
		30	0	59·7	448·6		151·1	
		35	0	60·1				
		40	0	58·6	448·2		151·6	
		45	0	58·9				
		50	0	58·7	444·1		151·5	

FEBRUARY 26 and 27, 1841.

d.	h.	m.	s.	Decl. Sc.-Div.	Hor. Force. Sc.-Div.	Ther.	Vert. Force. Sc.-Div.	Ther.
27	3	55	0	60·1				
	4	0	0	59·2	444·6	45	151·4	46
		5	0	59·6				
		10	0	58·8	442·3		151·5	
		15	0	58·0				
		20	0	57·1	440·0		151·7	
		25	0	56·6				
		30	0	55·1	437·5		151·8	
		35	0	54·8				
		40	0	54·4	437·4		152·0	
		45	0	54·7				
		50	0	54·1	432·5		152·1	
		55	0	52·9				
	5	0	0	52·3	432·1	45	152·2	46
		5	0	53·1				
		10	0	52·8	427·7		153·1	
		15	0	51·2				
		20	0	50·3	432·0		153·4	
		25	0	51·6				
		30	0	52·3	432·2		153·3	
		35	0	51·4				
		40	0	52·1	434·8		154·3	
		45	0	51·2				
		50	0	49·9	435·7		154·4	
		55	0	50·2				
	6	0	0	50·0	440·0	45	155·4	46
		5	0	48·3				
		10	0	48·2	439·2		155·7	
		15	0	49·7				
		20	0	48·6	442·9		155·9	
		25	0	48·5				
		30	0	48·3	442·3		155·8	
		35	0	48·6				
		40	0	47·6	439·9		156·2	
		45	0	47·0				
		50	0	46·6	437·6		156·4	
		55	0	47·1				
	7	0	0	47·3	438·9	45	156·4	46
		5	0	47·4				
		10	0	47·3	445·9		156·8	
		15	0	46·3				
		20	0	46·6	440·2		157·1	
		25	0	46·8				
		30	0	45·7	435·0		157·3	
		35	0	45·0				
		40	0	45·3	438·1		156·5	
		45	0	44·4				
		50	0	44·2	437·0		157·0	
		55	0	44·2				
	8	0	0	44·1	433·3	45	157·0	46
		5	0	44·9				
		10	0	46·1	435·0		156·7	
		15	0	47·0				
		20	0	46·5	444·0		156·8	
		25	0	46·7				
		30	0	46·9	446·3		157·6	
		35	0	47·5				
		40	0	47·2	449·0		157·0	
		45	0	47·1				
		50	0	47·4	449·7		156·5	
		55	0	48·4				
	9	0	0	48·7	447·7	45	156·2	46
		5	0	48·3				
		10	0	48·7	449·8		156·1	
		15	0	49·1				
		20	0	49·3	451·1		156·5	
		25	0	49·4				

G

FEBRUARY 26 and 27, 1841.

d.	h.	m.	s.	Decl. Sc.-Div.	Hor. Force Sc.-Div.	Ther.	Vert. Force Sc.-Div.	Ther.
27	9	30	0	49·2	455·3		157·0	
		35	0	49·3				
		40	0	49·3	448·7		156·9	
		45	0	49·6				
		50	0	49·7	458·1		157·3	
		55	0	49·5				
	10	0	0	49·5	446·2	49	150·1	49
	12	0	0	53·6	441·0	48	152·5	49
	14	0	0	60·6	448·9	47	152·5	48
	16	0	0	53·2	446·5	47	154·7	48

The Mean Positions at the usual hours of observation, are given in page 39.

ST. HELENA { Decl. 1 Scale Division = 0′·71 / H. F. k = ·0002; q = ·0003

Regular and term observations

The H. F. was observed at 2ᵐ·30ˢ· after the times specified.

d.	h.	m.	s.	Decl. Sc.-Div.	Hor. Force Sc.-Div.	Ther.
26	0	0	0	36·1	70·8	69
	2	0	0	35·8	64·8	70
	3	0	0	36·9	64·0	70
	4	0	0	34·0	65·0	70
	5	0	0	33·3	66·9	70
	6	0	0		60·1	70
	8	0	0	34·5	62·0	69
	10	0	0	37·9	62·8	69
		5	0	37·6		
		10	0	37·4	62·1	
		15	0	37·2		
		20	0	37·5	62·2	
		25	0	37·4		
		30	0	37·8	62·8	
		35	0	37·9		
		40	0	38·0	62·8	
		45	0	38·1		
		50	0	38·0	62·7	
		55	0	38·0		
	11	0	0	38·0	61·8	69
		5	0	38·0		
		10	0	37·9	61·0	
		15	0	37·9		
		20	0	37·9	60·6	
		25	0	38·0		
		30	0	37·6	60·4	
		35	0	37·2		
		40	0	36·9	59·1	
		45	0	36·8		
		50	0	36·2	59·0	
		55	0	36·0		
	12	0	0	35·9	58·9	69
		5	0	35·9		
		10	0	35·8	58·8	
		15	0	35·9		
		20	0	35·9	58·8	
		25	0	35·9		
		30	0	35·9	58·8	
		35	0	35·9		
		40	0	36·0	61·8	
		45	0	36·2		
		50	0	37·1	65·8	
		55	0	38·0		
	13	0	0	38·1	66·2	69
		5	0	38·9		
		10	0		65·0	

FEBRUARY 26 and 27, 1841.

d.	h.	m.	s.	Decl. Sc.-Div.	Hor. Force Sc.-Div.	Ther.
26	13	15	0	39·0		
		20	0	39·1	64·1	
		25	0	38·8		
		30	0	38·9	63·1	
		35	0	38·9		
		40	0	38·9	62·1	
		45	0	38·7		
		50	0	38·9	62·1	
		55	0	38·9		
	14	0	0	38·9	62·1	69
		5	0	38·1		
		10	0	38·0	62·5	
		15	0	37·9		
		20	0	37·9	61·5	
		25	0	37·9		
		30	0	37·9	61·3	
		35	0	37·8		
		40	0	37·5	61·0	
		45	0	37·5		
		50	0	37·4	61·2	
		55	0	37·5		
	15	0	0	37·5	61·5	69
		5	0	37·9		
		10	0	37·9	62·0	
		15	0	37·9		
		20	0	38·2	62·6	
		25	0	38·8		
		30	0	38·0	63·2	
		35	0	38·0		
		40	0	38·1	65·2	
		45	0	38·9		
		50	0	38·9	65·9	
		55	0	39·1		
	16	0	0	39·1	66·5	69
		5	0	39·0		
		10	0	38·9	64·9	
		15	0	38·8		
		20	0	38·4	63·8	
		25	0	38·2		
		30	0	38·1	63·0	
		35	0	38·1		
		40	0	38·1	62·9	
		45	0	38·0		
		50	0	38·0	62·1	
		55	0	38·0		
		0	0	38·0	62·1	69
		5	0	37·9		
		10	0	38·0	62·6	
		15	0	37·9		
		20	0	37·9	62·7	
		25	0	37·9		
		30	0	37·9	62·8	
		35	0	37·2		
		40	0	37·2	62·6	
		45	0	37·1		
		50	0	37·0	62·2	
		55	0	36·9		
	18	0	0	36·9	62·1	69
		5	0	36·6		
		10	0	36·4	62·0	
		15	0	37·0		
		20	0	36·9	62·1	
		25	0	36·9		
		30	0	37·1	62·3	
		35	0	37·3		
		40	0	37·3	62·7	
		45	0	37·1		

FEBRUARY 26 and 27, 1841.

d.	h.	m.	s.	Decl. Sc.-Div.	Hor. Force Sc.-Div.	Ther.
26	18	50	0	36·9	62·7	
		55	0	36·9		
	19	0	0	36·7	62·8	69
		5	0	36·8		
		10	0	36·9	62·3	
		15	0	36·6		
		20	0	36·4	62·0	
		25	0	36·1		
		30	0	35·9	62·4	
		35	0	35·5		
		40	0	35·1	63·2	
		45	0	35·6		
		50	0	34·8	63·6	
		55	0	34·2		
	20	0	0	34·4	64·0	69
		5	0	34·1		
		10	0	33·9	64·4	
		15	0	33·6		
		20	0	33·0	64·1	
		25	0			
		30	0	32·5	64·2	
		35	0			
		40	0	32·4	64·3	
		45	0	32·5		
		50	0	32·4	64·5	
		55	0	32·2		
	21	0	0	32·2	64·5	68
		5	0	32·4		
		10	0	32·3	64·5	
		15	0	32·8		
		20	0	32·9	65·1	
		25	0	33·0		
		30	0	33·0	65·1	
		35	0	33·1		
		40	0	33·1	66·1	
		45	0	33·2		
		50	0	33·5	67·0	
		55	0	33·8		
	22	0	0	33·8	68·0	68
		5	0	34·1		
		10	0	34·4	68·0	
		10	0	34·4		
		20	0	34·5	68·0	
		25	0	35·1		
		30	0	35·2	67·9	
		35	0	35·0		
		40	0	35·2	67·9	
		45	0	35·4		
		50	0	35·3	67·9	
		55	0	35·2		
	23	0	0	35·1	68·0	68
		5	0	35·2		
		10	0	35·4	68·2	
		15	0	35·7		
		20	0	35·9	68·5	
		25	0	36·1		
		30	0	36·8	68·2	
		35	0	37·0		
		40	0	37·2	67·9	
		45	0	37·2		
		50	0	38·0	67·8	
		55	0	38·2		
27	0	0	0	38·9	67·8	69
		5	0	39·0		
		10	0	39·3	67·8	
		15	0	39·9		
		20	0	40·0	67·9	

FEBRUARY 26 and 27, 1841.

M. Gött. Time.				Decl.	Hor. Force.		Vert. Force.	
d.	h.	m.	s.	Sc.-Div.	Sc.-Div.	Ther.	Sc.-Div.	Ther.
27	0	25	0	40·1				
		30	0	40·7	68·0			
		35	0	41·0				
		40	0	41·1	68·1			
		45	0	41·8				
		50	0	41·9	68·2			
		55	0	42·0				
	1	0	0	42·0	68·3	69		
		5	0	42·1				
		10	0	42·4	68·2			
		15	0	42·5				
		20	0	42·6	68·1			
		25	0	42·7				
		30	0	42·7	68·2			
		35	0	42·8				
		40	0	42·9	68·2			
		45	0	42·9				
		50	0	42·5	68·0			
		55	0	42·1				
	2	0	0	42·0	67·6	69		
		5	0	42·0				
		10	0	42·2	67·0			
		15	0	42·2				
		20	0	42·0	66·2			
		25	0	42·0				
		30	0	42·0	66·1			
		35	0	42·0				
		40	0	42·0	65·8			
		45	0	42·0				
		50	0	42·0	65·2			
		55	0	42·0				
	3	0	0	42·0	65·3	69		
		5	0	42·0				
		10	0	41·8	65·3			
		15	0	41·8				
		20	0	41·5	65·3			
		25	0	41·5				
		30	0	41·3	65·0			
		35	0	41·1				
		40	0	41·0	64·9			
		45	0	40·9				
		50	0	40·7	64·8			
		55	0	40·5				
	4	0	0	40·2	64·7	69		
		5	0	39·9				
		10	0	39·2	64·2			
		15	0	39·0				
		20	0	38·9	64·0			
		25	0	38·8				
		30	0	38·9	63·6			
		35	0	38·1				
		40	0		63·1			
		45	0	37·9				
		50	0	37·1	62·9			
		55	0	37·1				
	5	0	0	37·1	62·1	69		
		5	0	36·9				
		10	0	36·7	62·0			
		15	0	36·5				
		20	0	36·2	61·9			
		25	0	36·0				
		30	0	35·8	62·0			
		35	0	35·9				
		40	0	35·7	62·1			
		45	0	35·3				
		50	0	35·4	62·3			
		55	0	35·1				

FEBRUARY 26 and 27, 1841.

M. Gött. Time.				Decl.	Hor. Force.		Vert. Force.	
d.	h.	m.	s.	Sc.-Div.	Sc.-Div.	Ther.	Sc.-Div.	Ther.
27	6	0	0	34·9	62·1	69		
		5	0	34·9				
		10	0	35·0	62·4			
		15	0	35·0				
		20	0	35·0	62·2			
		25	0	35·0				
		30	0	34·9	62·1			
		35	0	34·9				
		40	0	34·9	61·9			
		45	0	34·9				
		50	0	34·9	61·9			
		55	0	35·1				
	7	0	0	35·1	61·9	69		
		5	0	35·9				
		10	0	35·9	61·9			
		15	0	36·0				
		20	0	36·0	61·9			
		25	0	36·0				
		30	0	36·0	62·1			
		35	0	36·0				
		40	0	36·1	62·1			
		45	0	36·1				
		50	0	36·1	62·1			
		55	0	36·1				
	8	0	0	36·1	62·4	69		
		5	0	36·9				
		10	0	37·0	62·8			
		15	0	37·2				
		20	0	37·3	63·1			
		25	0	37·3				
		30	0	37·3	63·4			
		35	0	37·2				
		40	0	37·3	63·7			
		45	0	37·6				
		50	0	37·7	64·0			
		55	0	37·8				
	9	0	0	37·8	64·1	69		
		5	0	37·9				
		10	0	37·9	64·1			
		15	0	37·9				
		20	0	37·8	64·0			
		25	0	37·9				
		30	0	38·0	64·0			
		35	0	37·9				
		40	0	37·4	63·8			
		45	0	37·4				
		50	0	37·4	64·1			
		55	0	37·5				

The Mean Positions at the usual hours of observation during the Month are given in page 37.

VAN DIEMEN ISLAND { Decl. 1 Scale Division = 0'·71 / H. F. k = ·0003 ; q = / V. F. k = ; q = }

Regular and extra observations.

The V. F. was observed at 2m. 30s. before, and the H. F. 2m. 30s. after the times specified.

M. Gött. Time.				Decl.	Hor. Force.		Vert. Force.	
d.	h.	m.	s.	Sc.-Div.	Sc.-Div.	Ther.	Sc.-Div.	Ther.
26	0	0	0	29·8	72·7	68	93·5	67
		15	0	27·6				
		20	0	25·8	71·8		91·6	
		25	0	29·0				
		30	0	33·5	71·3		92·3	
		35	0	33·1				
		40	0	32·7	70·3		92·8	

FEBRUARY 26 and 27, 1841.

M. Gött. Time.				Decl.	Hor. Force.		Vert. Force.	
d.	h.	m.	s.	Sc.-Div.	Sc.-Div.	Ther.	Sc.-Div.	Ther.
26	0	45	0	34·0				
		50	0	31·3	68·4		93·9	
	1	0	0	29·0	69·3		95·2	67
		10	0	28·1	68·1		93·2	
		15	0	29·5				
		20	0	29·6	67·7		93·9	
		25	0	29·5				
		30	0	28·3	66·7		93·7	
		35	0	28·0				
		40	0	28·6	65·4		93·8	
		45	0	29·1				
	2	0	0	32·6	65·3	67	94·5	67
		12	30	34·6				
		17	30	34·6	66·6		95·4	
		22	30	34·2				
		27	30	34·8	66·9		95·1	
		32	30	35·2				
		37	30	35·1	67·8		94·9	
		42	30	35·1				
		47	30	34·7	66·8		94·3	
	3	0	0	36·5	67·4		96·6	67
	4	0	0	49·2	69·1	65	95·3	64
		10	0	46·1	70·2			
		15	0	45·3				
		20	0	44·7	70·2		96·2	
		25	0	43·9				
		30	0	43·1	71·0		95·6	
		35	0	43·6				
		40	0	42·9	70·2		95·3	
		45	0	43·8				
		50	0	44·8			95·6	
	5	0	0	45·1	70·5		93·4	63
	6	0	0	49·9	70·4	64	91·9	62
	7	0	0	48·5	74·9		96·0	62
	8	0	0	44·9	69·7	64	94·9	61
	9	0	0	38·4	72·5		95·5	60

Term observations.

The Decl., H. F., and V. F. were observed simultaneously at the times specified.

M. Gött. Time.				Decl.	Hor. Force.		Vert. Force.	
d.	h.	m.	s.	Sc.-Div.	Sc.-Div.	Ther.	Sc.-Div.	Ther.
26	10	0	0	38·4	72·6	60	94·9	57
		2	30	37·9	71·9		95·5	
		5	0	37·4	72·1		95·6	
		7	30	37·3	71·3		95·4	
		10	0	39·4	71·4		95·3	
		12	30	38·8	71·3		95·8	
		15	0	39·5	71·2		95·9	
		17	30	38·7	71·4		96·1	
		20	0	39·3	71·2		96·0	
		22	30	40·6	71·3		96·2	
		25	0	39·4	71·1		96·2	
		27	30	38·9	70·7		96·2	
		30	0	39·4	71·3		96·2	
		32	30	39·6	71·3		96·3	
		35	0	39·0	71·2		95·9	
		37	30	39·7	71·3		95·9	
		40	0	39·4	71·0		95·8	
		42	30	39·2	71·1		96·0	
		5	0	38·9	70·9		95·8	
		47	30	38·9	71·2		95·8	
		50	0	39·2	71·4		95·8	
		52	30	39·6	71·4		95·9	
		55	0	39·8	71·3		95·7	
		57	30	39·8	71·5		95·6	
	11	0	0	40·7	71·7	61	96·0	60

FEBRUARY 26 and 27, 1841.

M. Gött. Time (d. h. m. s.)	Decl. (Sc.-Div.)	Hor. Force (Sc.-Div.)	Ther.	Vert. Force (Sc.-Div.)	Ther.
26 11 2 30	40·2	71·8		95·5	
5 0	39·9	72·0		95·3	
7 30	39·2	71·6		95·8	
10 0	39·1	71·5		95·9	
12 30	39·2	71·7		96·0	
15 0	38·8	71·4		96·0	
17 30	38·5	71·5		96·0	
20 0	39·2	71·8		95·9	
22 30	40·1	72·2		96·1	
25 0	40·0	72·1		95·7	
27 30	39·6	72·2		95·6	
30 0	39·7	72·3		95·7	
32 30	39·7	72·3		95·6	
35 0	39·9	71·7		95·3	
37 30	40·9	72·2		96·0	
40 0	41·2	71·8		95·7	
42 30	41·3	72·1		96·2	
45 0	39·2	72·1		96·0	
47 30	38·7	71·9		95·6	
50 0	37·9	71·3		95·1	
52 30	37·4	70·6		95·0	
55 0	37·8	70·7		95·2	
57 30	38·0	70·1		95·4	
12 0 0	37·8	69·7	60	95·9	59
2 30	38·1	69·5		95·1	
5 0	38·1	69·3		95·1	
7 30	38·0	69·1		96·1	
10 0	38·9	68·9		95·5	
12 30	37·2	68·6		96·2	
15 0	37·7	68·4		96·5	
17 30	37·0	68·4		96·4	
20 0	37·4	68·3		96·3	
22 30	36·9	68·2		96·0	
25 0	36·9	68·2		95·8	
27 30	37·8	68·2		96·2	
30 0	37·5	68·1		96·4	
32 30	37·7	67·6		96·4	
35 0	38·0	67·4		96·6	
37 30	37·3	67·2		96·6	
40 0	37·3	66·7		97·1	
42 30	37·5	66·1		97·3	
45 0	36·3	65·4		97·2	
47 30	35·9	64·9		97·6	
50 0	36·1	64·6		97·7	
52 30	36·4	64·5		98·3	
55 0	36·2	64·6		98·1	
57 30	36·9	64·3		98·5	
13 0 0	35·9	63·9	60	98·3	59
2 30	36·5	63·3		99·1	
5 0	35·2	63·2		99·3	
7 30	35·0	63·1		98·9	
10 0	35·0	63·2		99·5	
12 30	34·2	63·4		99·1	
15 0	34·4	63·6		99·0	
17 30	34·8	63·9		99·1	
20 0	34·7	64·3		98·6	
22 30	35·7	64·4		98·7	
25 0	35·4	64·8		99·0	
27 30	36·5	65·0		98·9	
30 0	37·4	65·2		98·6	
32 30	36·4	65·3		98·4	
35 0	36·9	65·4		98·5	
37 30	37·7	65·9		98·3	
40 0	37·9	66·0		98·3	
42 30	38·4	66·2		98·2	
45 0	39·1	66·3		98·5	
47 30	39·8	66·7		98·1	

FEBRUARY 26 and 27, 1841.

M. Gött. Time (d. h. m. s.)	Decl. (Sc.-Div.)	Hor. Force (Sc.-Div.)	Ther.	Vert. Force (Sc.-Div.)	Ther.
26 13 50 0	40·0	66·6		97·6	
52 30	40·4	66·9		97·9	
55 0	40·6	66·7		97·9	
57 30	41·1	67·0		97·8	
14 0 0	41·5	67·1	60	98·0	59
2 30	41·8	67·2		97·6	
5 0	41·9	67·3		97·5	
7 30	42·2	67·4		97·7	
10 0	42·0	67·3		97·7	
12 30	42·0	67·2		97·4	
15 0	42·2	67·1		97·4	
17 30	42·4	66·9		97·5	
20 0	42·5	66·9		97·7	
22 30	42·8	66·8		97·7	
25 0	42·6	66·8		98·0	
27 30	42·7	66·9		97·9	
30 0	43·1	67·0		98·1	
32 30	43·1	67·2		97·9	
35 0	43·3	67·4		98·0	
37 30	43·6	67·5		97·9	
40 0	43·7	67·6		97·8	
42 30	44·0	67·7		97·7	
45 0	43·2	67·7		97·5	
47 30	43·3	67·9		97·6	
50 0	43·5	68·2		97·6	
52 30	43·6	68·3		97·0	
55 0	43·8	68·4		97·3	
57 30	43·4	68·1		96·8	
15 0 0	43·8	68·2	60	97·4	60
2 30	43·8	67·8		97·2	
5 0	42·5	67·9		97·3	
7 30	42·6	67·4		96·9	
10 0	42·9	67·0		97·4	
12 30	44·0	66·7		98·0	
15 0	44·7	66·4		97·9	
17 30	44·7	66·2		98·7	
20 0	44·8	65·8		99·1	
22 30	44·5	65·7		98·5	
25 0	43·4	65·3		98·1	
27 30	41·7	65·3		98·0	
30 0	41·3	66·1		98·8	
32 30	41·6	66·9		98·6	
35 0	41·5	66·8		98·5	
37 30	41·8	66·9		98·7	
40 0	41·8	66·3		98·0	
42 30	42·5	66·7		98·4	
45 0	43·8	65·9		99·1	
47 30	45·1	66·6		99·2	
50 0	45·5	67·0		99·4	
52 30	45·7	67·4		99·6	
55 0	46·0	67·3		99·1	
16 0 0	46·4	67·7	60	99·0	59
2 30	46·7	67·6		100·3	
5 0	47·1	67·8		100·0	
7 30	47·6	68·9		100·0	
10 0	47·6	69·4		99·9	
12 30	47·6	69·5		100·0	
15 0	47·6	69·6		100·0	
17 30	47·7	70·3		100·0	
20 0	47·7	70·5		100·0	
22 30	47·3	70·6		99·8	
25 0	47·4	70·9		99·6	
27 30	47·4	71·2		99·4	
30 0	47·5	71·8		99·1	
32 30	47·5	72·3		98·8	
35 0	47·5	72·6		98·7	

FEBRUARY 26 and 27, 1841.

M. Gött. Time (d. h. m. s.)	Decl. (Sc.-Div.)	Hor. Force (Sc.-Div.)	Ther.	Vert. Force (Sc.-Div.)	Ther.
26 16 37 30	47·6	72·8		98·4	
40 0	47·6	72·7		98·4	
42 30	48·1	73·0		98·4	
45 0	48·1	73·1		98·3	
47 30	48·0	73·3		97·9	
50 0	48·0	73·6		98·0	
52 30	48·1	74·0		97·8	
55 0	48·0	74·2		97·6	
57 30	48·0	74·2		97·4	
17 0 0	48·4	74·2	60	97·4	58
2 30	48·1	74·2		97·4	
5 0	48·1	74·1		97·3	
7 30	48·2	74·2		97·2	
10 0	48·3	74·0		97·6	
12 30	48·5	Lost.		97·6	
15 0	48·2	73·5		97·4	
17 30	48·5	73·3		97·6	
20 0	48·5	73·4		98·0	
22 30	48·5	73·3		98·0	
25 0	48·7	73·4		98·2	
27 30	49·2	73·7		98·3	
30 0	49·2	73·6		98·4	
32 30	49·4	73·5		98·2	
35 0	49·5	73·3		98·3	
37 30	49·1	73·1		98·3	
40 0	49·1	73·1		98·2	
42 30	49·0	73·1		98·2	
45 0	49·3	73·3		98·4	
47 30	49·7	73·5		98·3	
50 0	49·6	73·5		98·3	
52 30	49·7	73·5		98·3	
55 0	49·5	73·3		98·2	
57 30	49·6	73·2		98·2	
18 0 0	49·9	73·3	60	98·1	58
2 30	49·9	73·2		98·2	
5 0	50·0	73·5		98·2	
7 30	50·0	73·6		98·1	
10 0	50·0	73·9		98·1	
12 30	49·9	73·9		97·8	
15 0	49·7	74·1		97·7	
17 30	49·6	74·5		97·7	
20 0	49·7	75·1		97·7	
22 30	49·7	75·8		97·7	
25 0	49·5	75·7		97·3	
27 30	49·6	75·9		97·2	
30 0	49·3	75·9		96·9	
32 30	49·4	75·9		97·2	
35 0	49·4	76·0		97·2	
37 30	49·3	75·9		97·2	
40 0	48·6	75·7		97·1	
42 30	48·5	75·7		97·1	
45 0	48·5	75·7		97·1	
47 30	48·5	75·5		97·3	
50 0	48·5	75·7		97·5	
52 30	48·4	75·4		97·4	
55 0	47·7	74·6		97·2	
57 30	48·0	75·0		97·7	
19 0 0	47·7	74·5	60	97·7	58
2 30	47·3	74·1		98·0	
5 0	46·7	73·6		98·3	
7 30	45·5	73·5		97·9	
10 0	45·3	73·6		97·8	
12 30	44·8	73·4		97·6	
15 0	43·9	73·4		97·2	
17 30	43·1	73·6		97·0	
20 0	42·4	73·6		96·7	
22 30	42·6	74·8		96·7	

FEBRUARY 26 and 27, 1841.

M. Gött. Time (d. h. m. s.)	Decl. (Sc.-Div.)	Hor. Force (Sc.-Div.)	Ther.	Vert. Force (Sc.-Div.)	Ther.
26 19 25 0	43·1	75·4		96·4	
27 30	43·1	75·9		96·3	
30 0	43·4	76·3		96·0	
32 30	43·9	76·7		96·0	
35 0	44·4	76·9		95·7	
37 30	45·0	77·2		95·7	
40 0	44·9	76·8		95·5	
42 30	44·3	76·5		95·3	
45 0	44·3	76·3		95·3	
47 30	44·0	76·2		95·4	
50 0	44·4	76·3		95·4	
52 30	44·9	76·4		95·7	
55 0	44·7	76·2		95·7	
57 30	44·2	75·8		95·9	
20 0 0	44·0	75·6	60	95·7	58
2 30	44·0	75·5		95·8	
5 0	43·9	75·5		96·0	
7 30	43·6	75·4		96·1	
10 0	43·8	75·4		96·3	
12 30	43·9	75·6		96·0	
15 0	43·4	75·5		96·0	
17 30	43·5	75·4		96·2	
20 0	43·5	75·6		96·3	
22 30	43·5	75·9		96·3	
25 0	43·6	76·2		96·3	
27 30	43·5	76·3		96·1	
30 0	43·8	75·9		96·2	
32 30	44·2	76·1		96·2	
35 0	44·2	75·8		96·3	
37 30	44·0	75·6		96·3	
40 0	43·7	75·3		96·3	
42 30	43·5	75·1		96·3	
45 0	43·5	74·9		96·4	
47 30	43·3	74·9		96·4	
50 0	43·0	74·9		96·5	
52 30	43·2	74·9		96·5	
55 0	43·0	74·4		96·7	
57 30	43·0	74·4		96·8	
21 0 0	42·2	73·8	60	96·8	59
2 30	41·9	73·7		96·8	
5 0	41·4	73·2		96·9	
7 30	40·9	72·8		97·3	
10 0	40·0	72·6		97·2	
12 30	39·5	72·8		97·3	
15 0	40·0	73·3		97·7	
17 30	39·3	73·3		97·3	
20 0	39·9	73·5		96·9	
22 30	40·7	74·0		96·9	
25 0	41·8	74·1		96·4	
27 30	42·1	73·8		96·1	
30 0	42·1	73·8		95·9	
32 30	42·0	73·8		95·8	
35 0	41·9	73·8		95·7	
37 30	41·6	74·0		95·5	
40 0	41·2	74·3		95·5	
42 30	41·1	74·4		95·0	
45 0	41·1	74·6		94·9	
47 30	41·3	74·9		94·7	
50 0	41·0	75·4		94·8	
52 30	41·0	75·4		94·6	
55 0	40·9	75·4		94·3	
57 30	40·9	75·3		94·3	
22 0 0	41·0	75·1	60	94·1	60
2 30	41·3	75·1		94·2	
5 0	41·4	75·0		94·2	
7 30	41·4	75·0		94·4	
10 0	41·5	74·9		94·1	

FEBRUARY 26 and 27, 1841.

M. Gött. Time (d. h. m. s.)	Decl. (Sc.-Div.)	Hor. Force (Sc.-Div.)	Ther.	Vert. Force (Sc.-Div.)	Ther.
26 22 12 30	41·5	74·7		94·3	
15 0	41·4	74·7		94·2	
17 30	41·3	74·6		94·3	
20 0	41·4	74·6		94·4	
22 30	41·4	74·4		94·7	
25 0	41·5	74·2		94·8	
27 30	41·5	74·0		94·9	
30 0	41·3	73·9		95·0	
32 30	61·1	73·9		95·0	
35 0	41·0	73·9		94·9	
37 30	40·7	74·0		95·0	
40 0	40·7	74·1		95·2	
42 30	40·8	74·2		95·2	
45 0	41·0	74·3		95·0	
47 30	41·0	74·4		94·9	
50 0	41·2	74·5		95·0	
52 30	41·2	74·6		94·8	
55 0	41·2	74·8		94·9	
57 30	41·1	74·9		94·8	
23 0 0	41·3	74·8	60	94·9	60
2 30	41·4	74·7		94·9	
5 0	41·4	74·7		94·8	
7 30	41·4	74·7		94·8	
10 0	41·2	74·8		94·6	
12 30	41·1	74·7		94·7	
15 0	40·8	74·8		94·6	
17 30	40·8	74·8		94·3	
20 0	40·9	74·8		94·8	
22 30	41·0	74·8		94·9	
25 0	40·9	74·7		95·1	
27 30	40·3	74·8		94·7	
30 0	40·1	74·9		94·9	
32 30	39·9	75·2		94·7	
35 0	39·3	75·2		94·4	
37 30	38·8	75·2		94·5	
40 0	37·7	75·6		94·6	
42 30	36·6	76·7		94·2	
45 0	36·5	77·4		93·9	
47 30	37·5	78·4		93·4	
50 0	38·1	78·7		93·2	
52 30	38·6	78·6		92·9	
55 0	38·4	78·3		92·7	
57 30	38·1	78·1		92·5	
27 0 0	38·1	77·8	60	92·4	59
2 30	37·9	77·5		92·4	
5 0	37·6	77·3		92·5	
7 30	37·2	77·1		92·3	
10 0	37·5	77·1		92·5	
12 30	37·7	77·0		92·6	
15 0	38·1	76·8		92·8	
17 30	38·6	76·7		93·0	
20 0	38·9	76·7		92·9	
22 30	39·1	76·6		93·2	
25 0	39·1	76·5		93·2	
27 30	39·2	76·4		93·3	
30 0	39·3	76·4		93·3	
32 30	39·6	76·2		93·6	
35 0	39·9	75·9		93·6	
37 30	39·2	75·7		93·7	
40 0	39·0	75·6		93·9	
42 30	38·9	75·4		93·9	
45 0	38·7	75·3		94·1	
47 30	38·9	75·3		93·9	
50 0	38·9	75·2		94·3	
52 30	38·9	75·2		94·3	
55 0	38·9	75·2		94·3	
57 30	39·0	75·3		94·6	

FEBRUARY 26 and 27, 1841.

M. Gött. Time (d. h. m. s.)	Decl. (Sc.-Div.)	Hor. Force (Sc.-Div.)	Ther.	Vert. Force (Sc.-Div.)	Ther.
27 1 0 0	38·9	75·2	60	94·5	59
2 30	38·8	75·2			
5 0	38·8	75·2			
7 30	38·7	75·3			
10 0	38·8	75·1			
12 30	38·8	75·1			
15 0	39·7	75·1			
17 30	38·9	75·5		96·7	
20 0	39·1	75·7		96 6	
22 30	39·3	75·7		96·6	
25 0	39·4	75·7		96·7	
27 30	39·1	75·7		96·7	
30 0	39·2	75·7		96·7	
32 30	38·9	75·7		96·7	
35 0	38·9	75·8		96·6	
37 30	38·6	75·7		96·5	
40 0	38·6	75·6		96·5	
42 30	38·4	75·5		96·6	
45 0	38·3	75·4		96·6	
47 30	38·2	75·4		96·6	
50 0	38·1	75·3		96·5	
52 30	38·1	75·3		96·6	
55 0	38·1	75·2		96·5	
57 30	38·6	75·3		96·8	
2 0 0	38·6	75·2	60	96·8	59
2 30	38·6	75·1		96·8	
5 0	38·6	75·2		96·9	
7 30	38·5	75·2		96·7	
10 0	38·3	75·3		96·8	
12 30	38·1	75·5		97·1	
15 0	38·2	75·6		96·9	
17 30	38·6	75·7		96·7	
20 0	38·5	75·6		96·7	
22 30	38·3	75·6		96·7	
25 0	38·1	75·7		96·7	
27 30	37·9	75·8		96·5	
30 0	37·3	75·8		96·5	
32 30	36·9	75·7		96·5	
35 0	36·3	75·3		96·2	
37 30	35·7	75·4		96·3	
40 0	35·4	75·1		96·4	
42 30	34·4	75·0		96·6	
45 0	34·1	75·0		96·1	
47 30	33·9	75·0		96·1	
50 0	34·0	75·0		96·3	
52 30	34·1	75·0		96·1	
55 0	34·3	75·0		96·0	
57 30	34·5	75·1		96·4	
3 0 0	35·0	75·2	60	96·5	60
2 30	35·3	75·1		96·5	
5 0	35·9	75·3		96·5	
7 30	36·1	75·3		96·5	
10 0	36·3	75·2		96·5	
12 30	36·8	75·0		96·7	
15 0	37·2	75·0		96·7	
17 30	37·7	75·0		96·9	
20 0	37·9	74·9		97·0	
22 30	38·0	74·9		97·0	
25 0	38·1	75·0		97·2	
27 30	38·5	75·1		97·4	
30 0	38·8	75·1		97·4	
32 30	39·1	75·1		97·1	
35 0	39·2	75·2		97·4	
37 30	39·5	75·3		97·5	
40 0	39·9	75·5		97·6	
42 30	40·1	75·6		97·4	
45 0	40·0	75·3		97·0	

FEBRUARY 26 and 27, 1841.

d.	h.	m.	s.	Decl. Sc.-Div.	Hor. Force Sc.-Div.	Ther.	Vert. Force Sc.-Div.	Ther.
27	3	47	30	39·9	75·2		97·2	
		50	0	39·9	75·3		97·3	
		52	30	40·0	75·4		97·5	
		55	0	40·5	75·6		97·0	
		57	30	40·7	75·6		97·1	
	4	0	0	40·5	75·5	60	97·0	60
		2	30	40·2	75·5			
		5	0	40·4	75·5		97·0	
		7	30	40·4	75·5		97·1	
		10	0	40·2	75·6		96·9	
		12	30	40·2	75·7		96·9	
		15	0	40·3	75·7		96·8	
		17	30	40·2	75·7		96·8	
		20	0	40·1	75·9		96·8	
		22	30	40·1	75·9		96·8	
		25	0	39·9	75·9		97·0	
		27	30	39·9	76·1		97·0	
		30	0	40·0	76·1		97·0	
		32	30	40·2	76·1		96·9	
		35	0	40·1	76·2		96·9	
		37	30	40·0	76·4		97·0	
		40	0	40·3	76·5		96·9	
		42	30	40·4	76·5		96·9	
		45	0	40·4	76·5		96·9	
		47	30	40·6	76·6		96·7	
		50	0	40·8	76·6		96·7	
		52	30	40·7	76·5		96·8	
		55	0	40·4	76·4		96·4	
		57	30	39·7	76·3		96·4	
	5	0	0	39·4	76·3	60	96·6	60
		2	30	39·5	76·5		96·6	
		5	0	39·9	76·5		96·8	
		7	30	39·9	76·5		96·4	
		10	0	39·5	76·5		96·6	
		12	30	39·3	76·5		96·4	
		15	0	39·1	76·4		96·5	
		17	30	38·9	76·2		96·5	
		20	0	39·4	75·7		96·8	
		22	30	40·2	75·8		97·0	
		25	0	40·9	75·9		97·1	
		27	30	41·5	76·1		97·3	
		30	0	41·8	76·1		97·3	
		32	30	42·0	76·2		97·4	
		35	0	42·5	75·8		97·3	
		37	30	42·6	75·7		97·4	
		40	0	42·7	75·8		97·4	
		42	30	42·5	76·0		97·2	
		45	0	42·5	76·2		97·2	
		47	30	42·7	76·2		97·1	
		50	0	43·0	76·1		97·0	
		52	30	43·5	75·9		97·0	
		55	0	44·0	75·9		97·1	
		57	30	44·4	76·0		97·1	
	6	0	0	44·2	75·8	60	97·1	60
		2	30	44·1	75·6		97·1	
		5	0	44·1	75·6		97·1	

FEBRUARY 26 and 27, 1841.

d.	h.	m.	s.	Decl. Sc.-Div.	Hor. Force Sc.-Div.	Ther.	Vert. Force Sc.-Div.	Ther.
27	6	7	30	44·1	75·7		96·9	
		10	0	44·7	76·0		97·0	
		12	30	45·0	76·5		96·9	
		15	0	45·8	76·7		96·8	
		17	30	46·2	76·7		96·9	
		20	0	46·4	76·8		96·7	
		22	30	47·0	76·8		96·7	
		25	0	47·0	77·0		96·7	
		27	30	48·0	77·0		96·7	
		30	0	48·3	77·4		96·8	
		32	30	48·0	77·6		96·7	
		35	0	48·5	77·9		96·5	
		37	30	48·5	77·4		96·3	
		40	0	47·5	77·4		96·1	
		42	30	46·4	77·5		95·5	
		45	0	46·2	77·7		95·8	
		47	30	45·8	77·9		95·7	
		50	0	45·3	77·9		95·6	
		52	30	45·2	78·2		95·3	
		55	0	45·1	78·1		95·3	
		57	30	44·7	77·8		95·3	
	7	0	0	44·6	77·4	60	95·4	58
		2	30	43·0	77·1		95·1	
		5	0	42·6	76·8		95·0	
		7	30	43·5	76·7		95·3	
		10	0	44·6	76·8		95·5	
		12	30	44·4	76·6		95·6	
		15	0	44·2	76·5		96·0	
		17	30	43·9	76·2		96·0	
		20	0	44·1	76·1		96·1	
		22	30	44·2	76·2		96·1	
		25	0	45·5	76·3		96·3	
		27	30	46·2	76·6		96·5	
		30	0	47·0	77·0		96·7	
		32	30	47·9	77·2		96·8	
		35	0	48·0	77·6		96·7	
		37	30	49·0	77·8		96·6	
		40	0	49·3	78·0		96·6	
		42	30	49·4	78·0		96·4	
		45	0	49·1	77·9		96·3	
		47	30	48·7	77·9		95·9	
		50	0	47·9	77·8		95·7	
		52	30	47·3	77·9		95·6	
		55	0	46·5	77·6		95·4	
		57	30	45·6	77·5		95·1	
	8	0	0	45·0	77·3	60	95·1	59
		2	30	44·8	77·3		94·9	
		5	0	44·0	76·9		94·9	
		7	30	44·1	76·8		94·9	
		10	0	43·9	76·6		95·0	
		12	30	44·0	76·7		95·3	
		15	0	43·8	76·5		95·4	
		17	30	43·0	76·3		95·5	
		20	0	43·1	76·4		95·7	
		22	30	43·0	76·2		95·8	
		25	0	43·0	76·3		95·9	

FEBRUARY 26 and 27, 1841.

d.	h.	m.	s.	Decl. Sc.-Div.	Hor. Force Sc.-Div.	Ther.	Vert. Force Sc.-Div.	Ther.
27	8	27	30	42·5	76·2		95·8	
		30	0	42·5	76·3		95·7	
		32	30	42·5	76·3		95·7	
		35	0	42·4	76·4		95·7	
		37	30	42·1	76·4		95·7	
		40	0	41·5	76·3		95·7	
		42	30	41·3	76·3		95·5	
		45	0	41·4	76·3		95·6	
		47	30	41·0	76·4		95·6	
		50	0	40·7	76·4		95·7	
		52	30	40·4	76·3		95·5	
		55	0	40·0	76·3		95·7	
		57	30	40·0	76·3		95·4	
	9	0	0	40·0	76·3	60	95·7	59
		2	30	40·0	76·3		95·7	
		5	0	39·8	76·2			
		7	30	39·0	76·1		95·9	
		10	0	39·2	76·3		95·9	
		12	30	38·4	76·0		95·8	
		15	0	38·8	76·1		95·7	
		17	30	39·0	76·2		95·8	
		20	0	39·0	76·1		95·9	
		22	30	39·1	76·2		95·8	
		25	0	39·1	76·2		95·8	
		27	30	38·9	76·3		96·0	
		30	0	38·5	76·2		96·1	
		32	30	37·9	76·2		96·1	
		35	0	38·2	76·2		96·3	
		37	30	38·9	76·2		96·1	
		40	0	38·9	76·1		96·1	
		42	30	38·8	75·9		96·0	
		45	0	38·5	75·9		96·0	
		47	30	38·7	75·8		96·1	
		50	0	38·1	75·5		96·2	
		52	30	37·1	75·5		96·2	
		55	0	37·0	75·6		96·1	
		57	30	37·6	75·8		96·3	

The Mean Positions at the usual hours of observation during the Month, are given in pages 36 and 37.

MARCH 14, 15, 16, and 17, 1841.

TORONTO[a] $\left\{\begin{array}{l}\text{Decl. 1 Scale Division} = 0'\cdot72 \\ \text{H. F. } k = \cdot000076\,;\ q = \cdot0002 \\ \text{V. F. } k = \cdot00009\,;\ q = \cdot00018\end{array}\right.$

Extra observations.

The V. F. was observed at 1m. 30t. before, and the H. F. 2m. after the times specified.

d.	h.	m.	s.	Decl. Sc.-Div.	Hor. Force Sc.-Div.	Ther.	Vert. Force Sc.-Div.	Ther.
14	18	20	0h	55·5	415·8	35	98·4	35
		25	0	55·7	411·6		100·3	
		30	0	59·5	415·8		100·4	
		35	0	62·9	431·2		101·0	

[a] TORONTO, March, 1841. — *Times of observation during the Month at which the Magnetometers were disturbed, but the mean readings were not materially changed.*

[*Extract from the Toronto Return of December, 1841.*—The H. F. magnet is said to be "considerably," or "very much" disturbed when it vibrates in an arc of 35 to 45 scale divisions; to be "much" disturbed when it vibrates in an arc of 20 to 35 divisions; "moderately," when in an arc of 10 to 20 divisions; and "slightly," when in an arc of 5 to 10 divisions.

The same terms are used for the Decl. magnet when it vibrates through half the above number of scale divisions.]

d.	h.	
1	20	Magnets much disturbed by irregular movements, or shocks.
2	2	Slightly disturbed; shocks.
3	0½	Much disturbed by shocks, especially the Decl. magnet.
?	2	Slightly disturbed by shocks; the Decl. magnet vibrating considerably at 1h. 52m. 30t. The copper rings prevent the bar from vibrating under ordinary circumstances.
4	22	Much disturbed with shocks; vibrating at 4d. 22h. and 5d. 0h.
5	0½	H. F. much disturbed by shocks.

MARCH 14, 15, 16, and 17, 1841.

M. Gött. Time.				Decl.	Hor. Force.		Vert. Force.	
d.	h.	m.	s.	Sc.-Div^m.	Sc.-Div^m.	Ther.	Sc.-Div^m.	Ther.
14	18	40	0	62·9	419·2		105·4	
		45	0	49·3	402·3		101·6	
		50	0	37·6	399·3		98·1	
		55	0	33·0	389·3		94·4	
19		5	0	34·6	405·3	35	86·4	36
		10	0	33·7	406·9		90·3	
		15	0	32·2	410·2		88·8	
		25	0	38·7	395·1		90·9	
		30	0	45·2	405·6		88·4	
		35	0	48·9	411·2		90·8	
		40	0	52·3	385·8		86·8	
		45	0	52·7	394·8		85·3	
		50	0	56·0	413·6		90·0	
		55	0	57·2	419·2		94·8	
15	0	25	0	59·8	463·3	32	114·4	33
		30	0	58·9	456·5		114·8	
		35	0	58·9	459·5		115·0	
		40	0	58·0	462·0		114·4	
		45	0	59·1	452·8		114·5	
		50	0	60·1	466·8			
		55	0	60·4	467·4			

MARCH 14, 15, 16, and 17, 1841.

M. Gött. Time.				Decl.	Hor. Force.		Vert. Force.	
d.	h.	m.	s.	Sc.-Div^m.	Sc.-Div^m.	Ther.	Sc.-Div^m.	Ther.
15	1	5	0	53·1	468·5	31	119·7	33
		15	0	53·0	454·0		119·8	
		20	0	60·7	466·0		118·5	
		25	0	59·6	468·9		118·8	
		30	0	61·8	474·6		119·0	
		35	0	58·6	475·8		120·3	
		40	0	58·1	473·5		119·7	
		45	0	57·0	474·1		119·7	
2		40	0	54·8	465·3	32	119·8	34
		45	0	55·5	469·0		119·5	
		50	0	57·1	469·4		119·0	
		55	0	58·1	469·8		118·8	
3		5	0	62·3	454·0	34	116·8	35
		10	0	59·1	464·1		120·2	
		15	0	56·9	467·5		119·4	
		20	0	52·0	467·3		119·2	
		25	0	54·5	461·8		119·8	
		30	0	59·1	457·4		118·3	
		35	0	57·8	460·0		118·3	
		40	0	57·1	458·2		118·5	
		45	0	57·0	450·4		118·3	

MARCH 14, 15, 16, and 17, 1841.

M. Gött. Time.				Decl.	Hor. Force.		Vert. Force.	
d.	h.	m.	s.	Sc.-Div^m.	Sc.-Div^m.	Ther.	Sc.-Div^m.	Ther.
15	3	50	0	56·0	442·6		117·9	
4		5	0	53·1	432·0	36	118·4	36
		10	0	51·0	427·0		118·2	
		15	0	49·5	416·6		117·7	
		20	0	47·3	410·9		116·1	
		25	0	48·3	407·7		116·5	
		30	0	45·2	401·5		116·4	
		35	0	39·7	402·1		116·6	
		40	0	34·0	401·9		120·3	
		45	0	39·9	401·5		121·4	
		50	0	41·4	400·0		121·0	
		55	0	40·9	410·2		121·2	
5		5	0	38·7	417·0	37	123·1	37
		10	0	37·2	413·3		122·5	
		15	0	37·3	408·1		122·7	
		20	0	38·6	403·3		122·6	
		25	0	40·0	401·2		122·8	
		30	0	40·6	403·7		123·8	
		35	0	41·9	405·4		124·4	
		40	0	41·9	420·6		125·5	
		45	0	38·8	416·6		126·2	

d. h.	
10 0 } 20 }	H. F. magnet much disturbed; shocks.
11 0 } 11 22 }	Decl. aud V. F. much disturbed; H. F. vibrating moderately at 11d. 0h.
12 0 }	Much disturbed; vibrating with shocks at 12d. 0h.
14 18 to 15 6 }	Magnets much disturbed; vibrating with shocks.
16 20 17 0 } 14 }	Moderate disturbance; vibrating with shocks.
18 0 }	Moderate vibrations; shocks.
21 18 20 }	Much disturbed; vibrating with shocks.
22 0 } 2 } 4 } 12 }	Much disturbed; shocks.
29 0 2 }	Much disturbed; vibrating with shocks. Decl. magnet only not vibrating at 29d. 0h.
30 18 to 31 0 }	Moderate disturbance; vibrating with shocks.

Notices of the Aurora Australis seen from Her Majesty's Ship Terror, during the month of March, 1841, transmitted by Captain Crozier, R.N.

Date. Time at Ship.	Ship's Approximate Position. Lat. S.	Long. E.	
d. h. m.	o	o	
1 11 0	68·8	168·1	Aurora in the N.W., the rays extending to 10° from the zenith from the base of a dark cloud.
30			The Aurora extended from N.W. by W., to N.E. by E.
2 12 0	68·3	168·0	The horizon to the South was illuminated by a pale light.
5 9 57	65·7	167·0	Two pale streaks of light in the E.N.E. which lasted about two minutes, commencing at an altitude of thirty degrees.
17 12 0	64·4	152·0	A faint Aurora seen in the S.W. towards midnight.
21 11 0	64·7	140	Observed an Aurora in the N.W. shooting rays of light at an altitude of 6° obliquely towards the horizon.
21 12 0 } to 14 30 }			The Aurora in the N.W. shooting rays of light behind dark clouds.
22 9 0	62·8	139	Observed the Aurora in the North, shooting rays of light towards the zenith.
22 11 20			The Aurora became very brilliant, extending from East to West; the colour was a pale white, shooting pencils of rays of a red tinge to the South and West.
11 40			The Aurora illuminated the whole heavens, giving a light equal to the moon.
12 0			The Aurora visible in the zenith till 12h. 30m. Wind unsteady.

Date. Time at Ship	Ship's Approximate Position. Lat. S.	Long. E.	
d. h. m.			
23 7 50 } to 8 40 }	62	135	Observed the Aurora extending across the zenith in a N.N.W. and S.S.E. direction: between 6m. 20s. and 8m. 40s. the Aurora was very brilliant, shooting pencil rays to the zenith, and forming a canopy of pale bright light from an altitude of 30° over the mast-heads, latterly diffusing, from its radiated appearance, and then disappearing during some of the most brilliant coruscations; stars were visible through it.
24 10 0	61	133	An Aurora occasionally very bright from N.W. to E.N.E
25 0 0	60·5	131·5	The Aurora occasionally seen from N.W. to S.E.
7 0			Observed a bright Aurora from North to East during the night; it occasionally became very bright in the zenith.
27 8 0	58	128·0	An Aurora was observed to form due West, which at 8m. 30s. extended across the zenith in an Easterly direction, until at 9m. 30s., it reached within three degrees of the Eastern horizon, where it appeared to rest upon a massy cloud; it became very brilliant, shooting streams of light in a horizontal direction easterly, showing various colours, bright red, purple, and deep blue, giving a moonlight appearance; at 11m. 20s. it entirely disappeared.
24 8 0 } to 11 0 }	57	127	The Aurora was shooting pencils of rays of pale light from behind dark clouds.
30			The Aurora extended from a thick massy cloud in the N.W. across the zenith, to within 6° of the horizon in the S.E., giving a moonlight appearance.
29 0 0	56	129	A brilliant Aurora in the zenith, at times exhibiting rays of crimson, blue, purple, and green light.
7 30			The Aurora extended from East to West in an arch, shooting bright variegated streams of light from behind a bank of dark clouds which skirted the horizon in that direction; at times very brilliant, and reflected by the sea.
30 0 0	55	132	The Aurora West at an altitude of 4°, extending in a narrow arch through the zenith, and meeting the Eastern horizon; the part of the arch to the Westward being very bright.
10 0			The Aurora commenced soon after dark, from East, in a curved line terminating in the West.
11 30			From 10m. to 11m. 30s. the Aurora was very bright.

b The extra observations commenced at midnight of Sunday at Toronto, at which time the Aurora was visible, and is described as follows:—

" 14 19 Calm, clear, and unclouded; bright auroral light in the North; a number of faint streamers shooting up from it.

" 14 20 Wind N. by W. light; clear and unclouded; since 19h. the features of the Aurora have been continually changing from arches of light, to detached banks and patches; some very vivid pulsations occurred about 19h.; nothing now appearing except two very faint streamers in the N. W.

" 14 22 Calm, clear, and unclouded; a bright bank of light remaining in the North."

MARCH 14, 15, 16, and 17, 1841.

M. Gött. Time. d. h. m. s.	Decl. Sc.-Div^ns	Hor. Force. Sc.-Div^ns	Ther.	Vert. Force. Sc.-Div^ns	Ther.
15 5 50 0	42·5	424·7		126·2	
55 0	41·4	437·8		126·2	
6 5 0	42·5	444·7	39	126·8	38
10 0	40·9	449·4		126·8	
15 0	43·5	459·7		127·0	
20 0	41·8	467·5		127·5	
25 0	43·2	466·7		127·7	
30 0	43·4	473·9		127·7	
35 0	42·8	476·5		127·0	
40 0	43·8	487·0		127·0	
45 0	44·9	488·6		127·1	
50 0	44·0	488·8		127·3	
8 30 0	45·4	483·3	40	125·4	40
9 0 0	49·7	476·0	41	125·0	40
30 0	56·1	499·0	41	129·5	40

Positions at the usual hours of observation, from March 14th, 18h. to March 17th, 6h.

M. Gött. Time. d. h. m. s.	Decl. Sc.-Div^ns	Hor. Force. Sc.-Div^ns	Ther.	Vert. Force. Sc.-Div^ns	Ther.
14 18 0 0	57·2	402·4	35	94·2	35
20 0 0	57·7	437·2	34	98·9	35
22 0 0	59·4	414·3	33	99·1	35
15 0 0 0	58·5	467·7	32	112·4	33
2 0 0	57·7	457·5	32	116·9	34
4 0 0	53·7	427·0	35	111·7	36
6 0 0	43·2	423·5	39	115·3	39
8 0 0	45·9	467·0	40	111·8	40
10 0 0	51·0	460·0	41	112·0	41
12 0 0	54·5	443·8	40	115·3	40
14 0 0	48·3	458·0	40	114·3	40
16 0 0	63·4	450·0	38	109·3	39
18 0 0	44·5	450·2	38	101·1	38
20 0 0	65·2	436·6	37	100·9	37
22 0 0	46·1	460·8	37	107·6	38
16 0 0 0	59·8	474·8	36	112·3	37
2 0 0	64·5	475·7	36	111·9	37
4					
6		a			
8					
10					
12 0 0					
14 0 0	54·2	442·3			
16 0 0	52·4	460·5	41	104·1	42
18 0 0	57·5	461·9	38	101·3	39
20 0 0	55·2	472·5	37	103·3	38
22 0 0	53·6	470·7	37	102·3	37
17 0 0 0	49·8	459·7	35	100·1	36
2 0 0	58·4	474·7	36	102·2	36
4 0 0	52·7	449·7	39	100·9	39
6 0 0	47·9	450·4	41	98·9	41

Mean Positions at the usual hours of observation during the Month.[b]

M. Gött. Time. d. h. m. s.	Decl. Sc.-Div^ns	Hor. Force. Sc.-Div^ns	Ther.	Vert. Force. Sc.-Div^ns	Ther.
0 0 0	55·7	473·3	38	109·2	42
2 0 0	57·9	468·0	38	110·2	42
4 0 0	55·0	455·5	39	107·1	43
6 0 0	47·3	453·0	41	106·3	44
8 0 0	45·1	461·3	42	106·0	45

MARCH 14, 15, 16 and 17, 1841.

M. Gött. Time. h. m. s.	Decl. Sc.-Div^ns	Hor. Force. Sc.-Div^ns	Ther.	Vert. Force. Sc.-Div^ns	Ther.
10 0 0	47·8	463·6	43	105·9	46
12 0 0	51·0	459·3	43	106·5	46
14 0 0	53·0	462·8	42	107·3	46
16 0 0	54·0	460·9	41	107·7	45
18 0 0	54·2	461·4	40	106·7	44
20 0 0	53·3	465·1	39	107·2	43
22 0 0	55·2	467·1	39	107·3	42

St. Helena { Decl. 1. Scale Division = 0'·71 / H.F. k = ·000180; q = ·00025 }

Positions at the usual hours of observation, from March 14th, 14h. to March 17th, 6h.

M. Gött. Time. d. h. m. s.	Decl. Sc.-Div^ns	Hor. Force. Sc.-Div^ns	Ther.
14 14 0 0c	35·0	d	
15 0 0	35·8		
16 0 0	35·8		
18 0 0	35·1		
19 30 0	36·4		
20 0 0	35·5		
20 30 0	33·0		
22 0 0	31·2		
23 0 0	31·3		
15 0 0 0	34·4		
2 0 0	37·4		
3 0 0	38·7		
4 0 0	36·5		
5 0 0	34·1		
6 0 0	36·3		
8 0 0	37·0		
10 0 0	37·4		
11 0 0	38·1		
12 0 0	38·0		
13 0 0	38·0		
14 0 0	38·2	36·3	72
15 0 0	39·1	39·0	71
16 0 0	40·0	40·0	71
18 0 0	39·3	49·2	71
19 30 0	41·4	50·8	71
20 0 0	41·6	53·1	71
20 30 0	41·4	52·3	70
22 0 0	37·6	57·5	70
23 0 0	38·0	61·6	71
16 0 0 0	38·0	65·0	71
2 0 0	36·1	63·9	71
3 0 0	35·0	62·1	72
4 0 0	33·9	56·0	72
5 0 0	32·7	46·6	72
6 0 0	33·2	43·0	72
8 0 0	38·0	47·4	71
10 0 0	37·0	47·0	71
11 0 0	38·8	52·6	71
12 0 0	38·3	50·0	71
13 0 0	38·0	49·0	71
14 0 0	37·3	48·5	71
15 0 0	37·4	49·4	71
16 0 0	37·6	50·5	71
18 0 0	38·0	51·1	70
19 30 0	40·0	54·0	70
20 0 0	38·1	55·2	70

MARCH 14, 15, 16 and 17, 1841.

M. Gött. Time. d. h. m. s.	Decl. Sc.-Div^ns	Hor. Force. Sc.-Div^ns	Ther.	Vert. Force. Sc.-Div^ns	Ther.
16 20 30 0	36·8	57·1	70		
22 0 0	33·8	62·0	70		
23 0 0	35·2	62·0	70		
17 0 0 0	37·8	63·8	70		
2 0 0	37·8	62·4	71		
3 0 0	36·1	60·0	71		
4 0 0	33·8	55·8	71		
5 0 0	35·0	55·0	71		
6 0 0	37·6	54·9	71		

Mean Positions at the same hours during the Month.[e]

M. Gött. Time. d. h. m. s.	Decl. Sc.-Div^ns	Hor. Force. Sc.-Div^ns	Ther.
0 0 0	38·3	65·5	70
2 0 0	40·2	63·7	71
3 0 0	39·1	59·8	71
4 0 0	37·6	56·3	71
5 0 0	39·1	54·5	72
6 0 0	37·2	52·9	72
8 0 0	39·0	52·8	71
10 0 0	39·0	52 7	70
11 0 0	39·3	53·1	70
12 0 0	39·4	54·5	70
13 0 0	39·1	55·2	70
14 0 0	38·9	54·7	70
15 0 0	38·9	54·4	70
16 0 0	39·0	54·8	70
18 0 0	39·1	56·0	69
19 30 0	39·6	57·0	69
20 0 0	38·3	58·0	69
20 30 0	37·1	58·9	69
22 0 0	35·3	62·7	69
23 0 0	36·5	64·5	69

Van Diemen Island { Decl. 1 Scale Division = 0'·71 / H.F. k = ·0003; q = / V.F. k = ; q = }

Extra observations.

The V.F. was observed at 2m. 30s. before, and the H.F. 2m. 30s. after the times specified.

M. Gött. Time. d. h. m. s.	Decl. Sc.-Div^ns	Hor. Force. Sc.-Div^ns	Ther.	Vert. Force. Sc.-Div^ns	Ther.
15 5 10 0	42·5	78·5	68	88·3	67
15 0	44·2				
20 0	46·7	77·4		88·8	
25 0	49·9				
30 0	44·7	69·1		88·8	
35 0	37·7				
40 0	32·4	71·2		89·4	
45 0	31·1				
50 0	34·9	76·9		89·1	
6 10 0f	43·4	75·2	66	88·5	66
15 0	41·7				
20 0	41·7	74·6		88·6	
25 0	42·4				
30 0	42·1	73·7		89·5	
35 0	42·4				
40 0	43·4	72·4		90·3	
45 0	44·2				
50 0	44·6	71·2		92·2	

[a] During these hours the instruments were employed in observations of the absolute horizontal intensity.
[b] The mean positions of the H. F. magnet are from the 1st to the 15th of March, inclusive; and of the V. F. magnet from the 2nd to the 15th, inclusive.
[c] Commencing after midnight on Sunday at St. Helena.
[d] The H. F. magnet was undergoing adjustment until 15d. 14h.

[e] The mean positions of the Decl. magnet are from the 1st to the 22nd of March, inclusive; and those of the H. F. magnet from the 16th to the 31st of March, inclusive; the connexion of the daily observations with the mean positions of the H. F. magnet, cannot be regarded as strictly determined, for the reasons assigned in page 37.
[f] "Calm; clear moonlight; heavy dew; no appearance of Aurora from the top of the hill near the Observatory."

MARCH 14, 15, 16, and 17, 1841.

d. h. m. s.	Decl. Sc.-Div.	Hor. Force. Sc.-Div.	Ther.	Vert. Force. Sc.-Div.	Ther.
16 6 10 0	43·2	78·7	66		66
15 0	46·7				
20 0	49·2	77·5		92·7	
25 0	53·5				
30 0	54·9	74·1		94·4	
35 0	54·9				
40 0	55·7	72·7		96·3	
45 0	57·0				
50 0	56·9	72·7		97·2	
7 15 0	53·5		66		65
20 0	51·4	74·6		94·1	
25 0	50·3				
30 0	47·5	74·2		93·5	
35 0	48·6				
40 0	47·4			93·3	

Positions at the usual hours of observation, from March 14th, 10h, to March 17th, 6h.

d. h. m. s.	Decl. Sc.-Div.	Hor. Force. Sc.-Div.	Ther.	Vert. Force. Sc.-Div.	Ther.
14 10 0 0	46·9	83·4	61	94·0	61
11 0 0	38·4	79·9		93·2	
12 0 0	41·1	75·7	63	95·8	62
13 0 0	47·8	71·8		96·7	
14 0 0	53·8	69·7	67	96·6	67
15 0 0	58·1	69·9		95·2	
16 0 0	62·8	67·9	71	95·6	71
17 0 0	63·0	67·0		95·7	
18 0 0	67·2	72·2	74	94·8	72
19 0 0	64·5	67·6		96·5	
20 0 0	52·2	65·7	75	96·2	72
21 0 0	50·5	65·9		97·1	
22 0 0	49·0	65·5	72	97·8	69
23 0 0	45·6	68·0		97·7	
15 0 0 0	44·4	78·3	69	92·4	69
1 0 0	45·0	70·8		92·6	
2 0 0	50·5	73·5	69	93·8	69
3 0 0	44·9	73·6		92·2	
4 0 0	43·6	74·9	68	91·7	68
5 0 0	38·3	78·4		88·3	
6 0 0	40·5	77·1	66	88·5	66
7 0 0	46·8	81·1		93·5	
8 0 0	44·3	72·7	64	93·5	63
9 0 0	49·6	72·2		96·3	
10 0 0	46·1	74·0	63	94·7	62
11 0 0	48·2	70·9		95·9	
12 0 0	47·0	72·5	64	93·7	64
13 0 0	45·0	72·6		91·6	
14 0 0	46·5	71·4	68	91·2	69
15 0 0	48·6	68·1		93·0	
16 0 0	53·5	68·0	73	93·2	72
17 0 0	56·5	68·7		92·7	
18 0 0	53·4	71·0	75	92·1	73
19 0 0	55·1	69·3		94·6	
20 0 0	55·6	69·3	75	95·1	73
21 0 0	50·6	70·4		94·8	
22 0 0	47·1	72·3	73	93·8	71
23 0 0	49·4	74·5		93·9	
16 0 0 0	49·6	75·1	71	93·9	69
1 0 0	47·6	76·0		93·9	
2 0 0	47·3	75·8	68	94·5	67
3 0 0	48·4	75·8		95·1	
4 0 0	47·5	77·6	67	95·9	66
5 0 0	43·7	83·1		90·2	
6 0 0	40·9	78·7	66	92·1	65

* Commencing at midnight of Sunday, at Toronto.

MARCH 14, 15, 16, and 17, 1841.

m. s.	Decl. Sc.-Div.	Hor. Force. Sc.-Div.	Ther.	Vert. Force. Sc.-Div.	Ther.
16 7 0 0	57·1	73·2		96·3	
8 0 0	48·3	74·5	65	92·3	65
9 0 0	46·7	76·6		93·3	
10 0 0	45·9	75·5	63	94·7	63
11 0 0	49·6	71·3		97·2	
12 0 0	46·5	73·6	64	94·2	65
13 0 0	47·2	73·8		93·0	
14 0 0	50·9	70·1	69	93·7	70
15 0 0	54·4	69·5		91·8	
16 0 0	57·3	69·8	74	91·0	73
17 0 0	58·3	70·9		90·9	
18 0 0	54·1	69·5	76	92·3	73
19 0 0	54·6	71·0		91·7	
20 0 0	52·6	71·2	76	92·0	72
21 0 0	52·1	72·5		91·7	
22 0 0	50·6	73·1	73	92·6	71
23 0 0	44·5	74·3		91·0	
17 0 0 0	43·5	75·6	71	91·7	69
1 0 0	44·3	76·5		91·8	
2 0 0	47·4	75·4	69	93·2	
3 0 0	48·6	74·3		94·0	
4 0 0	52·0	75·9	69	94·4	67
5 0 0	50·7	76·2		93·4	
6 0 0	49·6	76·8	67	93·9	66

Mean Positions at the same hours during the Month.

m. s.	Decl. Sc.-Div.	Hor. Force. Sc.-Div.	Ther.	Vert. Force. Sc.-Div.	Ther.
0 0 0	48·0	76·2	68	93·3	66
1 0 0	47·4	76·5		93·4	
2 0 0	46·4	76·4	67	93·4	65
3 0 0	47·4	76·8		93·5	
4 0 0	47·6	77·1	65	94·1	64
5 0 0	47·5	77·6		93·7	
6 0 0	48·3	77·7	64	94·1	63
7 0 0	48·7	78·2		94·6	
8 0 0	48·0	78·1	63	94·3	62
9 0 0	47·0	78·0		94·7	
10 0 0	45·7	77·7	62	95·0	61
11 0 0	43·4	76·6		94·3	
12 0 0	43·4	75·4	63	94·2	63
13 0 0	45·9	74·0		94·4	
14 0 0	50·0	72·9	66	94·2	66
15 0 0	53·9	72·7		94·0	
16 0 0	56·7	72·9	69	93·4	69
17 0 0	57·6	73·8		93·2	
18 0 0	57·1	74·4	72	92·9	69
19 0 0	54·1	74·0		93·2	
20 0 0	51·8	74·2	72	93·1	69
21 0 0	50·8	74·1		93·2	
22 0 0	50·2	75·0	70	93·3	67
23 0 0	48·9	75·8		93·3	

MARCH 22 and 23, 1841.

TORONTO { Decl. 1 Scale Division = 0'·72 ; H. F. k = ·000076; q = ·0002 ; V. F. k = ·00009; q = ·00018

Extra observations.

The V. F. was observed at 1m. 30s. before, and the H. F. 2m. after the times specified.

	Decl.	Hor. Force.	Ther.	Vert. Force.	Ther.
22 4 30 0	44·9	432·2	42	96·1	43
35 0	48·3	425·2		95·9	
40 0	42·1	422·6		97·1	

MARCH 22 and 23, 1841.

d. h. m. s.	Decl. Sc.-Div.	Hor. Force. Sc.-Div.	Ther.	Vert. Force. Sc.-Div.	Ther.
22 4 45 0	40·7	420·0		96·3	
50 0	39·0	438·1		96·5	
55 0	40·8	445·4		95·7	
5 5 0	38·8	445·3	42	97·1	44
10 0	37·9	434·7		96·7	
15 0	38·0	434·7		95·1	
20 0	42·0	443·7		95·0	
25 0	42·5	448·7	43	95·1	44
30 0	41·1	452·4		95·2	
35 0	44·1	437·6		95·2	
40 0	43·2	439·5		95·2	
45 0	42·5	440·5		95·3	
50 0	43·0	429·6		96·8	
55 0	41·3	421·9		94·6	
6 30 0	41·3	459·9	44	98·2	44
7 0 0	37·5	449·3	44	100·1	44
7 30 0	39·2	454·1	44	97·5	44
8 30 0	40·2	478·7	44	98·5	44
9 0 0	37·7	484·5	45	100·3	45
9 30 0	43·1	450·3	45	106·1	45
10 30 0	46·9	430·7	45	112·6	45

Positions at the usual hours of observation, from March 21st, 18h, to March 23rd, 18h.

d. h. m. s.	Decl. Sc.-Div.	Hor. Force. Sc.-Div.	Ther.	Vert. Force. Sc.-Div.	Ther.
22 18 0 0*	58·6	455·2	44	92·0	44
20 0 0	69·3	454·4	44	86·4	44
22 0 0	58·9	474·0	44	94·6	44
22 0 0 0	63·8	474·9	43	93·8	44
2 0 0	73·1	458·1	42	93·8	43
4 0 0	39·2	411·4	42	93·4	43
6 0 0	37·6	418·0	44	98·0	44
8 0 0	41·7	488·5	44	97·1	44
10 0 0	44·4	441·1	45	108·2	45
12 0 0	68·2	460·0	45	102·2	45
14 0 0	57·5	437·8	45	98·2	45
16 0 0	68·7	474·5	45	91·2	45
18 0 0	47·8	435·8	44	90·3	45
20 0 0	44·5	396·0	44	67·2	45
22 0 0	61·6	447·7	45	81·6	45
23 0 0 0	58·6	449·0	45	87·3	46
2 0 0	52·6	435·0	46	86·7	46
4 0 0	56·8	437·6	46	88·9	46
6 0 0	45·4	423·4	47	89·0	47
8 0 0	44·9	443·3	48	90·3	47
10 0 0	42·8	450·2	49	90·2	48
12 0 0	53·1	440·5	49	92·4	49
14 0 0	56·3	443·7	48	90·7	49
16 0 0	52·0	419·4	48	88·8	48
18 0 0	56·6	450·6	48	89·9	48

Mean Positions at the same hours during the Month.[b]

d. h. m. s.	Decl. Sc.-Div.	Hor. Force. Sc.-Div.	Ther.	Vert. Force. Sc.-Div.	Ther.
0 0 0	55·7	458·4	45	88·4	42
2 0 0	57·9	453·4	45	89·6	42
4 0 0	55·0	436·9	46	87·8	43
6 0 0	47·3	430·3	48	87·1	44
8 0 0	45·1	447·3	49	87·4	45
10 0 0	47·8	448·9	50	88·7	46
12 0 0	51·0	447·4	50	88·9	46
14 0 0	53·0	441·5	49	89·8	46
16 0 0	54·0	447·4	48	87·7	45
18 0 0	54·2	450·5	46	88·6	44
20 0 0	53·3	458·1	46	84·9	43
22 0 0	55·2	455·1	45	87·9	43

[b] The mean positions of the H. F. and V. F. magnets, are from the 17th to the 31st of March, inclusive.

H

MARCH 22 and 23, 1841.

St. Helena { Decl. 1 Scale Division = 0'·71 ; H. F. k = ·00018; q = ·00025

Extra observations.

The H. F. was observed at 0m. 30s. after the times specified.

M. Gött. Time. d. h. m. s.	Decl. Sc.-Div⁰ˢ	Hor. Force. Sc.-Div⁰ˢ	Ther.°	Vert. Force. Sc.-Div⁰ˢ	Ther.°
22 2 13 36		68·0	70		
18 36	40·9	67·0			
23 36	41·5	67·2			
28 36	41·9	67·1			
33 36	41·1	65·6			
38 36	40·9	63·1			
43 36	39·2	61·0			
48 36	38·2	58·9			
53 36	37·2	55·8			
58 36	37·2	54·1			
3 3 36	37·2	52·3	70		
8 36	37·2	51·0			
13 36	37·2	49·8			
18 36	37·2	48·0			
23 36	37·8	47·3			
28 36	37·8	46·0			
33 36	37·8	47·9			
38 36	35·2	47·9			
43 36	35·2	48·0			
48 36	35·0	48·0			
53 36	34·8	47·5			
58 36	34·2	47·0			
4 3 36	33·8	46·6	70		
8 36	33·8	45·8			
13 36	34·0	44·3			
18 36	34·2	43·2			
23 36	34·2	43·0			
28 36	33·2	43·1			
33 36	32·8	42·8			
38 36	32·9	41·5			
43 36	33·2	40·0			
48 36	33·2	40·4			
53 36	32·5	40·3			
58 36	32·4	40·3			
5 3 36	32·2	40·4	70		
8 36	32·5	40·7			
13 36	32·8	40·0			
18 36	33·0	40·1			
23 36	32·8	40·9			
28 36	32·2	41·1			
33 36	32·3	40·0			
38 36	32·4	38·4			
43 36	32·1	37·0			
48 36	32·4	36·0			
53 36	32·3	35·0			
58 36	32·2	34·8			
6 3 36	32·2	34·7	71		
8 36	32·4	35·2			
13 36	32·4	36·0			
18 36	32·8	38·0			
23 36	33·6	39·8			
28 36	33·8	40·0			
33 36	33·9	40·0			
38 36	34·1	40·0			
43 36	34·6	39·8			
48 36	35·0	39·4			
53 36	35·2	39·0			
58 36	35·4	38·4			
7 3 36	35·5	38·4	71		
8 36	35·5	38·1			

MARCH 22 and 23, 1841.

M. Gött. Time. m. s.	Decl. Sc.-Div⁰ˢ	Hor. Force. Sc.-Div⁰ˢ	Ther.°	Vert. Force. Sc.-Div⁰ˢ	Ther.°
22 7 13 36	35·5	38·8			
18 36	35·6	38·9			
23 36	35·9	39·0			
28 36	36·0	39·5			
33 36	36·1	40·0			
38 36	36·1	40·2			
43 36	36·2	40·5			
48 36	36·2	40·6			
53 36	36·1	41·5			
58 36	36·0	42·0			
8 3 36	36·0	42·0	70		
8 36	36·1	40·9			
13 36	36·1	39·9			
18 36	36·0	39·1			
23 36	36·0	38·3			
28 36	35·8	38·8			
33 36	35·3	39·0			
38 36	34·9	39·1			
43 36	34·8	39·9			
48 36	34·8	39·1			
53 36	35·1	39·5			
58 36	35·9	39·9			
9 3 36	36·0	40·8	70		·
8 36	36·0	41·2			
13 36	36·0	42·0			
18 36	36·0	43·1			
23 36	36·0	43·9			
28 36	36·0	44·5			
33 36	35·8	44·9			
38 36	35·2	44·9			
43 36	35·0	44·3			
48 36	35·0	44·0			
53 36	35·1	43·1			
58 36	35·4	42·8			
10 3 36	35·9	41·9	70		
8 36	36·0	41·0·			
13 36	36·0	40·0			
18 36	36·0	39·8			
23 36	36·0	39·1			
28 36	36·1	39·0			
33 36	36·1	39·0			
38 36	36·0	39·1			
43 36	36·1	39·8			
48 36	36·1	39·8			
53 36	36·0	39·9			
58 36	36·0	39·9			
11 3 36	36·1	40·0	70		
8 36	36·2	40·9			
13 36	36·5	41·1			
18 36	36·9	43·9			
23 36		46·2			
28 36	37·0				
33 36	37·8	48·9			
38 36	38·1	51·1			
43 36	38·9	52·0			
48 36	38·9	52·5			
53 36	38·9	52·8			
58 36	38·9	52·8			
12 3 36	39·0	53·0	70		
8 36	39·0	53·1			
18 36	39·1	52·9			
23 36	39·1	52·0			
33 36	39·1	50·2			
38 36	38·5	50·9			
43 36	38·2	50·9			
48 36	38·3	50·9			
53 36	38·1	50·9			

MARCH 22 and 23, 1841.

M. Gött. Time. d. h. m. s.	Decl. Sc.-Div⁰ˢ	Hor. Force. Sc.-Div⁰ˢ	Ther.°	Vert. Force. Sc.-Div⁰ˢ	Ther.°
22 12 58 36	38·1	50·8			
13 3 36	38·3	50·7	69		
8 36	38·4	50·2			
13 36	38·4	50·2			
18 36	38·4	50·2			
23 36	38·1	50·2			
28 36	37·9	50·2			
33 36	37·9	50·3			
38 36	37·9	50·3			
43 36	38·0	50·5			
48 36	38·3	50·8			
53 36	38·5	51·2			
58 36	38·6	51·8			
14 3 36	38·8	52·1	69		
8 36	38·9	52·1			
13 36	38·9	52·1			
18 36	38·8	52·2			
23 36	38·6	52·1			
28 36	38·4	52·3			
33 36	38·5	52·4			
38 36	38·4	52·3			
43 36	38·4	52·0			
48 36	38·5	52·0			
53 36	38·6	51·8			
58 36	38·5	51·4			
15 3 36	38·5	51·6	69		
8 36	38·5	51·6			
13 36	38·6	51·9			
18 36	38·6	52·1			
23 36	39·0	52·8			
28 36	39·3	53·3			
33 36	39·4	53·7			
38 36	39·5	54·1			
43 36	39·9	54·4			
48 36	39·9	54·8			
53 36	40·1	54·7			
58 36	39·9	54·2			
16 3 36	39·6	54·3	69		
8 36	39·4	54·2			
13 36	39·1	54·2			
18 36	38·9	54·1			
23 36	38·8	54·0			
28 36	38·7	53·9			
33 36	38·5	53·9			
38 36	38·5	53·3			
43 36	38·5	53·4			
48 36	38·5	53·3			
53 36	38·6	53·3			
58 36	38·7	53·7			
17 3 36	38·9	53·6	69		
8 36	39·1	53·9			·
13 36	39·1	54·0			
18 36	39·1	54·1			
23 36	39·1	54·5			
28 36	39·1	54·9			
33 36	39·1	54·9			
38 36	39·1	54·8			
43 36	39·0	54·1			
48 36	39·0	54·3			
53 36	39·0	54·0			
58 36	39·0	54·1			
18 3 36	38·9	54·4	69		
8 36	39·0	54·8			
13 36	39·0	54·9			
18 36	39·0	54·8			
23 36	39·0	54·4			
28 36	39·0	54·0			

MARCH 22 and 23, 1841.

M. Gött. Time (d. h. m. s.)	Decl. (Sc.-Div)	Hor. Force (Sc.-Div)	Hor. Force (Ther.)	Vert. Force (Sc.-Div)	Vert. Force (Ther.)
22 18 33 36	39·0	52·9			
38 36	38·9	53·5			
43 36	38·9	53·0			
48 36	38·9	54·2			
53 36	38·9	53·9			
58 36	39·0	54·3			
19 3 36	39·1	54·9	69		
8 36	39·2	55·1			
13 36	39·6	55·8			
18 36	39·9	56·9			
23 36	40·1	55·9			
28 36	40·2	55·8			
33 36	40·0	55·2			
38 36	40·0	54·5			
43 36	39·5	54·5			
48 36	38·0	54·0			
53 36	38·9	53·9			
58 36	38·1	53·7			
20 3 36		53·5	69		
8 36	37·9	53·4			
13 36	37·8	53·7			
28 36	37·1	52·9			
33 36		52·5			
38 36	37·1	52·8			
43 36	37·1	52·5			
48 36	37·1	52·8			
53 36	37·1	52·9			
58 36	37·0	53·4			
21 3 36	37·0	53·2	69		
8 36	37·0	54·0			
13 36	36·9	54·4			
18 36	36·9	54·9			
23 36	37·0	55·6			
28 36	37·0	55·9			
33 36	37·0	56·0			
38 36	37·0	56·0			
43 36	36·9	56·1			
48 36	36·6	56·1			
53 36	36·4	56·2			
58 36	36·1	56·1			
22 3 36	36·0	56·3	69		
8 36	36·1	56·2			
13 36	36·3	56·1			
18 36	36·5	56·5			
23 36	36·9	57·0			
28 36	37·0	57·0			
33 31	37·1	57·0			
38 36	37·7	57·2			
43 36	38·0	57·3			

Positions at the usual hours of observation, from March 21st, 14ʰ, to March 23rd, 23ʰ.

M. Gött. Time (d. h. m. s.)	Decl. (Sc.-Div)	Hor. Force (Sc.-Div)	Ther.
21 14 0 0	39·8	62·2	68
15 0 0	39·5	59·2	69
16 0 0	39·3	60·0	68
18 0 0	40·0	62·3	68
19 30 0	41·2	70·0	69
20 0 0	39·6	68·3	69
20 30 0	37·1	68·2	69
22 0 0	35·1	67·0	68
23 0 0	36·8	69·6	68
22 0 0 0	39·4	73·9	69

MARCH 22 and 23, 1841.

M. Gött. Time (d. h. m. s.)	Decl. (Sc.-Div)	Hor. Force (Sc.-Div)	Ther.	Vert. Force (Sc.-Div)	Ther.
22 2 0 0	42·6	75·0	70		
3 0 0	37·2	52·3	70		
4 0 0	34·2	46·8	70		
5 0 0	32·4	40·3	70		
6 0 0	32·2	34·5	71		
8 0 0	36·0	42·1	70		
10 0 0	35·4	42·3	70		
11 0 0	36·1	39·9	70		
12 0 0	38·9	52·8	70		
13 0 0	38·1	50·9	69		
14 0 0	38·6	52·2	69		
15 0 0	38·5	51·4	69		
16 0 0	39·9	54·2	69		
18 0 0	39·0	54·1	69		
19 30 0	40·2		69		
20 0 0	38·2	53·5	69		
20 30 0	37·1	53·0	69		
22 0 0	36·1	56·1	69		
23 0 0	38·4	57·0	69		
23 0 0 0ᵃ	41·0	56·8	70		
2 0 0		53·0	71		
3 0 0	33·9	51·5	71		
4 0 0	33·0	51·8	71		
5 0 0	32·0	51·2	72		
6 0 0	32·0	51·3	72		
8 0 0	35·0	52·1	72		
10 0 0	33·8	50·1	71		
11 0 0	36·0	54·0	71		
12 0 0	35·4	51·4	71		
13 0 0	35·8	51·9	71		
14 0 0	36·0	54·3	70		
15 0 0	37·0	55·9	70		
16 0 0	37·2	55·2	70		
18 0 0	37·1	53·9	69		
19 30 0	36·1	55·2	69		
20 0 0	35·0	56·1	69		
20 30 0	33·2	57·9	69		
22 0 0	32·9	61·9	69		
23 0 0	35·0	63·0	69		

The Mean Positions of the H. F. magnet at the same hours during the Month are given in page 48, as are also the Mean Positions of the Decl. magnet up to March 23rd, 0ʰ, at which time the connexion of the series was broken.

VAN DIEMEN ISLAND { Decl. 1 Scale Division = 0'·71 ; H. F. k = ·0003; q = ; V. F. k = ; q = }

Extra observations.

The V. F. was observed at 2ᵐ. 30ˢ. before, and the H. F. 2ᵐ. 30ˢ. after the times specified.

M. Gött. Time (d. h. m. s.)	Decl. (Sc.-Div)	Hor. Force (Sc.-Div)	Ther.	Vert. Force (Sc.-Div)	Ther.
22 2 12 30	41·3				
17 30	45·3	74·1		92·8	
22 30	49·3			92·8	
27 30	49·9	73·3			
32 30	50·3			92·5	
37 30	52·4	75·6			
42 30	57·0		68	91·5	66
47 30	57·3	78·8			

MARCH 22 and 23, 1840.

M. Gött. Time (d. h. m. s.)	Decl. (Sc.-Div)	Hor. Force (Sc.-Div)	Ther.	Vert. Force (Sc.-Div)	Ther.
22 3 10 0	28·4	73 8		78·9	
15 0	28·8				
20 0	28·5	71·9		79·6	
25 0	29·3				
30 0	36·0	73·2		82·1	
35 0	40·2				
40 0	42·9	72·4		84·8	
45 0	44·2				
50 0	42·2	70·5		85·6	
4 10 0	37·1	72·0	69	85·2	69
15 0	32·2				
20 0	28·4	71·7		82·7	
25 0	28·6				
30 0	29·1	71·1		82·5	
35 0	30·2				
40 0	28·9	70·9		83·2	
45 0	27·3				
50 0	28·5	70·6		82·7	
5 10 0	29·5	68·8		85·0	
15 0	27·3				
20 0	27·9	68·5		84·1	
25 0	28·3				
30 0	30·3	66·9		84·4	
35 0	34·4				
40 0	40·5	66·3		87·9	
45 0	44·3				
50 0	55·5	70·2		89·5	
6 10 0	59·3	74·6	70		69
15 0	54·1				
20 0	51·7	74·0		82·2	
25 0	48·9				
30 0	46·4	74·0		81·7	
35 0	43·1				
40 0	41·1	73·8		80·5	
45 0	39·7				
50 0	39·9	75·1		80·5	
7 10 0	38·7	75·9			
15 0	40·6				
20 0	40·7	73·7			
25 0	41·6				
30 0	43·7	72·4			
35 0	44·8				
40 0	45·1	70·9			
45 0	47·9				
50 0	48·8	72·2			
10 10 0		69·5	67		67
15 0	62·5				
20 0	62·8	72·1		93·6	
25 0	60·1				
30 0	58·8	73·4		90·4	
35 0	55·9				
40 0	54·8	72·7		88·4	
45 0	53·7				
50 0	54·1	73·0		88·2	
11 10 0		72·1			
15 0	50·4				
20 0	50·1	71·0		87·6	
25 0	48·5				
20 10 0	43·6	69·2	81	87·7	76
15 0	42·3				
20 0	43·2	69·1		88·5	
25 0	43·9				
30 0	44·1	65·6		89·1	
35 0	43·1				

* At 23ᵈ. 2ʰ. the Decl. magnet was employed in observations on the absolute horizontal intensity, whereby the connexion of the differential series was broken; the subsequent observations with this instrument are therefore unconnected with the preceding ones.

H 2

.OBSERVATIONS WITH THE MAGNETOMETERS ON DAYS OF UNUSUAL MAGNETIC DISTURBANCE, 1840-1841.

MARCH 22 and 23, 1841.

M. Gött. Time. d. h. m. s.	Decl. Sc.-Div.	Hor. Force. Sc.-Div.	Ther.	Vert. Force. Sc.-Div.	Ther.
22 20 40 0	43·9	66·8		90·5	
45 0	45·9				
50 0	47·2	66·8		91·4	

Positions at the usual hours of observation, from March 21st, 14ʰ. to March 23rd, 23ʰ.

M. Gött. Time. d. h. m. s.	Decl. Sc.-Div.	Hor. Force. Sc.-Div.	Ther.	Vert. Force. Sc.-Div.	Ther.
21 14 0 0	51·4	75·7	63	93·8	62
15 0 0	55·6	76 4		92·3	
16 0 0	56·1	76·5	66	91·7	64
17 0 0	58·6	77·3		91·7	
18 0 0	60·3	79·2	68	92·8	65
19 0 0	57·8	73·4		94·8	
20 0 0	60·2	74·6	67	94·2	67
21 0 0	54·7	74·6		92·7	
22 0 0	53·3	77·7	69	90·5	67
23 0 0	52·6	76·6		91·5	
22 0 0 0	51·3	76·8	68	91·7	67
1 0 0	49·1	77·6		91·3	
2 0 0	42·1	76·4	67	89·1	66
3 0 0	39·6	74·7		80·7	
4 0 0	39·1	71·0	69	86·0	69
5 0 0	32·1	69·7		84·6	
6 0 0	68·2	73·7	70	89·3	69
7 0 0	41·0	74·0			
8 0 0	50·5	71·9	69	89·3	69
9 0 0	52·2	72·5		90·2	
10 0 0	59·6	68·5	67	94·7	67
11 0 0	52·5	73·0		87·3	
12 0 0	46·0	67·9	68	90·0	67
13 0 0	50·7	68·7		91·2	
14 0 0	50·9	69·6	69	89·9	67
15 0 0	52·6	69·6		89·5	
16 0 0	54·8	68·0	74	90·2	73
17 0 0	54·8	70·1		87·6	
18 0 0	56·4	69·1	79	88·5	75
19 0 0	48·3	67·3		89·7	
20 0 0	40·6	71·7	81	87·9	76
21 0 0	47·7	67·1		91·2	
22 0 0	49·6	66·9	80	91·2	76
23 0 0	48·8	69·2		89·7	
23 0 0 0	46·0	68·1	77	91·0	74
1 0 0	42·6	69·7		89·1	
2 0 0	34·1	67·7	74	88·9	73
3 0 0	48·4	69·3		90·0	
4 0 0	55·2	70·7	73	90·6	71
5 0 0	54·3	73·5		87·7	
6 0 0	47·8	72·8	71	89·0	70
7 0 0	48·5	74·4		90·4	
8 0 0	49·3	74·9	68	91·2	67
9 0 0	49·4	74·4		91·9	
10 0 0	49·1	75·5	66	93·3	65
11 0 0	47·1	73·6		93·3	
12 0 0	47·6	73·6	66	92·1	66
13 0 0	46·2	73·4			
14 0 0	48·1	71·6	69	90·3	69
15 0 0	51·5	67·7		92·0	
16 0 0	53·9	66·8	72	92·0	72
17 0 0	55·8	68·2		91·3	
18 0 0	55·2	71·5	74	89·1	73
19 0 0	59·2	69·2		91·3	
20 0 0	51·3	71·0	76	90·7	72

MARCH 22 and 23, 1841.

M. Gött. Time. d. h. m. s.	Decl. Sc.-Div.	Hor. Force. Sc.-Div.	Ther.	Vert. Force. Sc.-Div.	Ther.
23 21 0 0	51·1	71·7		99·7	
22 0 0	49·9	72·9	73	91·0	71
23 0 0	46·5	70·4		92·0	

The Mean Positions at the same hours during the Month are given in page 49.

APRIL 18, 19, 20, and 21, 1841.

TORONTO { Decl. 1 Scale Division = 0'·72 ; H. F. k = ·000076; q = ·0002 ; V. F. k = ·00009; q = ·00018 }

Positions at the usual hours of observation, April 18, 19, 20, and 21.

M. Gött. Time. d. h. m. s.	Decl. Sc.-Div.	Hor. Force. Sc.-Div.	Ther.	Vert. Force. Sc.-Div.	Ther.
18 18 0 0[a]	48·2	49·1	44	84·7	45
20 0 0	55·4	42·1	44	79·8	44
22 0 0	52·7	68·4	43	85·0	44
19 0 0 0	55·2	60·1	44	86·6	45
2 0 0	57·2	65·4	44	87·0	45
4 0 0	50·1	32·5	45	85·1	45
6 0 0	44·1	50·1	46	85·1	46
8 0 0	41·1	55·6	49	83·1	48
10 0 0	39·8	63·8	50	86·1	49
12 0 0	40·8	53·4	50	91·5	50
14 0 0	42·9	53·7	50	86·2	50
16 0 0	47·9	41·7	49	86·4	49
18 0 0	52·6	31·1	48	81·3	48
20 0 0	54·0	53·6	47	79·5	48
22 0 0	45·6	45·4	47	67·1	47
20 0 0 0	57·4	65·4	46	75·9	47
2 0 0	56·5	50·7	46	83·5	46
4 0 0	52·6	38·5	46	84·4	46
6 0 0	44·0	22·2	46	87·8	46
8 0 0	40·8	67·5	47	91·1	47
10 0 0	45·9	61·4	47	93·8	47
12 0 0	44·1	53·1	47	92·7	47
14 0 0	57·8	57·0	47	87·0	47
16 0 0	61·2	45·0	46	86·5	46
18 0 0	66·2	28·1	45	71·6	46
20 0 0	53·9	64·5	45	86·5	46
22 0 0	42·2	43·0	44	84·0	45
21 0 0 0	57·0	68·3	45	84·0	45
2 0 0	57·3	50·0	46	84·7	46
4 0 0	49·9	37·2	48	81·9	48
6 0 0	47·5	44·5	49	81·3	49
8 0 0	44·8	54·9	51	81·2	50
10 0 0	46·6	59·4	52	80·9	52
12 0 0	52·4	52·4	54	81·3	53
14 0 0	55·2	36·0	53	79·8	54
16 0 0	52·6	54·8	52	71·8	53
18 0 0	54·2	45·2	51	76·5	52
20 0 0	53·6	44·0	50	77·8	51
22 0 0	58·5	53·6	49	77·6	49

Mean Positions at the same hours during the Month.

M. Gött. Time. d. h. m. s.	Decl. Sc.-Div.	Hor. Force. Sc.-Div.	Ther.	Vert. Force. Sc.-Div.	Ther.
0 0 0	55·1	62·8	48	82·0	48
2 0 0	56·1	52·8	48	82·0	48
4 0 0	51·9	37·7	49	80·2	49
6 0 0	43·4	34·1	50	79·6	50
8 0 0	40·7	50·6	51	80·4	51

APRIL 18, 19, 20, and 21, 1841.

M. Gött. Time. h. m. s.	Decl. Sc.-Div.	Hor. Force. Sc.-Div.	Ther.	Vert. Force. Sc.-Div.	Ther.
10 0 0	44·2	55·8	52	80·7	51
12 0 0	48·3	54·8	53	81·2	52
14 0 0	51·0	47·9	52	80·6	52
16 0 0	52·9	50·7	51	80·4	51
18 0 0	52·1	54·2	49	80·6	49
20 0 0	53·0	58·5	48	81·4	49
22 0 0	53·6	61·2	48	81·3	48

ST. HELENA { Decl. 1 Scale Division = 0'·71 ; H. F. k = ·00018; q = ·00025 }

Positions at the usual hours of observation, April 18, 19, 20, and 21.

M. Gött. Time. h. m. s.	Decl. Sc.-Div.	Hor. Force. Sc.-Div.	Ther.
18 14 0 0[b]	27·1	64·9	70
15 0 0	27·1	64·2	70
16 0 0	28·0	65·9	70
18 0 0	27·6	64·9	70
20 0 0	29·9	66·9	70
20 30 0	28·0	65·0	70
22 0 0	28·0	69·2	70
23 0 0	29·8	71·6	70
19 0 0 0	31·2	71·0	71
2 0 0	27·8	70·8	71
3 0 0	26·8	68·9	71
4 0 0	26·3	63·9	72
5 0 0	26·0	62·0	72
6 0 0	26·8	65·1	72
7 0 0	28·0	65·0	72
10 0 0	28·2	63·0	71
11 0 0	27·8	62·9	71
12 0 0	28·9	61·0	71
13 0 0	30·0	65·9	71
14 0 0	30·0	65·8	71
15 0 0	30·1	66·4	70
16 0 0	30·0	65·9	70
18 0 0	31·3	67·0	70
19 30 0	31·9	68·0	70
20 0 0	31·9	68·1	70
20 30 0	31·0	69·1	70
22 0 0	30·0	69·1	70
23 0 0	31·0	68·8	70
20 0 0 0	28·4	71·1	71
2 0 0	27·1		
3 0 0	23·9	68·0	72
4 0 0	26·1	66·2	72
5 0 0	25·8	63·7	72
6 0 0	26·0	61·1	72
8 0 0	26·8	59·9	71
10 0 0	27·0	62·0	71
11 0 0	28·1	64·1	71
12 0 0	29·0	64·1	71
13 0 0	28·9	63·9	71
14 0 0	29·1	65·1	72
15 0 0	29·0	66·9	70
16 0 0	30·2	67·6	70
18 0 0	31·2	72·0	70
19 30 0	31·8	68·9	70
20 0 0	31·4	68·0	70
20 30 0	30·5	68·9	70
22 0 0	29·6	72·4	70
23 0 0	30·2	72·9	70
21 0 0 0	31·9	76·0	70

[a] Commencing after midnight of Sunday at Toronto.   [b] Commencing after midnight of Sunday at St. Helena.

April 18, 19, 20, and 21, 1841.

M. Gött. Time. d. h. m. s.	Decl. Sc.-Div^ns	Hor. Force. Sc.-Div^ns	Ther.	Vert. Force. Sc.-Div^ns	Ther.
21 2 0 0	29·0	74·9	71		
3 0 0	26·6	70·9	71		
4 0 0	26·9	66·9	72		
5 0 0	27·9	65·6	72		
6 0 0	28·9	66·0	72		
8 0 0	28·8	66·0	71		
10 0 0	29·0	67·1	71		
11 0 0	30·1	70·2	71		
12 0 0	30·4	68·9	70		
13 0 0	30·5	68·0	70		
14 0 0	30·4	66·9	70		
15 0 0	30·8	67·2	70		
16 0 0	31·0	69·1	70		
18 0 0	29·9	68·1	69		
19 30 0	31·1	69·7	70		
20 0 0	30·8	69·2	70		
20 30 0	30·1	68·0	70		
22 0 0	28·9	73·5	70		
23 0 0	29·2	74·8	69		

Mean Positions at the same hours during the Month.

M. Gött. Time. d. h. m. s.	Decl. Sc.-Div^ns	Hor. Force. Sc.-Div^ns	Ther.	Vert. Force. Sc.-Div^ns	Ther.
0 0 0	30·5	78·8	70		
2 0 0	29·8	77·6	71		
3 0 0	28·9	74·8	71		
4 0 0	29·0	72·0	71		
5 0 0	29·0	70·1	71		
6 0 0	28·7	68·8	71		
8 0 0	29·2	67·5	71		
10 0 0	29·6	67·9	70		
11 0 0	29·7	68·4	70		
12 0 0	29·8	67·9	70		
13 0 0	30·1	68·5	70		
14 0 0	29·8	68·9	70		
15 0 0	30·1	69·3	70		
16 0 0	30·4	69·8	70		
18 0 0	30·7	71·0	69		
19 30 0	31·7	71·8	69		
20 0 0	31·0	71·9	69		
20 30 0	29·7	72·7	69		
22 0 0	27·8	76·5	69		
23 0 0	28·8	77·9	69		

CAPE OF GOOD HOPE { Decl. 1 Scale Division = 0'·75
H. F. k = ·00018; q = ·0003

Positions at the usual hours of observation, April 18, 19, 20, and 21.

M. Gött. Time. d. h. m. s.	Decl. Sc.-Div^ns	Hor. Force. Sc.-Div^ns	Ther.
18 12 0 0[a]	54·7	45·9	67
14 0 0	51·0	45·8	67
16 0 0	52·8	42·3	66
18 0 0	50·0	43·8	66
20 0 0	51·2	45·9	66
22 0 0	47·1	46·9	66
19 0 0 0	50·2	45·0	67
2 0 0	49·9	43·8	68
4 0 0	47·5	40·0	69
6 0 0	48·8	40·7	69
8 0 0	50·2	40·8	69
10 0 0	48·5	41·7	69

April 18, 19, 20, and 21, 1841.

M. Gött. Time. d. h. m. s.	Decl. Sc.-Div^ns	Hor. Force. Sc.-Div^ns	Ther.	Vert. Force. Sc.-Div^ns	Ther.
19 12 0 0	47·8	43·0	68		
14 0 0	51·6	45·1	68		
16 0 0	52·1	45·1	67		
18 0 0	53·6	43·6	66		
20 0 0	50·2	47·0	66		
22 0 0	48·2	42·3	67		
20 0 0 0	51·0	41·8	67		
2 0 0	49·8	39·8	68		
4 0 0	49·6	38·9	69		
6 0 0	46·3	37·5	69		
8 0 0	47·8	36·9	69		
10 0 0	47·9	41·7	69		
12 0 0	50·1	43·6	68		
14 0 0	51·9	44·0	68		
16 0 0	54·2	45·3	68		
18 0 0	56·3	49·1	68		
20 0 0	50·5	48·1	67		
22 0 0	48·2	44·4	67		
21 0 0 0	51·4	44·6	68		
2 0 0	51·6	42·1	69		
4 0 0	50·5	48·8	70		
6 0 0	50·0	43·8	69		
8 0 0	50·1	43·4	68		
10 0 0	50·8	45·7	68		
12 0 0	51·8	47·6	68		
14 0 0	52·1	45·9	69		
16 0 0	54·5	45·5	69		
18 0 0	51·9	46·4	68		
20 0 0	50·7	48·3	68		
22 0 0	50·5	45·9	67		

Mean Positions at the same hours during the Month. [b]

M. Gött. Time. d. h. m. s.	Decl. Sc.-Div^ns	Hor. Force. Sc.-Div^ns	Ther.
0 0 0	51·1	40·3	66
2 0 0	52·7	39·4	67
4 0 0	51·8	39·4	68
6 0 0	50·9	38·6	68
8 0 0	51·1	38·2	67
10 0 0	50·7	39·8	67
12 0 0	51·2	39·8	67
14 0 0	51·6	40·7	67
16 0 0	52·2	41·0	66
18 0 0	51·6	42·1	66
20 0 0	47·4	44·3	66
22 0 0	47·4	42·7	66

VAN DIEMEN ISLAND { Decl. 1 Scale Division = 0'·71
H. F. k = ·0003; q =
V. F. k = ; q =

Extra observations.

The V. F. was observed at 2m. 30s. before, and the H. F. 2m. 30s. after the times specified.

M. Gött. Time. d. h. m. s.	Decl. Sc.-Div^ns	Hor. Force. Sc.-Div^ns	Ther.	Vert. Force. Sc.-Div^ns	Ther.
18 5 5 0	60·3				
10 0	62·2	208·9		93·3	
15 0	63·8				
20 0	63·9	210·4		92·4	
25 0	62·8				
30 0	60·6	211·3		90·4	

April 18, 19, 20, and 21, 1841.

M. Gött. Time. d. h. m. s.	Decl. Sc.-Div^ns	Hor. Force. Sc.-Div^ns	Ther.	Vert. Force. Sc.-Div^ns	Ther.
18 5 35 0	59·0				
40 0	57·8	211·2		89·0	
45 0	56·1				
50 0	54·8	210·1		88·3	
6 10 0	54·6	208·0	54		54
15 0	54·2				
20 0	53·6	206·9		89·3	
25 0	52·8				
30 0	51·5	207·2		89·5	
35 0	50·0				
40 0	48·9	207·9		88·9	
45 0	48·5				
50 0	48·1	208·8		88·1	
8 10 0	57·9	207·4	54		53
15 0	56·6				
20 0	54·1	208·3		89·8	
25 0	53·4				
30 0	53·6	207·9		88·6	
35 0	54·0				
40 0	54·8	206·4		88·8	
45 0	56·0				
50 0	56·8	205·1		90·0	
9 10 0	56·0	204·4			
15 0	54·9				
20 0	54·4	203·4		90·8	
25 0	52·8				
30 0	51·3	203·4		90·6	
35 0	51·0				
40 0	51·5	203·4		90·6	
45 0	50·5				
50 0	49·3	204·1		90·4	
10 10 0	50·6	204·5	54		5
15 0	49·2				
20 0	49·2			89·7	
19 21 10 0	36·9	197·4	64		6
15 0	34·7				
20 0	33·0	198·4		90·9	
25 0	33·4				
30 0	33·8	199·1		90·3	
35 0	34·9				
40 0	34·9	200·3		89·9	
45 0	35·7				
50 0	36·8	200·4		89·4	
22 10 0	38·4	200·4	65		6
15 0	39·6				
20 0	39·0	198·5		89·8	
25 0	37·5				
30 0	37·3	198·1		90·3	
35 0	36·8				
40 0	35·8	197·9		90·3	
45 0	35·3				
50 0	36·2	197·7		90·4	
23 10 0	38·8	197·0	65		
15 0	39·6				
20 0	40·4	196·9		92·0	65
25 0	41·0				
30 0	41·4	195·7		92·1	
35 0	41·9				
40 0	41·1	195·9		92·8	
45 0	41·1				
50 0	40·8	196·3		92·8	
55 0	40·9				
20 0 10 0	40·6	196·5[c]	66	c	65

[a] Commencing after midnight of Sunday at the Cape of Good Hope.

[b] The mean positions of the H. F. magnet are not strictly comparable with the positions of the 18th to the 21st, owing to the progressive stretching of the wires, which continued throughout the month.

[c] From 20d. 0h. 10m. to 20d. 1h. 45m., the V. F. was observed at 5m. before, and the H. F. 5m. after the times specified.

APRIL 18, 19, 20, and 21, 1841.

M. Gött. Time.		Decl.	Hor. Force.		Vert. Force.	
m.	s.	Sc.-Div⁸⁸.	Sc.-Div⁸⁸.	Ther.	Sc.-Div⁸⁸.	Ther.
		°		°		°
20 0 20	0	43·0				
30	0	45·5	197·8		91·8	
40	0	40·8			·	
50	0	38·0	198·0		90·9	
1 5	0	44·1	197·2			
15	0	42·7				
25	0	42·9	197·6		91·9	
35	0	43·2				
45	0	44·5			91·9	
10 5	0	58·0				
10	0	58·0	201·0	58	94·2	57
15	0	56·9				
20	0	56·2	202·8		94·1	
25	0	57·6				
30	0	58·9	203·8		93·5	
35	0	59·9				
40	0	60·1	204·2		95·4	
45	0	60·5				
50	0	60·1	203·9		92·6	
55	0	60·3				
11 5	0	57·3				
10	0	55·8	204·2	58	91·4	58
15	0	54·6				
20	0	53·5	203·8		90·9	
25	0	52·8				
30	0	52·5	203·8		90·8	
35	0	51·9				
40	0	51·9	203·1		90·5	
45	0	51·2				
50	0	51·2	202·8		90·7	

Positions at the usual hours of observation,
April 18, 19, 20, and 21.

18 3 0 0ᵃ	47·3	207·0		92·9	
4 0 0	50·5	208·4	53	92·7	53
5 0 0	59·7	208·2		93·3	
6 0 0	54·1	210·1	54	88·6	54
7 0 0	47·7	209·1		87·5	
8 0 0	59·4	206·7	54	92·6	53
9 0 0	57·9	204·8		91·1	
10 0 0	49·3	204·2	54	90·2	54
11 0 0	48·3	204·4		90·0	
12 0 0	48·0	208·1	54	94·2	54
13 0 0	48·3	198·5		93·7	
14 0 0	49·3	199·8	55	92·5	55
15 0 0	49·8	202·4		90·3	
16 0 0	50·8	198·7	57	92·9	57
17 0 0	52·9	201·3		92·9	
18 0 0	54·6	202·2	58	93·6	57
19 0 0	44·8	202·9		92·4	
20 0 0	51·0	200·8	57	94·7	57
21 0 0	45·5	204·5		93·4	
22 0 0	49·3	205·1	57	92·7	56
23 0 0	48·3	205·8		92·4	
19 0 0 0	46·1	202·4	56	93·7	55
1 0 0	42·8	207·0		91·8	

APRIL 18, 19, 20, and 21, 1841.

M. Gött. Time.		Decl.	Hor. Force.		Vert. Force.	
m.	s.	Sc.-Div⁸⁸.	Sc.-Div⁸⁸.	Ther.	Sc.-Div⁸⁸.	Ther.
		°		°		°
19 2 0	0	45·1	205·3	56	92·8	55
3 0	0	45·6	204·4		93·5	
4 0	0	46·1	208·5	55	90·4	55
5 0	0	47·5	205·2		92·8	
6 0	0	49·5	205·3	55	93·1	55
7 0	0	48·4	205·2		93·1	
8 0	0	48·7	206·7	54	93·1	54
9 0	0	50·4	206·3		93·4	
10 0	0	50·9	206·9	56	93·8	54
11 0	0	48·2	205·4		94·4	
12 0	0	49·2	203·1	55	93·9	55
13 0	0	47·0	202·3		92·6	
14 0	0	45·9	203·1	57	89·9	56
15 0	0	48·4	201·7		90·4	
16 0	0	51·1	197·8	60	92·6	59
17 0	0	55·5	199·3		91·9	
18 0	0	52·6	198·7	62	94·3	60
19 0	0	51·4	200·0		94·1	
20 0	0	50·5	200·7	63	92·7	61
21 0	0	40·7	196·9		91·9	
22 0	0	37·1	201·0	65	89·3	64
23 0	0	37·3	196·5		91·6	
20 0 0	0	40·8	196·3	66	92·7	65
1 0	0	40·9	198·1		90·9	
2 0	0	45·8	201·8	65	90·5	65
3 0	0	48·7	199·8		91·8	
4 0	0	48·0	202·3	63	90·2	62
5 0	0	50·3	203·5		92·7	
6 0	0	50·4	205·7	61	89·1	60
7 0	0	50·6	203·4		90·6	
8 0	0	52·2	203·6	59	91·4	58
9 0	0	51·6	202·1		93·0	
10 0	0	56·6	201·2	57	94·9	57
11 0	0	58·7	204·4		91·8	
12 0	0	51·2	202·5	59	91·0	60
13 0	0	45·1	201·9		89·8	
14 0	0	47·6			90·3	60
15 0	0	48·1	199·8		90·0	
16 0	0	51·7	198·8	62	92·1	60
17 0	0	52·2	197·6		93·6	
18 0	0	52·6	199·5	61	94·8	59
19 0	0	54·2	202·1		93·9	
20 0	0	52·2	203·8	61	92·6	60
21 0	0	50·0	202·8		91·6	
22 0	0	43·4	204·2	59	91·9	59
23 0	0	42·4	202·7		92·2	
21 0 0	0	45·0	201·6	57	95·3	56
1 0	0	49·5	205·6		92·8	
2 0	0	42·2	203·9	56	92·3	55
3 0	0	45·9	204·7		93·5	
4 0	0	48·4	206·5	55	93·4	54
5 0	0	49·1	206·0		94·2	
6 0	0	54·0	204·3	53	95·8	52
7 0	0	52·0	204·6		94·1	
8 0	0	50·7	204·0	52	94·5	52
9 0	0	50·1	204·1		95·0	
10 0	0	48·3	205·2	50	94·7	50
11 0	0	50·6	204·8		96·0	

APRIL 18, 19, 20, and 21, 1841.

M. Gött. Time.			Decl.	Hor. Force.		Vert. Force.	
d. h.	m.	s.	Sc.-Div⁸⁸.	Sc.-Div⁸⁸.	Ther.	Sc.-Div⁸⁸.	Ther.
			°		°		°
21 12	0	0	50·7	204·9	51	95·8	53
13	0	0	47·6	203·2		94·2	
14	0	0	50·3	200·4	55	93·3	57
15	0	0	54·5	199·5		93·7	
16	0	0	54·5	200·4	59	92·0	60
17	0	0	56·0	202·5		91·1	
18	0	0	56·0	201·7	61	93·0	61
19	0	0	53·4	101·7		92·8	
20	0	0	49·7	201·3	62	92·4	62
21	0	0	44·4	202·2		91·5	
22	0	0	49·6	201·1	61	94·6	62
23	0	0	47·5	202·0		93·1	

Mean Positions at the same hours during the Month.

0 0	0	47·4	203·6	59	93·4	58
1 0	0	46·4	204·7		92·7	
2 0	0	46·3	204·6	58	92·8	57
3 0	0	47·0	204·3		93·3	
4 0	0	48·0	204·8	57	93·2	56
5 0	0	48·9	204·8		93·7	
6 0	0	49·3	205·0	56	93·4	55
7 0	0	48·6	205·1		93·4	
8 0	0	48·8	205·4	55	93·4	55
9 0	0	49·0	205·5		93·6	
10 0	0	47·4	205·7	54	93·5	54
11 0	0	46·3	205·4		93·4	
12 0	0	45·1	204·6	54	93·4	54
13 0	0	45·5	203·1		93·3	
14 0	0	47·8	202·3	56	93·2	56
15 0	0	50·6	201·6		93·1	
16 0	0	53·8	201·3	59	93·4	58
17 0	0	55·0	201·9		93·3	
18 0	0	54·3	202·4	61	93·1	59
19 0	0	52·8	203·0		92·7	
20 0	0	51·6	203·2	61	92·7	59
21 0	0	49·5	203·2		92·3	
22 0	0	48·8	203·5	60	92·4	59
23 0	0	47·7	203·4		92·7	

MAY 9 and 10, 1841.

Toronto [c]
$\left\{\begin{array}{l}\text{Decl. 1 Scale Division} = \cdot 0'\cdot 72 \\ \text{H. F. } k = \cdot 000076\,;\, q = \cdot 0002 \\ \text{V. F. } k = \cdot 00009\,;\, q = \cdot 00018\end{array}\right.$

Extra observations.

The V. F. was observed at 1ᵐ. 30ˢ. before, and the H. F. 2ᵐ. after the times specified. [b]

10 0 25	0	64·7	401·0	53	58·8	53
30	0	67·4	401·7		57·3	
35	0	67·4	402·0		60·5	

ᵃ Commencing after midnight of Sunday at Van Diemen Island. .
ᵇ Weather densely clouded, with haze and drizzling rain.

ᶜ TORONTO, May, 1841.—*Times of observation at which the Magnetometers were disturbed, but the mean readings of the Instruments were not materially changed.*

2 18   Decl., H. F., and V. F., much disturbed, vibrating.
20   Decl., H. F., and V. F., very much disturbed, vibrating, with shocks.
22   Decl., H. F., and V. F., very much disturbed, vibrating, with shocks.

d. h.
3 0   Decl., H. F., and V. F., moderate vibrations; Decl. much disturbed with shocks.
5 0   H. F. and V. F. slight vibrations; Decl. very slight vibrations and shocks.
16   Decl., H. F., and V. F., moderate vibrations and shocks.
6 0   V. F. vibrating much.
2   V. F. slight vibrations; Decl. and H. F. moderate shocks.
7 23   Decl. vibrating much, with shocks.
8 0   Decl. and H. F. vibrating much, with shocks.
10 23   H. F. slight shocks.

MAY 9 and 10, 1841.

M. Gött. Time	Decl.	Hor. Force		Vert. Force	
d. h. m. s.	Sc.-Div.	Sc.-Div.	Ther.	Sc.-Div.	Ther.
10 0 40 0	62·2	406·6		61·3	
45 0	56·4	409·1		60·7	
50 0	54·5	404·8		61·6	
55 0	55·6	405·3		62·3	
1 0 0	54·7	435·6	53	60·6	53
5 0	45·9	443·2		62·3	
10 0	44·8	443·1		61·6	
15 0	42·0	429·1		60·5	
20 0	36·2	427·5		60·8	
25 0	35·1	422·7		61·0	
30 0	34·8	423·5		61·4	
35 0	35·0	424·8		63·2	
40 0	36·7	224·2		65·1	
45 0	36·9	419·6		65·0	
2 40 0	47·4	414·2	53	71·2	53
45 0	46·3	405·7		71·5	
50 0	48·1	409·3		72·8	
55 0	48·6	404·9		73·8	
3 0 0	46·4	403·7	52	73·8	53
5 0	47·0	405·5		74·2	
10 0	46·1	411·7		74·2	
25 0	47·9	409·4		74·3	
30 0	45·8	407·4		75·6	
35 0	46·3	405·3		75·3	

MAY 9 and 10, 1841.

M. Gött. Time	Decl.	Hor. Force		Vert. Force	
d. h. m. s.	Sc.-Div.	Sc.-Div.	Ther.	Sc.-Div.	Ther.
10 3 40 0	44·8	406·0		75·0	
45 0	43·5	404·0		75·2	
50 0	46·4	406·8		75·1	
55 0	48·5	412·8		76·6	
4 10 0	49·5	418·3	54	77·7	54
15 0	50·3	423·7		78·7	
20 0	51·4	429·0		79·5	
25 0	50·3	433·2		80·1	
30 0	50·0	437·0		80·8	
35 0	48·9	438·7		81·3	
40 0	50·7	441·4		82·8	
45 0	49·2	438·7		83·2	
50 0	47·0	438·7		84·2	
55 0	46·9	432·8		84·2	
5 0 0	47·9	435·5	54	85·5	54
5 0	45·3	431·9		85·0	
10 0	45·9	425·7		84·7	
15 0	45·1	423·3		84·7	
20 0	44·4	428·8		85·2	
25 0	45·7	435·6		85·6	
30 0	48·0	445·0		87·0	
35 0	48·0	449·0		87·1	
40 0	48·6	455·6		87·5	
45 0	49·9	462·6		88·7	

MAY 9 and 10, 1841.

M. Gött. Time	Decl.	Hor. Force		Vert. Force	
d. h. m. s.	Sc.-Div.	Sc.-Div.	Ther.	Sc.-Div.	Ther.
10 5 50 0	52·6	470·1		89·9	
55 0	51·1	476·6		91·8	
6 5 0	47·4	478·9	54	92·4	54
10 0	47·2	484·3		93·4	
15 0	46·5	491·7		93·4	

Positions at the usual hours of observation, from May 9th, 18h. to May 10th, 22h.

M. Gött. Time	Decl.	Hor. Force		Vert. Force	
9 18 0 0	51·9	442·8	54	64·5	54
20 0 0	61·4	432·3	53	59·2	54
22 0 0	55·5	437·5	53	57·5	54
10 0 0 0	71·5	400·7	53	49·0	53
2 0 0	34·9	410·3	53	65·3	53
4 0 0	50·6	415·5	54	76·5	54
6 0 0	50·4	475·1	54	92·6	54
8 0 0	44·7	527·5	54	92·4	54
10 0 0	43·5	516·0	54	90·2	54
12 0 0	48·0	485·5	55	91·6	54
14 0 0	42·7	483·3	55	86·0	55
16 0 0	45·3	462·7	54	80·9	55
18 0 0	46·0	463·9	54	79·5	55
20 0 0	50·0	463·3	54	79·0	54
22 0 0	51·2	461·8	55	78·8	54

d. h.
11 0 H. F. slight shocks.
2 Decl. and H. F. moderate shocks.
16 Decl. and H. F. vibrating much.
12 18 Decl., H. F., and V. F., vibrating much, with shocks.
22 Decl. and H. F. vibrating very much.
13 0 Decl. vibrating very much.
2 Decl. vibrating very much.
8 Decl. and H. F. moderate shocks.
14 18 Decl. vibrating very much, and H. F. moderately.
20 Decl. and H. F. vibrating very much.
22 Decl., H. F., and V. F., vibrating moderately.
15 0 Decl. moderate vibrations; H. F. slight vibrations and shocks; V. F. vibrating very much.
2 H. F. strong shocks at 1h. 55m.
16 18 Decl. vibrating very much.
17 2 Decl. vibrating slightly; H. F. moderate shocks.
4 Decl., H. F., and V. F., slight vibrations.
20 Decl. vibrating much.
22 Decl. and H. F. vibrating much.
18 0 Decl. and H. F. moderate vibrations; V. F. vibrating very much.
20 Decl. vibrating much.
19 0 Decl. vibrating much, with shocks; H. F. slight shocks.
6 H. F. slight shocks.
8 H. F. slight shocks.
20 Decl. and H. F. vibrating much, with shocks.
22 Decl. and H. F. vibrating much, with shocks.
20 0 Decl. vibrating much; H. F. slight vibrations and shocks.
21 2 Decl. and H. F. slight shocks.
4 H. F. strong shocks.
6 Decl. and H. F. strong shocks.
12 Decl. and H. F. slight vibrations and shocks.
14 Decl. and H. F. slight vibrations and shocks.
22 10 H. F. moderate vibrations and shocks.
12 H. F. slight vibrations and shocks.
14 H. F. very slight vibrations and shocks.
24 2 H. F. slight shocks.
25 0 Decl. slight vibrations and shocks.
18 H. F. slight shocks.
20 H. F. moderate shocks.
22 H.F. moderate shocks.
27 2 H. F. vibrating much, with shocks.
4 H. F. vibrating much, with shocks.
6 H. F. vibrating much, with shocks.
30 22 Decl. and H. F. vibrating much.
31 0 Decl., H. F., and V. F., vibrating much, with shocks.
2 Decl. and H. F. slight shocks.

TORONTO, June, 1841.—*Times of observation at which the Magnetometers were disturbed, but the mean readings of the Instruments were not materially changed.*
d. h.
1 18 Decl. and H. F. slightly disturbed, vibrating, with shocks.
2 22 Decl. and H. F. considerable vibrations.
3 0 Decl. and H. F. slight vibrations.
2 Decl. and H. F. moderate vibrations.
4 22 V. F. considerable vibrations.
5 4 V. F. considerable vibrations.
6 22 H. F. slight shocks.
9 0 Decl. moderate shocks.
2 Decl. moderate shocks.
4 Decl. and H. F. moderate shocks.
10 0 Decl. vibrating much, H. F. moderately, V. F. considerably, with shocks.
2 Decl. and H. F. moderate vibrations and shocks.
11 20 Decl. and H. F. moderate vibrations and shocks.
22 Decl. and H. F. moderate vibrations and shocks.
12 0 Decl., H. F., and V. F., moderate vibrations, with shocks.
10 Decl. and H. F. slight vibrations and shocks.
12 H. F. moderate shocks.
14 Decl. and H. F. moderate shocks.
15 0 Decl. and H. F. slight vibrations; H. F. vibrating much, with shocks.
2 Decl. slight shocks; H. F. vibrating much, with shocks.
4 Decl. moderate shocks.
10 H. F. moderate shocks.
16 0 Decl. and H. F. slight vibrations.
10 H. F. slight vibrations.
18 H. F. slight vibrations.
22 Decl. and H. F. slight vibrations and shock
17 0 Decl. and H. F. moderate vibrations and shocks.
2 Decl. and H. F. slight vibrations and shocks.
4 H. F. moderate shocks.
20 H. F. moderate shocks.
22 H. F. moderate shocks.
18 2 H. F. slight shocks.
21 18 V. F. slight vibrations.
23 2 Decl. and H. F. moderate shocks.
23 4 Decl. and H. F. moderate vibrations and shocks.
8 H. F. moderate shocks.
25 0 Decl. much disturbed by shocks; H. F. much, and vibrating.
2 Decl. much disturbed by shocks; H. F. vibrating very much, with shocks.
8 Decl. and H. F. moderate shocks.
28 20 V. F. slightly vibrating.
29 0 H. F. moderate vibrations.
8 H. F. slight shocks.
10 H. F. moderate shocks.

* Commencing after midnight of Sunday at Toronto.

MAY 9 and 10, 1841.

Mean Positions at the same hours during the Month.[*]

M. Gött. Time.			Decl.	Hor. Force.		Vert. Force.	
h.	m.	s.	Sc.-Div⁹ˢ	Sc.-Div⁹ˢ	Ther.	Sc.-Div⁹ˢ	Ther.
0	0	0	54·9	477·8	56	73·0	56
2	0	0	55·7	468·4	57	72·3	57
4	0	0	50·4	454·3	58	70·8	58
6	0	0	42·6	458·9	59	71·7	58
8	0	0	40·8	470·2	60	72·4	59
10	0	0	43·7	477·1	60	73·3	60
12	0	0	48·0	472·6	61	72·7	60
14	0	0	50·1	467·7	60	71·8	60
16	0	0	50·4	467·9	59	72·0	59
18	0	0	49·8	470·4	58	72·0	58
20	0	0	49·3	472·9	57	72·7	57
22	0	0	50·6	475·3	56	72·5	56

ST. HELENA { Decl. 1 Scale Division = 0′·71
H. F. $k = ·00018$; $q = ·00025$

Regular and extra observations.

The H. F. was observed at 1ᵐ. after the times specified.

h.	m.	s.	Sc.-Div⁹ˢ	Sc.-Div⁹ˢ	Ther.	Sc.-Div⁹ˢ	Ther.
9 14	0	0[b]	29·8	68·2	65		
15	0	0	27·0	63·9	66		
16	0	0	31·1	63·0	66		
18	0	0	29·1	66·8	66		
19 30	0	0	33·6	66·5	66		
20	0	0	32·1	63·0	66		
20 30	0	0	32·0	60·0	66		
22	0	0	28·0	61·3	66		
23	0	0	27·1	63·4	66		
10 0	0	0	26·0	64·1	67		
2	0	0	22·5	55·0	68		
12	28		22·0	54·8			
17	28		21·9	54·8			
22	28		21·9	54·3			
27	28		21·9	54·0			
32	28		22·1	53·9			
37	28		22·3	53·9			
42	28		22·1	53·2			
47	28		22·1	53·1			
52	28		22·0				
57	28		22·0	53·1	68		
3 0	0	0	22·0				
2	28		21·9	53·0			
7	28		21·8	52·5			
12	28		21·7	52·1			
17	28		21·5	51·6			
22	28		21·8	51·0			
27	28		21·9	50·9			
32	28		21·3	50·0			
37	28		21·9	50·0			
42	28		22·0	50·0			
47	28		22·7	49·9			
52	28		23·0	49·9			
57	28		22·9	50·0			
4 0	0	0	22·9	50·0	68		
2	28		23·0	50·0			
7	28		23·1	50·0			
12	28		23·1	50·0			
17	28		23·2	50·0			
22	28		23·5	49·0			
27	28		23·6	48·8			

MAY 9 and 10, 1841.

M. Gött. Time.			Decl.	Hor. Force.		Vert. Force.	
.	h.	m. s.	Sc.-Div⁹ˢ	Sc.-Div⁹ˢ	Ther.	Sc.-Div⁹ˢ	Ther.
10	4 32	28	23·4	48·2			
	37	28	23·1	47·9			
	42	28	23·0	47·2			
	47	28	23·0	47·0			
	52	28	23·0	47·0			
	57	28	22·9	47·0			
5	0	0	22·9	47·0	68		
	2	28	22·9	47·0			
	7	28	22·7	47·1			
	12	28	22·6	47·9			
	17	28		48·9			
	22	28	22·3	47·9			
	27	28	22·8	48·2			
	32	28	23·1	48·9			
	37	28	23·5	49·0			
	42	28	24·1	49·9			
	47	28	24·3	50·1			
	52	28	24·2	50·1			
	57	28	24·1				
6	0	0	24·1	50·1	68		
	2	28	24·1	50·1			
	7	28	23·9	50·0			
	12	28	23·9	50·0			
	17	28	23·9	50·1			
	22	28	23·8	50·2			
	27	28	23·8	51·0			
	32	28	23·8	51·2			
	37	28	23·0	51·5			
	42	28	23·0	51·3			
	47	28	22·6	51·1			
	52	28	22·0	51·0			
	57	28	22·0	51·0			
7	2	28	22·0	50·9			
	7	28	22·0	50·2			
	12	28	22·0	50·1			
	17	28	22·0	50·0			
	22	28	21·9	49·8			
	27	28	22·0	49·1			
	32	28	22·2	48·9			
	37	28	22·7	48·8			
	42	28	23·0	48·7			
	47	28	23·0	48·8			
	52	28	23·2	48·9			
	57	28	23·6	49·0			
8	0	0	23·8	49·0	68		
	2	28	23·8	49·0			
	7	28	23·5	49·5			
	12	28	23·1	49 8			
	17	28	23·0	49·9			
	22	28	23·0	49·9			
	27	28	23·0	50·0			
	32	28	23·1	50·6			
	37	28	23·4	50·8			
	42	28	23·8	50·9			
	47	28	23·8	50·9			
	52	28	23·9	51·0			
	57	28	24·0	51·3			
9	2	28	24·0	51·7	68		
	7	28	24·1	51·9			
	12	28	24·3	52·2			
	17	28	24·4	52·4			
	22	28	24·5	52·4			
	27	28	24·6	52·3			
	28		24 6	52·4			

MAY 9 and 10, 1841.

M. Gött. Time.			Decl.	Hor. Force.		Vert. Force.	
d.	h.	m. s.	Sc.-Div⁹ˢ	Sc.-Div⁹ˢ	Ther.	Sc.-Div⁹ˢ	Ther.
10	9 37	28	24·6	52·2			
	42	28	24·8	52·3			
	47	28	25·0	53·1			
	52	28	25·2	53·5			
	57	28	25·1	53·7			
10	0	0	25·1	53·4	69		
	2	28	25·0	53·4			
	7	28	24 4	53·4			
	12	28	24·5	53·3			
	17	28	24·9	53·2			
	22	28	25·9	53·4			
	27	28	24·9	53·0			
	32	28	24·9	53·0			
	37	28	25·0	53·0			
	42	28	25·0	53·0			
	47	28	25·1	53·1			
	52	28	25·1	53·5			
	57	28	25·1	53·3			
11	0	0	25·1	53·3	67		
	2	28	25·6	53·2			
	7	28	25·8	53·2			
	12	28	25·8	53·4			
	17	28	25·2	53·8			
	22	28	25·3	54·0			
	27	28	25·4	54·1			
	32	28	25·5	54·1			
	37	28	25·7	54·9			
	42	28	25·9	55·1			
	47	28	25·9	55·5			
	52	28	26·1	56·0			
	57	28	26·7	56·7			
12	0	0	26·8	57·0	67		
	2	28	27·0	57·0			
	7	28	27·1	58·0			
	12	28	27·1	58·1			
	17	28	27·9	58·8			
	22	28	28·1	59·1			
	27	28	28·1	59·2			
	32	28	28·2	59·7			
	37	28	28·9	59·8			
	42	28	28·9	60·0			
	47	28	28·9	59·9			
	52	28	29·0	59·1			
	57	28	29·0	59·1			
13	0	0	29·0	·59·1	67		
	2	28	29·0	59·1			
	7	28	29·0	59·0			
	12	28	29·0	58·7			
	17	28	29·1	58·7			
	22	28	29·1	58·5			
	27	28	29·1	58·2			
	32	28	29·2	58·1			
	37	28	29·8	58·2			
	42	28	30·1	58·8			
	47	28	30·3	58·9			
	52	28	30·6	58·9			
	57	28	30·2	58·9			
14	0	0		57·0	67		
	2	28	30·1	58·9			
	7	28	30·2	58·9			
	12	28	30·4	59·1			
	17	28	30·8	59·2			
	22	28	30·2	59·9			
	27	28	30·2	60·0			

---

[*] The mean positions of the Decl. magnet are from the 1st to the 24th of May, inclusive; and those of the V. F. magnet from the 1st to the 25th, inclusive.

[b] Commencing after midnight of Sunday at St. Helena.

MAY 9 and 10, 1841.

M. Gött. Time	Decl. (Sc.-Div)	Hor. Force (Sc.-Div)	Ther. (Vert. Force)
10 14 32 28	30·5	60·9	
37 28	30·2	61·0	
42 28	30·1	61·0	
47 28	30·1	61·1	
52 28	30·0	61·2	
57 28	30·0	61·0	
15 0 0	30·0	61·1	67
2 28	30·0	61·1	
7 28	30·0	61·5	
12 28	30·0	61·9	
17 28	30·0	62·0	
22 28	30·0	62·1	
27 28	30·0	62·5	
32 28	29·2	62·4	
37 28	29·0	62·4	
42 28	29·4	62·1	
47 28	29·1	62·1	
52 28	29·1	62·0	
57 28	29·0	62·0	
16 0 0	29·0	62·0	67
2 28	29·0		
7 28	29·0	62·0	
12 28		62·0	
17 28	29·1	62·1	
22 28	29·2	62·1	
27 28	29·3	62·1	
32 28	29·5	62·2	
37 28	29·6	62·2	
42 28	29·7	62·0	
47 28	30·0	62·0	
52 28	30·0	62·0	
57 28	30·1	62·0	
17 2 28	30·2	62·0	
7 28	30·2	62·0	
12 28	30·1	62·1	
17 28	30·1	62·2	
22 28	30·1	62·4	
27 28	30·0	62·4	
32 28	30·0	62·5	
37 28	30·0	62·6	
42 28	30·0	62·7	
47 28	30·0	62·7	
52 28	30·1	62·7	
57 28	30·2	62·7	
18 0 0	30·3	62·7	67
2 28	30·3	62·7	
7 28	30·2	62·8	
12 28	30·2	62·9	
17 28	30·1	63·0	
22 28	30·1	63·2	
27 28	30·0	63·5	
32 28	30·0	63·9	
37 28	30·0	64·0	
42 28	30·0	64·2	
47 28	30·1	64·4	
52 28	30·2	64·5	
57 28	30·2	64·6	
19 2 28	30·2	64·7	67
7 28	30·3	64·7	
12 28	30·4	64·7	
17 28	30·5	64·8	
22 28	30·5	64·8	

MAY 9 and 10, 1841.

M. Gött. Time	Decl. (Sc.-Div)	Hor. Force (Sc.-Div)	Ther.	Vert. Force (Sc.-Div)	Ther.
10 19 27 28	30·5	64·9			
32 28	30·4	64·9			
37 28	30·4	65·0			
42 28	30·3	65·2			
47 28	30·1	65·4			
52 28	30·0	65·8			
57 28	30·0	66·0			
20 0 0	29·9	65·9	67		
20 30 0	28·9	67·0	66		
22 0 0	25·0	71·7	66		
23 0 0	23·1	74·1	66		

Mean Positions at the usual hours of observation during the Month.

M. Gött. Time	Decl. (Sc.-Div)	Hor. Force (Sc.-Div)	Ther.	Vert. Force (Sc.-Div)	Ther.
0 0 0	27·5	78·7	67		
2 0 0	27·7	75·0	68		
3 0 0	27·7	72·9	68		
4 0 0	28·0	70·7	68		
5 0 0	27·7	68·7	68		
6 0 0	27·3	68·0	68		
8 0 0	27·8	67·3	68		
10 0 0	28·2	67·7	67		
11 0 0	28·2	67·9	67		
12 0 0	28·6	68·1	67		
13 0 0	28·8	68·9	67		
14 0 0	29·3	69·6	67		
15 0 0	29·3	69·9	67		
16 0 0	29·5	70·1	67		
18 0 0	30·2	70·5	66		
19 30 0	31·2	71·3	66		
20 0 0	31·3	71·4	66		
20 30 0	30·6	72·1	66		
22 0 0	27·8	75·9	66		
23 0 0	27·3	78·0	66		

CAPE OF GOOD HOPE { Decl. 1 Scale Division = 0'·75 ; H. F. k = ·000185; q. = ·0003

Positions at the usual hours of observation, from May 9th, 12ʰ., to May 10th, 22ʰ.

M. Gött. Time	Decl. (Sc.-Div)	Hor. Force (Sc.-Div)	Ther.	Vert. Force (Sc.-Div)	Ther.
9 12 0 0ᵃ	53·4	74·4	62		
14 0 0	50·2	76·4	61		
16 0 0	55·2	65·8	61		
18 0 0	55·4	67·5	60		
20 0 0	51·8	63·3	60		
22 0 0	48·0	57·3	60		
10 0 0 0	50·3	57·7	61		
2 0 0	47·1	46·8	62		
4 0 0		43·1	63		
6 0 0	43·9	47·4	63		
8 0 0	43·2	49·2	64		
10 0 0	45·3	55·3	64		
12 0 0	49·6	60·3	63		
14 0 0	53·1	59·7	62		
16 0 0	51·3	63·0	62		
18 0 0	51·5	62·8	63		
20 0 0	49·1	67·0	63		
22 0 0	43·8	65·5	62		

MAY 9 and 10, 1841.

Mean Positions at the same hours during the Month.[b]

M. Gött. Time	Decl. (Sc.-Div)	Hor. Force (Sc.-Div)	Ther.	Vert. Force (Sc.-Div)	Ther.
0 0 0	50·1	73·6	61		
2 0 0	51·5	70·3	61		
4 0 0	51·2	70·9	62		
6 0 0	49·9	70·3	62		
8 0 0	49·7	70·0	62		
10 0 0	49·9	71·0	61		
12 0 0	51·0	72·5	61		
14 0 0	51·9	73·3	61		
16 0 0	52·2	73·8	61		
18 0 0	57·9	74·4	60		
20 0 0	50·7	77·1	60		
22 0 0	47·2	75·5	60		

VAN DIEMEN ISLAND { Decl. 1 Scale Division = 0'·71 ; H. F. k = ·0003; q = ; V. F. k = ; q =

Extra observations.

The V. F. was observed at 2ᵐ. 30ˢ. before, and the H. F. 2ᵐ. 30ˢ. after the times specified.

M. Gött. Time	Decl. (Sc.-Div)	Hor. Force (Sc.-Div)	Ther.	Vert. Force (Sc.-Div)	Ther.
9 21 10 0	48·0	14·7	53	100·4	53
15 0	58·7			100·7	
20 0	52·6	8·2		100·7	
25 0	47·1				
30 0	49·3	8·9		101·0	
35 0	47·8				
40 0	39·4	8·2		100·3	
45 0	44·1				
50 0	44·1	4·9		100·9	
55 0	39·0				
22 10 0	41·6	6·1	53	100·6	53
15 0	38·1				
20 0	36·6	7·3		100·0	
25 0	38·1				
30 0	38·5	7·4		99·8	
35 0	36·5				
40 0	36·5	7·7		99·5	
45 0	38·9				
50 0	34·6	7·5		99·1	
23 5 0	30·3				
10 0	31·3	8·0	53	98·8	53
15 0	32·8				
20 0	32·6	9·1		97·9	
25 0	30·5				
30 0	29·3	9·2		96·6	
35 0	30·6				
40 0	32·5				
45 0	36·9	10·5			
50 0	42·0	11·2		96·8	
10 1 20 0	31·6	9·9			
25 0	34·7		52		52
30 0	37·7	0·0		93·9	
35 0	38·9				
40 0	38·9	1·1		94·9	
45 0	39·6				
2 15 0	40·1	1·1	52		

---

a. Commencing after Sunday midnight at Cape Town.
b. The mean positions of the Decl. magnet are from the 5th to the 21st, inclusive; those of the H. F. are not strictly comparable with the observations of the 9th and 10th, in consequence of the stretching of the suspension wires in the early portion of the month; the true mean positions corresponding with those of the 9th and 10th of May would, it is probable, have been higher numbers than those which result from the mean of the readings during the month.

I

MAY 9 and 10, 1841.

M. Gött. Time (d. h. m. s.)	Decl. Sc.-Div	Hor. Force Sc.-Div	Ther.	Vert. Force Sc.-Div	Ther.
10 2 20 0	38·5	13·9		95·4	52
25 0	36·3				
30 0	35·4	14·8		94·3	
35 0	31·8				
40 0	25·9	12·5		92·4	
45 0	25·1				
50 0	25·4	11·4		92·4	
3 10 0	33·9	6·9	52	94·7	
15 0	36·4				
20 0	40·1	9·2		95·8	52
25 0	36·5				
30 0	31·9	14·2		94·5	
35 0	30·2				
40 0	28·7	17·3		91·5	
45 0	32·3				
50 0	34·2	18·4		89·9	
4 15 0	37·7	18·6	52	88·7	52
20 0	38·5	20·3			
25 0	37·7				
30 0	38·2	21·4		87·8	
35 0	44·6				
40 0	49·8	21·1		88·9	
45 0	53·2				
50 0	55·5	21·7		89·3	
5 5 0	54·9		52		
10 0	52·6	21·0		88·5	52
15 0	51·8				
20 0	53·0	21·5		87·7	
25 0	50·1				
30 0	48·0	18·8		86·6	
35 0	45·1				
40 0	43·4	19·7		86·4	
45 0	46·9				
50 0	50·7	21·4		86·9	
55 0	54·6				
6 5 0	59·7		52		
10 0	62·1	21·7		88·3	52
15 0	62·7				
20 0	63·5	22·0		88·3	52
25 0	64·0				
30 0	67·1	21·8		89·0	
35 0	70·2				
40 0	72·2	21·9		90·4	
45 0	75·5				
50 0	78·5			92·4	
55 0	79·7				
7 10 0	74·9	23·3	52		
15 0	74·2				
20 0	72·4	23·4		92·4	52
25 0	68·4				

MAY 9 and 10, 1841.

M. Gött. Time (d. h. m. s.)	Decl. Sc.-Div	Hor. Force Sc.-Div	Ther.	Vert. Force Sc.-Div	Ther.
10 7 30 0	66·2	23·5		92·5	
35 0	63·8				
40 0	61·0	22·7		92·6	
45 0	57·8				
50 0	55·8			92·2	
8 10 0	53·9		51		51
15 0	53·9				

Positions at the usual hours of observation, from May 9th, 3h. to May 10th, 23h.

M. Gött. Time (d. h. m. s.)	Decl. Sc.-Div	Hor. Force Sc.-Div	Ther.	Vert. Force Sc.-Div	Ther.
9 3 0 0h	45·7	21·8		91·8	
4 0 0	49·7	23·0	56	92·1	55
5 0 0	50·2	23·8		91·8	
6 0 0	52·1	22·8	55	92·8	54
7 0 0	52·3	23·2		92·6	
8 0 0	51·1	23·8	54	92·5	53
9 0 0	49·9	24·5		92·7	
10 0 0	48·9	24·8	52	92·7	52
11 0 0	47·6	25·1		92·5	
12 0 0	46·3	26·2	52	92·1	53
13 0 0	44·7	20·9		92·9	
14 0 0	50·6	12·6	53	97·4	53
15 0 0	57·0	10·4		95·8	
16 0 0	56·7	16·0	53	94·8	53
17 0 0	58·5	12·8		96·7	
18 0 0	59·6	16·1	52	96·5	52
19 0 0	60·6	15·2		97·6	
20 0 0	55·5	13·4	52	98·0	52
21 0 0	47·2	8·8		100·7	
22 0 0	39·1	7·6	53	100·7	54
23 0 0	31·3	7·1		98·6	
10 0 0 0	46·0	17·2	53	97·0	54
1 0 0	46·2	1·4		91·3	
2 0 0	40·0	10·0	52	95·4	52
3 0 0	27·1	9·3		92·5	
4 0 0	35·0	18·4	52	88·7	52
5 0 0	60·0	20·3			
6 0 0	56·6	21·9	52	87·8	52
7 0 0	79·0	23·2		91·0	
8 0 0	54·6	22·6	51	91·1	51
9 0 0	56·6	21·7		91·6	
10 0 0	58·3	20·7	50	92·2	50
11 0 0	57·6	20·6		92·6	
12 0 0	54·5	20·1	50	92·3	50
13 0 0	54·5	20·0		92·8	
14 0 0	51·5	20·5	51	92·2	50
15 0 0	53·0	20·5		92·1	
16 0 0	54·1	19·9	52	92·6	52
17 0 0	53·1	19·9		92·6	

MAY 9 and 10, 1841.

M. Gött. Time (d. h. m. s.)	Decl. Sc.-Div	Hor. Force Sc.-Div	Ther.	Vert. Force Sc.-Div	Ther.
10 18 0 0	52·5	19·6	53	92·8	53
19 0 0	53·2	20·6		92·7	
20 0 0	52·3	21·0	54	92·3	53
21 0 0	52·0	21·2		92·0	
22 0 0	51·6	21·4	54	92·0	53
23 0 0	51·3	21·7		92·0	

Mean Positions at the same hours during the Month.

M. Gött. Time (d. h. m. s.)	Decl. Sc.-Div	Hor. Force Sc.-Div	Ther.	Vert. Force Sc.-Div	Ther.
0 0 0	49·4	22·0	55	92·7	55
1 0 0	49·3	21·6		92·8	
2 0 0	49·0	22·4	54	92·7	54
3 0 0	48·5	22·3		92·6	
4 0 0	49·8	22·8	54	92·6	54
5 0 0	51·0	23·2		92·4	
6 0 0	51·7	23·6	53	92·6	53
7 0 0	52·8	23·8		92·9	
8 0 0	51·5	24·3	52	92·7	52
9 0 0	51·6	24·8		92·6	
10 0 0	52·2	24·9	52	92·8	51
11 0 0	51·3	25·1		92·5	
12 0 0	49·7	24·6	52	92·5	52
13 0 0	49·2	23·4		92·4	
14 0 0	49·6	21·8	54	92·6	53
15 0 0	52·0	20·7		92·6	
16 0 0	54·7	20·7	55	92·6	55
17 0 0	55·8	20·2		93·1	
18 0 0	56·1	20·8	57	92·9	56
19 0 0	55·3	21·5		92·6	
20 0 0	53·6	21·8	57	92·3	56
21 0 0	52·4	21·4		92·4	
22 0 0	50·8	21·4	56	92·5	55
23 0 0	49·9	21·4		92·7	

JULY 18, 19, and 20, 1841.

TORONTO * $\begin{cases} \text{Decl. 1 Scale Division} = 0'\cdot72 \\ \text{H. F. } k = \cdot000076 \text{ ; } q = \cdot0002 \\ \text{V. F. } k = \cdot000093 \text{ ; } q = \cdot00018 \end{cases}$

Extra observations.

The V. F. was observed at 1m. 30s. before, and the H. F. 2m. 30s. after the times specified.

d. h. m. s.	Decl.	Hor. Force	Ther.	Vert. Force	Ther.
19 15 15 0h	174·5	337·6	75	82·5	75
20 0	173·8	343·5		77·4	
25 0	174·3	349·1		82·1	
30 0	190·7	334·4			

---

* Commencing after midnight of Sunday at Van Diemen Island.

d. h. m.	Wind.	
19 15 0	Calm.	Clear; bright bank of auroral light in North, with a few faint streamers shooting from it.
30	Calm.	Clear; bank of light still remaining.
35		A bright semi-circular narrow strip of light extending from the N. W. to N.E. horizons, the centre reaching to an altitude of 85°; low bank still remaining in North horizon.
16 0	Calm.	Clear; bank of light growing fainter, and semi-circular form lost. No streamers visible.
30	N.N.E. very light	Clear; Aurora growing fainter.
17 0	N.N.E. very light	No apparent alteration of the Aurora since 16h. 30m.
18 0	Calm.	Clear; bank of auroral light in North, disappeared soon after 19h.

* Toronto, July, 1841.—*Times of observation at which the magnets were disturbed, but the mean readings were not materially changed.*

d. h.
1 14 Decl. and H. F. moderately disturbed, vibrating with shocks.
2 20 Decl. and V.F. moderately, H. F. much disturbed by vibrations and shocks.
22 Decl. and V. F. moderately; H. F. much disturbed by vibrations and shocks.
3 0 Decl. moderate shocks; H. F. vibrating much, with shocks.
5 16 Decl. moderate shocks; V. F. slightly vibrating.
20 Decl. moderate shocks; (22h.) Decl. and H. F. moderate shocks.
6 2 H. F. moderate shocks.
20 H. F. slightly disturbed; vibrations and shocks.
7 0 Decl. and H. F. moderate shocks.
2 Decl. and H. F. much disturbed by shocks.
9 2 Decl. slightly; H. F. moderate shocks.
14 22 Decl. and H. F. slight shocks; V. F. slight vibrations.
15 0 H. F. moderate vibrations and shocks.
18 Decl. and H. F. and V. F. vibrating considerably; H. F. much disturbed by shocks.
20 Decl., H. F., and V. F. vibrating moderate; H. F. moderate shocks.

July 18, 19, and 20, 1841.

M. Gött. Time.				Decl.	Hor. Force.		Vert. Force.	
d.	h.	m.	s.	Sc.-Div⁸ⁱ.	Sc.-Div⁸ⁱ.	Ther.	Sc.-Div⁸ⁱ.	Ther.
19	15	35	0	207·5	332·6		91·5	
		40	0	224·7	347·7		90·1	
		45	0	216·3	350·0		86·9	
		50	0	199·4	354·9		80·3	
		55	0	180·0	367·1		75·8	
16	5	0	150·8	362·8	75	80·5	76
	10	0	151·2	366·9		81·0	
	15	0	146·3	362·0		79·9	
	20	0	144·5	367·2		78·4	
	25	0	142·6	370·9		79·0	
	30	0	139·3	375·7		78·6	
	35	0	136·8	377·0		79·1	
	40	0	136·1	374·0		79·8	
	45	0	137·1	373·9		79·3	
	50	0	139·0	375·6		79·1	
	55	0	140·3	378·5		79·1	

Positions at the usual hours of observation, from
July 18th, 18ʰ. to July 20th, 22ʰ.

d.	h.	m.	s.	Decl.	Hor. Force	Ther.	Vert. Force	Ther.
18	18	0	0ʰ	131·9	412·3	71	83·1	72
	20	0	0	133·9	414·1	71	83·8	71
	22	0	0	135·3	414·0	70	84·1	70
19	0	0	0	143·4	416·8	69	86·5	71
	2	0	0	145·1	410·8	70	82·6	71
	4	0	0	142·5	396·6	72	80·6	72
	6	0	0	121·9	379·2	74	79·6	73
	8	0	0	117·6	405·0	74	78·8	74
	10	0	0	125·5	406·6	75	78·7	75
	12	0	0	134·5	418·6	76	80·5	76
	14	0	0	131·0	375·1	76	87·8	75
	16	0	0	161·6	377·6	75	77·1	76
	18	0	0	144·9	374·7	73	71·8	74
	20	0	0	142·0	389·9	72	72·6	73
	22	0	0	137·4	399·3	71	83·5	72
20	0	0	0	130·3	392·0	70	73·3	71
	2	0	0	141·4	394·1	71	79·8	72
	4	0	0	139·7	393·4	73	81·4	74
	6	0	0	129·1	361·1	74	80·0	75
	8	0	0	128·2	386·4	76	82·4	75
	10	0	0	124·0	357·8	76	83·7	75
	12	0	0	132·2	392·4	76	81·3	76
	14	0	0	134·8	387·0	76	79·3	75
	16	0	0	134·1	396·4	74	75·1	75
	18	0	0	134·7	394·3	73	80·6	74
	20	0	0	126·4	397·1	71	80·7	73
	22	0	0	132·2	397·5	70	83·8	71

Mean Positions at the same hours during the Month.

d.	h.	m.	s.	Decl.	Hor. Force	Ther.	Vert. Force	Ther.
	0	0	0	140·9	404·1	69	84·9	70
	2	0	0	143·1	400·2	70	85·0	70
	4	0	0	136·8	389·6	71	83·7	71

July 18, 19, and 20, 1841.

M. Gött. Time.				Decl.	Hor. Force.		Vert. Force.	
d.	h.	m.	s.	Sc.-Div⁸ⁱ.	Sc.-Div⁸ⁱ.	Ther.	Sc.-Div⁸ⁱ.	Ther.
	6	0	0	127·3	388·8	71	83·2	72
	8	0	0	125·3	403·2	72	83·8	72
	10	0	0	128·7	407·1	73	84·7	73
	12	0	0	133·1	402·7	73	84·8	73
	14	0	0	135·2	393·1	73	84·7	73
	16	0	0	136·2	394·4	72	82·8	72
	18	0	0	138·0	396·8	71	81·1	72
	20	0	0	136·2	398·3	70	81·8	71
	22	0	0	136·3	401·3	70	84·0	70

ST. HELENA $\left\{ \begin{array}{l} \text{Decl. 1 Scale Division} = 0'·71 \\ \text{H. F. } k = ·00018; \ q = ·00025 \\ \text{V. F. } k = ·0002; \ q = \end{array} \right.$

Extra observations.

The V. F. was observed at 2ᵐ. 30ˢ. before, and the H. F.
2ᵐ. 30ˢ. after the times specified.

d.	h.	m.	s.	Decl.	Hor. Force	Ther.	Vert. Force	Ther.
20	0	10	0	29·1	42·2	62		61
		15	0	29·8				
		20	0	29·9	43·0		50·7	
		25	0	29·8				
		30	0	29·8	43·3		50·7	
		35	0	29·9				
		40	0	30·1	43·0		50·3	
		45	0	30·1				
		50	0	29·9	42·0		50·1	
		55	0	29·9				
	1	0	0	29·3	41·7	62	49·9	61
		5	0	28·6				
		10	0	28·1	41·1		49·3	
		15	0	28·3				
		20	0	28·2	41·2		49·8	
		25	0	28·0				
		30	0	27·1	39·0		48·9	
		35	0	26·9				
		40	0	26·9	38·0		48·5	
		45	0	26·8				
		50	0	26·9	39·0		48·4	
		55	0	27·1				
	2	0	0	27·0	40·0	62	48·9	62
		5	0	27·0				
		10	0	27·1	40·0		48·6	
		15	0	27·0				
		20	0	26·9	39·9		48·7	
		25	0	27·0				
		30	0	26·7	39·1		48·8	
		35	0	26·1				
		40	0	25·2	38·0		48·6	
		45	0	25·1				
		50	0	25·0	37·6		48·4	
		55	0	25·0				

July 18, 19, and 20, 1841.

M. Gött. Time.				Decl.	Hor. Force.		Vert. Force.	
d.	h.	m.	s.	Sc.-Div⁸ⁱ.	Sc.-Div⁸ⁱ.	Ther.	Sc.-Div⁸ⁱ.	Ther.
20	3	0	0	25·0	37·2	63	48·4	62
	5	0		25·2				
	10	0		25·3	37·1		48·4	
	15	0		25·4				
	20	0		25·5	36·8		48·7	
	25	0		25·6				
	30	0		25·5	36·2		48·5	
	35	0		25·2				
	40	0		25·0	36·0		48·5	
	45	0		24·8				
	50	0		24·5	36·1		47·9	
	55	0		24·3				
	4	0	0	24·1	35·8	63	47·8	62
	5	0		23·9				
	10	0		23·8	35·2		47·8	
	15	0		23·6				
	20	0		23·7	34·9		47·7	
	25	0		23·6				
	30	0		23·8	34·5		47·7	
	35	0		23·8				
	40	0		23·9	34·1		47·7	
	45	0		24·0				
	50	0		24·0	34·9		47·5	
	55	0		24·1				
	5	0	0	24·1	34·2	63	47·3	62
	5	0		24·2				
	10	0		24·5	34·0		47·1	
	15	0		24·2				
	20	0		24·3	34·0		47·1	
	25	0		24·0				
	30	0		24·0	33·9		46·8	
	35	0		24·0				
	40	0		24·3	33·9		46·8	
	45	0		24·6				
	50	0		24·9	33·2		46·9	
	55	0		25·1				

Positions at the usual hours of observation,
July 18, 19, and 20.

d.	h.	m.	s.	Decl.	Hor. Force	Ther.	Vert. Force	Ther.
18	14	0	0ʰ	28·1	44·1	61	48·0	61
	15	0	0	27·7	41·7	61	48·4	61
	16	0	0	27·8	41·2	61	49·6	61
	18	0	0	28·2	42·2	61	48·8	60
19	30	0		30·0	41·3	61	48·8	60
	20	0	0	30·8	42·8	61	48·6	60
	20	30	0	30·2	42·1	61	48·6	60
	22	0	0	26·3	45·0	60	48·3	60
	23	0	0	24·9	45·9	60	46·3	60
19	0	0	0	26·9	47·0	61	45·4	61
	2	0	0	24·5	44·1	62	46·6	62
	3	0	0	27·1	43·9	62	45·8	62
	4	0	0	29·0	43·9	63	44·7	62

d. h.
15 22 Decl. and H. F. moderate vibrations and shocks.
16 0 Decl. and H. F. moderate vibrations and shocks.
20 0 Decl. slight, H. F. moderate, vibrations and shocks.
22 0 Decl. slight, H. F. moderate, vibrations and shocks.
17 0 H. F. and V. F. moderate vibrations.
18 22 H. F. much disturbed by shocks.
19 0 H. F. moderate shocks.
2 0 Decl. and H. F. moderate shocks.
20 0 H. F. much disturbed by shocks.
2 0 Decl. and H. F. much disturbed by shocks.
4 0 Decl. vibrating slightly ; H. F. vibrating much.
10 0 H. F. moderate vibrations and shocks.
20 0 Decl. and H. F. vibrating slightly ; V. F. vibrating much.
21 0 H. F. moderate vibrations.

d. h.
21 4 H. F. moderate vibrations and shocks.
22 20 H. F. strong shocks.
22 H. F. strong shocks.
20 0 V. F. vibrating much.
22 H. F. moderate shocks.
26 20 Decl. and H. F. slight, V. F. considerable, vibrations.
22 Decl. and H. F. slight vibrations.
28 0 Decl. moderate vibrations and shocks ; H. F. moderate shocks.
2 0 Decl. and H. F. slight shocks.
29 2 Decl. much disturbed by vibrations and shocks.
22 H. F. slight vibrations.

* Commencing after Sunday midnight at Toronto.
ᵇ Commencing after Sunday midnight at St. Helena.

I 2

JULY 18, 19, and 20, 1841.

d.	h.	m.	s.	Decl. Sc.-Div.	Hor. Force Sc.-Div.	Ther.	Vert. Force Sc.-Div.	Ther.
19	5	0	0	27·2	43·2	62	44·1	63
	6	0	0	27·0	a		43·9	62
	8	0	0	26·0			43·7	62
	10	0	0	27·0	41·3	62	45·4	61
	11	0	0	27·6	41·4	62	45·5	61
	12	0	0	27·2	41·9	62	45·9	61
	13	0	0	26·2	40·2	62	45·5	62
	14	0	0	27·0	35·7	62	50·9	62
	15	0	0	25·5	37·9	62	50·0	60
	16	0	0	27·6	43·2	61	50·8	61
	18	0	0	26·0	38·7	61	50·4	61
	19	30	0	27·1	39·9	61	51·1	61
	20	0	0	30·9	39·0	61	51·0	61
	20	30	0	31·0	38·0	61	51·1	61
	22	0	0	29·8	43·4	61	51·9	60
	23	0	0	28·1	44·2	61	51·0	61
20	0	0	0	28·1	41·6	61	50·8	61
	2	0	0	27·0	40·0	62	48·9	62
	3	0	0	25·0	37·2	62	48·4	62
	4	0	0	24·1	35·8	63	47·8	62
	5	0	0	24·1	34·2	63	47·3	72
	6	0.	0	25·0	33·0	63	47·2	62
	8	0	0	25·5	34·1	62	48·2	62
	10	0	0	24·7	32·2	62	48·3	61
	11	0	0	27·7	35·6	61	49·0	61
	12	0	0	27·2	35·1	61	48·7	61
	13	0	0	27·2	36·0	61	47·8	61
	14	0	0	27·5	37·5	61	47·8	60
	15	0	0	28·0	39·0	61	48·1	60
	16	0	0	28·5	39·1	61	48·0	60
	18	0	0	28·2	39·9	61	48·0	60
	19	30	0	30·0	39·9	61	48·0	60
	20	0	0	31·0	40·0	61	48·0	60
	20	30	0	30·7	41·5	61	47·9	60
	22	0	0	27·0	44·9	61	47·0	60
	23	0	0	25·0	44·7	61	45·2	61

Mean Positions at the same hours during the Month.b

d.	h.	m.	s.	Decl. Sc.-Div.	Hor. Force Sc.-Div.	Ther.
	0	0	0	25·0	46·2	62
	2	0	0	25·4	44·0	63
	3	0	0	25·7	41·6	63
	4	0	0	25·5	40·0	64
	5	0	0	24·9	39·0	63
	6	0	0	24·2	38·1	63
	8	0	0	24·7	37·8	63
	10	0	0	25·3	38·0	63
	11	0	0	25·6	38·1	64
	12	0	0	25·7	38·2	64
	13	0	0	25·8	37·9	63
	14	0	0	26·1	40·3	62
	15	0	0	26·1	40·1	62
	16	0	0	26·4	41·0	62
	18	0	0	26·7	41·2	62
	19	30	0	28·3	41·5	62
	20	0	0	29·3	41·2	62
	20	30	0	29·1	42·4	62
	22	0	0	25·9	45·4	62
	23	0	0	24·9	46·3	62

JULY 18, 19, and 20, 1841.

CAPE OF GOOD HOPE { Decl. 1 Scale Division = 0'·75; H.F. k = ·00018; q = ·0003

Positions at the usual hours of observation, July 18th, 12h. to July 20th, 22h.

d.	h.	m.	s.	Decl. Sc.-Div.	Hor. Force Sc.-Div.	Ther.	Vert. Force Sc.-Div.	Ther.
18	12	0	0e	53·4	69·1	60		
	14	0	0	53·9	71·7	60		
	16	0	0	53·7	68·9	60		
	18	0	0	54·6	65·2	60		
	20	0	0	57·0	72·9	60		
	22	0	0	49·6	73·1	60		
19	0	0	0	49·6	70·4	60		
	2	0	0	51·2	68·9	60		
	4	0	0	54·2	72·1	60		
	6	0	0	52·7	67·3	60		
	8	0	0	52·6	68·5	60		
	10	0	0	53·0	68·5	60		
	12	0	0	53·1	71·1	60		
	14	0	0	50·7	66·7	60		
	16	0	0	56·6	74·4	60		
	18	0	0	50·9	67·5	60		
	20	0	0	55·1	68·6	60		
	22	0	0	52·5	68·6	60		
20	0	0	0	52·1	61·0	60		
	2	0	0	50·8	61·3	60		
	4	0	0	50·7	62·0	60		
	6	0	0	50·4	59·3	60		
	8	0	0	52·0	59·3	60		
	10	0	0	46·6	61·1	59		
	12	0	0	52·7	62·5	59		
	14	0	0	53·0	62·0	59		
	16	0	0	55·2	64·9	59		
	18	0	0	53·7	68·3	58		
	20	0	0	52·0	70·3	58		
	22	0	0	50·0	69·2	58		

Mean Positions at the same hours during the Month.

d.	h.	m.	s.	Decl. Sc.-Div.	Hor. Force Sc.-Div.	Ther.
	0	0	0	51·0	67·3	58
	2	0	0	52·7	65·8	58
	4	0	0	52·4	66·1	59
	6	0	0	51·6	65·1	59
	8	0	0	51·9	65·0	59
	10	0	0	52·1	65·5	59
	12	0	0	52·7	66·7	58
	14	0	0	53·0	67·4	58
	16	0	0	53·7	68·2	58
	18	0	0	53·9	68·8	58
	20	0	0	53·6	70·9	57
	22	0	0	50·0	70·3	57

VAN DIEMEN ISLAND { Decl. 1 Scale Division = 0'·71; H. F. k = ·0003; q = ; V. F. k = ; q =

Extra observations.

The V. F. was observed at 2m. 30s. before, and the H. F. 2m. 30s. after the times specified.

d.	h.	m.	s.	Decl. Sc.-Div.	Hor. Force Sc.-Div.	Ther.	Vert. Force Sc.-Div.	Ther.
20	0	25	0d	56·0				
		30	0	55·3	17·5	46	13·4	45
		35	0	54·0				
		50	0	52·7	15·1		14·2	

JULY 18, 19, and 20, 1841.

d.	h.	m.	s.	Decl. Sc.-Div.	Hor. Force Sc.-Div.	Ther.	Vert. Force Sc.-Div.	Ther.
20	0	45	0	51·8				
		50	0	52·1	12·8		15·7	
		55	0	52·2				
	1	10	0e	58·2	18·1	46		46
		15	0	65·1				
		20	0	67·6	23·0		19·4	
		25	0	65·4				
		30	0	70·2	27·0		11·7	
		35	0	76·2				
		40	0	79·0	22·4		11·7	
		45	0	74·3				
		57	30	63·1	17·6		11·6	
	2	7	30	58·7	16·6	47	12·5	47
		15	0f	55·8				
		20	0	57·1	18·0		13·3	
		25	0	57·4				
		30	0	56·5	17·5		12·9	
		35	0g	58·3				
		40	0	59·1	17·6		14·5	
		45	0	60·0				
		50	0	60·6	17·6		15·7	
	7	5	0	70·6				
		10	0	70·0	21·6	43	19·2	43
		15	0	69·4				
	10	10	0	75·1	23·1	42		42
		15	0	74·4				
		20	0	73·9	23·5		20·7	
		25	0	71·4				
		30	0	69·1	23·3		17·8	
		35	0	68·1				
		40	0	68·1			15·7	

Positions at the usual hours of observation, July 18, 19, and 20.

d.	h.	m.	s.	Decl. Sc.-Div.	Hor. Force Sc.-Div.	Ther.	Vert. Force Sc.-Div.	Ther.
18	3	0	0h	21·0			8·4	
	4	0	0	64·8	22·0	51	8·2	50
	5	0	0		22·4		9·2	
	6	0	0	65·3	23·1	50	9·8	48
	7	0	0		23·2		10·3	
	8	0	0	765·5	24·3	48	10·6	47
	9	0	0		23·7		11·1	
	10	0	0	64·7	23·8	48	11·3	46
	11	0	0		24·6		12·0	
	12	0	0	61·4	24·6	46	11·3	46
	13	0	0		24·2		10·9	
	14	0	0	59·9	23·8	47	9·2	47
	15	0	0		20·7		8·6	
	16	0	0	65·5	19·6	51	8·8	50
	17	0	0		20·1		8·2	
	18	0	0	67·6	21·5	52	8·0	51
	19	0	0		22·2		9·6	
	20	0	0	65·5	22·6	51	9·3	49
	21	0	0		22·4		10·4	
	22	0	0	64·8	22·1	50	11·2	48
	23	0	0		22·4		11·9	
19	0	0	0	63·8	21·9	49	13·3	47
	1	0	0		22·1		14·3	
	2	0	0	62·5	21·3	47	14·2	46
	3	0	0		22·1		14·1	
	4	0	0	63·7	23·9	47	14·6	46
	5	0	0		24·4		14·0	

* Connexion broken; the readings of the H. F. magnet from 18d. 14h. to 19d. 5h. are not comparable with the mean positions during the month given below.
b The mean positions of the H. F. are from July 20th to July 30th, inclusive; there are no mean positions of the V. F. in consequence of frequent breaks in the series.
c Commencing after Sunday midnight at the Cape of Good Hope.
d "No Aurora visible."
e "The Aurora plainly visible in the S.E; altitude about 20°."
f "Aurora scarcely visible."
g "Aurora not visible."
h Commencing after Sunday midnight at Van Diemen Island.

JULY 18, 19, and 20, 1841.

M. Gött. Time (m. s.)	Decl. Sc.-Div^m.	Hor. Force Sc.-Div^m.	Ther.	Vert. Force Sc.-Div^m.	Ther.
19 6 0 0	64·9	25·9	46	11·7	45
7 0 0		25·0		13·9	
8 0 0	65·0	25·2	45	15·1	44
9 0 0		25·1		15·5	
10 0 0	64·4	25·2	45	14·8	44
11 0 0		27·1		13·0	
12 0 0	61·3	24·2	45	14·2	44
13 0 0		21·9		21·6	
14 0 0	63·9	22·9	45	15·3	44
15 0 0		18·7		19·7	
16 0 0	73·7	15·4	46	23·7	45
17 0 0		21·3		13·3	
18 0 0	74·4	21·1	47	17·3	46
19 0 0		17·5		24·6	
20 0 0	72·2	19·3	47	20·5	46
21 0 0		20·8		15·4	
22 0 0	67·3	20·7	47	16·2	45
23 0 0		18·2		19·0	
20 0 0 0	52·4	18·8	46	15·9	45
1 0 0		12·4		20·4	
2 0 0	61·8	17·6	47	12·1	47
3 0 0		17·8		15·7	
4 0 0	64·5	19·3	45	19·6	44
5 0 0		22·2		17·6	
6 0 0	64·9	22·0	43	19·8	43
7 0 0		21·9		20·5	
8 0 0	66·4	22·6	43	16·8	43
9 0 0		23·1		18·1	
10 0 0	76·5	24·1	42	22·8	42
11 0 0		22·5		14·1	
12 0 0	63·2	22·7	42	17·4	42
13 0 0		21·4		18·8	
14 0 0	63·3	21·1	43	15·8	43
15 0 0		19·3		16·1	
16 0 0	66·1	19·9	46	13·3	46
17 0 0		20·5		12·1	
18 0 0	67·9	20·9	47	13·4	46
19 0 0		21·0		12·0	
20 0 0	64·8	20·6	48	13·3	47
21 0 0		21·0		13·2	
22 0 0	64·4	20·9	47	13·7	46
23 0 0		17·9		16·8	

Mean Positions at the same hours during the Month.

M. Gött. Time (m. s.)	Decl. Sc.-Div^m.	Hor. Force Sc.-Div^m.	Ther.	Vert. Force Sc.-Div^m.	Ther.
0 0 0	61·6	20·8	48	12·6	47
1 0 0	61·4	20·9		12·9	
2 0 0	61·6	20·9	47	13·1	47
3 0 0	62·2	21·7		13·2	
4 0 0	62·9	22·3	46	14·1	46
5 0 0	63·4	23·0		14·6	
6 0 0	64·0	23·4	45	15·1	45
7 0 0	65·1	23·7		15·3	
8 0 0	64·3	23·9	44	15·1	44
9 0 0	64·0	24·3		15·1	
10 0 0	64·4	24·5	44	15·3	44
11 0 0	63·6	24·8		14·9	
12 0 0	62·0	24·5	44	14·1	44
13 0 0	61·4	23·5		14·0	
14 0 0	62·7	21·7	46	13·7	46
15 0 0	64·6	20·0		13·1	
16 0 0	66·7	19·2	48	13·5	48
17 0 0	68·1	19·3		12·6	
18 0 0	68·4	19·9	50	12·2	49

JULY 18, 19, and 20, 1841.

M. Gött. Time (d. h. m. s.)	Decl. Sc.-Div^m.	Hor. Force Sc.-Div^m.	Ther.	Vert. Force Sc.-Div^m.	Ther.
19 0 0	67·0	20·1		12·1	
20 0 0	65·4	20·3	50	11·2	49
21 0 0	64·2	20·4		11·4	
22 0 0	63·6	20·7	49	11·7	48
23 0 0	62·6	20·8		12·1	

ANTARCTIC EXPEDITION AT GARDEN ISLAND, SYDNEY, NEW SOUTH WALES.

Decl. 1 Scale Division $= 0'·73$
H. F. $k = ·000186$; $q =$
V. F. $k =$; $q =$

Positions at the usual hours of observation, July 18, 19, and 20.

M. Gött. Time (d. h. m. s.)	Decl. Sc.-Div^m.	Hor. Force Sc.-Div^m.	Ther.	Vert. Force Sc.-Div^m.	Ther.
18 3 0 0[a]	38·1	47·0	52	48·7	55
4 0 0	38·8	48·0	52	49·4	54
5 0 0	38·6	48·8	52	48·4	54
6 0 0	39·9	49·5	52	49·8	53
7 0 0	38·8	48·0	52	50·9	54
8 0 0	38·8	47·8	52	49·7	54
9 0 0	38·5	48·9	52	49·0	54
10 0 0	38·7	50·5	52	50·2	53
11 0 0	37·8	51·4	52	50·2	54
12 0 0	35·9	50·3	52	50·2	54
13 0 0	33·9	49·0	53	49·2	55
14 0 0	34·3	48·8	54	48·7	55
15 0 0	37·5	43·4	57	47·3	59
16 0 0	40·3	41·7	58	48·1	59
17 0 0	40·5	41·2	59	46·4	59
18 0 0	40·2	42·0	60	46·1	61
19 0 0	39·2	43·9	61	45·6	61
20 0 0	38·0	42·4	60	46·3	61
21 0 0	37·8	42·3	59	46·3	60
22 0 0	37·6	43·0	58	47·2	59
23 0 0	37·6	44·1	57	48·2	58
19 0 0	36·8	43·9	57	50·3	56
1 0 0	36·7	46·3	55	50·3	56
2 0 0	36·2	44·0	57	50·1	56
3 0 0	36·8	45·4	57	49·9	56
4 0 0	37·4	46·9	55	49·9	56
5 0 0	36·5	48·6	56	49·9	56
6 0 0	37·0	51·9	53	50·1	57
7 0 0	36·0	51·7	52	49·9	56
8 0 0	38·0	53·5	52	50·0	56
9 0 0	38·4	56·1	51	50·1	56
10 0 0	37·5	58·5	50	50·0	56
11 0 0	35·7	55·7	51	49·8	56
12 0 0	36·0	51·4	51	50·4	56
13 0 0	35·4	51·0	54	47·4	56
14 0 0	39·2	43·9	56	48·0	57
15 0 0	43·1	35·7	58	48·0	59
16 0 0	41·1	40·5	60	45·6	62
17 0 0	44·2	38·1	61	45·9	62
18 0 0	42·5	32·5	60	45·0	64
19 0 0	41·6	36·2	61	45·0	64
20 0 0	38·5	39·0	59	46·0	60
21 0 0	37·9	40·3	58	47·1	59
22 0 0	37·5	38·5	56	48·8	57
23 0 0	30·4	43·1	54	48·1	56
20 0 0	29·0	36·4	53	50·3	56
1 0 0	35·0	46·8	53	49·3	57
2 0 0	33·8	44·5	52	50·2	57

JULY 18, 19, and 20, 1841.

M. Gött. Time (d. h. m. s.)	Decl. Sc.-Div^m.	Hor. Force Sc.-Div^m.	Ther.	Vert. Force Sc.-Div^m.	Ther.
20 4 0 0	35·0	46·3	52	51·6	57
5 0 0	36·8	48·5	53	51·6	53
6 0 0	35·9	49·7	51	51·0	53
7 0 0	40·2	50·4	50	51·4	53
8 0 0	37·0	52·6	50	50·8	53
9 0 0	36·4	53·2	50	50·5	53
10 0 0	42·9	54·7	49	53·1	50
11 0 0	35·1	55·5	50	51·0	49
12 0 0	36·1	55·6	49	49·2	53
13 0 0	34·9	53·5	50	49·4	53
14 0 0	36·1	53·7	50	49·7	52
15 0 0	38·9	52·7	50	50·3	53
16 0 0	38·6	52·2	50	50·0	58
17 0 0	39·7	53·4	51	50·5	53
18 0 0	39·6	52·7	52	50·7	53
19 0 0	38·7	51·7	52	50·3	53
20 0 0	36·9	51·7	52	49·1	54
21 0 0	36·1	52·3	52	49·1	54
22 0 0	36·1	52·3	51	50·0	54
23 0 0	35·5	49·5	51	50·5	54

Mean Positions at the same hours, from July 19 to July 31 inclusive.

M. Gött. Time (d. h. m. s.)	Decl. Sc.-Div^m.	Hor. Force Sc.-Div^m.	Ther.	Vert. Force Sc.-Div^m.	Ther.
0 0 0	33·1	51·1	56	47·9	57
1 0 0	32·6	52·0		48·9	
2 0 0	33·1	53·7	56	49·0	56
3 0 0	34·0	53·0		48·5	
4 0 0	34·6	54·1	54	48·8	56
5 0 0	35·0	55·2		49·3	
6 0 0	35·6	56·3	53	50·3	55
7 0 0	36·4	57·3		50·2	
8 0 0	35·9	58·3	52	50·0	54
9 0 0	35·3	55·0		49·4	
10 0 0	35·9	60·2	51	49·9	54
11 0 0	34·6	61·2		49·4	
12 0 0	33·8	60·6	52	48·8	54
13 0 0	32·7	58·5		47·8	
14 0 0	33·4	55·6	54	47·0	56
15 0 0	35·5	51·7		47·3	
16 0 0	37·0	48·6	58	46·8	59
17 0 0	38·1	48·2		46·6	
18 0 0	38·5	48·1	60	46·4	61
19 0 0	37·0	47·6		46·3	
20 0 0	35·7	48·4	59	46·3	60
21 0 0	34·2	49·2		46·5	
22 0 0	34·8	49·6	57	47·5	59
23 0 0	33·9	50·7		47·2	

JULY 23 and 24, 1841.

TORONTO { Decl. 1 Scale Division $= 0'·72$
H. F. $k = ·000076$; $q = ·0002$
V. F. $k = ·000303$; $q = ·00018$

Positions at the usual hours of observation, from July 23d, 0^h. to July 24th, 16^h.

M. Gött. Time (d. h. m. s.)	Decl. Sc.-Div^m.	Hor. Force Sc.-Div^m.	Ther.	Vert. Force Sc.-Div^m.	Ther.
23 0 0 0	43·9	91·5	76	72·6	77
2 0 0	42·0	86·1	76	76·2	77
4 0 0	37·9	53·0	77	72·1	77
6 0 0	28·0	57·9	78	71·4	78
8 0 0	24·1	80·6	79	73·6	79
10 0 0	26·2	90·4	79	74·3	79

* Commencing after Sunday midnight at Sydney.

JULY 23 and 24, 1841.

M. Gött. Time				Decl.	Hor. Force.		Vert. Force.	
d.	h.	m.	s.	Sc.-Div	Sc.-Div	Ther.	Sc.-Div	Ther.
23	12	0	0	29.3	79.9	81	77.4	80
	14	0	0	39.8	60.0	80	74.2	80
	16	0	-0	42.8	45.0	79	67.6	80
	18	0	0	40.0	71.4	78	62.4	79
	20	0	0	33.4	59.9	78	62.5	79
	22	0	0	31.0	71.2	77	54.8	78
24	0	0	0	32.0	56.1	76	44.7	77
	2	0	0		28.4	77	60.6	77
	4	0	0	43.4	61.7	78	68.8	79
	6	0	0	31.2	63.5	79	71.9	80
	8	0	0	25.1	81.3	81	73.8	81
	10	0	0	35.8	30.1	81	84.2	81
	12	0	0	30.0	77.0	80	81.7	81
	14	0	0	35.8	67.9	80	80.0	80
	16	0	0[a]	46.2	64.3	79	75.5	79

The Mean Positions during the Month are given in page 59.

St. Helena $\left\{ \begin{array}{l} \text{Decl. 1 Scale Division} = 0'.71 \\ \text{H. F. } k = .00018; q = .00025 \\ \text{V. F. } k = .0002; q = \end{array} \right.$

Extra observations.
The H. F. was observed at 2m. 30s. after the times specified.

M. Gött. Time				Decl.	Hor. Force.	Vert. Force.
d.	h.	m.	s.	Sc.-Div	Sc.-Div	Ther.
23	23	40	0	29.1	40.0	62
		45	0	29.8		
		50	0	29.8	39.8	
		55	0	29.3		
24	0	5	0	29.0		
		10	0	29.0	39.0	62
		15	0	28.9		
		20	0	28.9	39.0	
		25	0	28.8		
		30	0	28.4	37.9	
		35	0	28.1		
		40	0	28.1	37.1	
		45	0	28.2		
		50	0	28.3	37.1	
		55	0	28.6		
	1	5	0	28.7		
		10	0	28.0	37.0	63
		15	0	28.0		
		20	0	28.0	37.0	
		25	0	28.1		
		30	0	28.6	38.7	
		35	0	28.8		
		40	0	28.6	38.0	
		45	0	28.1		
		50	0	28.1	38.2	
		55	0	28.1		
	2	5	0	28.0		
		10	0	28.0	38.0	63
		15	0	27.7		
		20	0	27.2	38.0	
		25	0	27.0		
		30	0	27.0	38.0	
		35	0	27.0		
		40	0	27.0	37.0	
		45	0	27.0		
		50	0	26.9	36.1	
		55	0	27.1		
	3	5	0	28.0		

JULY 23 and 24, 1841.

M. Gött. Time				Decl.	Hor. Force.		Vert. Force.	
d.	h.	m.	s.	Sc.-Div	Sc.-Div	Ther.	Sc.-Div	Ther.
24	3	10	0	28.0	35.8	63		
		15	0	27.8				
		20	0	27.5	35.9			
		25	0	27.3				
		30	0	27.0	35.3			
		35	0	26.7				
		40	0	26.5	35.5			
		45	0	26.5				
		50	0	26.3	35.0			
		55	0	26.2				
	4	5	0	25.9		63		
		10	0	25.4	35.4			
		15	0	25.4				
		20	0	25.9	34.8			
		25	0	26.0				
		30	0	26.0	34.0			
		35	0	26.0				
		40	0	26.0	34.7			
		45	0	25.9				
		50	0	25.2	34.8			
		55	0	25.1				
	5	30	0	23.2	34.6	64		
	8	5	0	23.9		62		
		10	0	23.9	29.5			
		15	0	23.9				
		20	0	24.0	29.1			
		25	0	24.1				
		30	0	24.1	28.8			
		35	0	24.2				
		40	0	24.2	28.4			
		45	0	24.1				
		50	0	24.4	28.1			
		55	0	24.6				
	9	5	0	24.5				
		10	0	24.2	27.3			
		15	0	24.2				
		20	0	24.0	27.1			
		25	0	23.9				
		30	0	23.9	27.1			
		35	0	23.9				
		40	0	23.9	28.6			
		45	0	24.1				
		50	0	24.7	31.8			
		55	0	25.4				
	10	5	0	25.1		63		
		10	0	24.9	35.4			
		15	0	24.8				
		20	0	24.7	36.6			
		25	0	24.9				
		30	0	24.9	38.0			
		35	0	24.8				
		40	0	25.0	37.2			
		45	0	25.0				
		50	0	25.0	35.7			
		55	0	24.9				
	11	5	0	25.1		62		
		10	0	25.0	33.1			
		15	0	24.9				
		20	0	24.7	32.2			
		25	0	24.7				
		30	0	24.7	31.8			
		35	0	24.7				
		40	0	24.8	32.0			
		45	0	24.9				

JULY 23 and 24, 1841.

M. Gött. Time				Decl.	Hor. Force.		Vert. Force.	
d.	h.	m.	s.	Sc.-Div	Sc.-Div	Ther.	Sc.-Div	Ther.
24	11	50	0	24.9	32.2			
		55	0	24.9				
	12	5	0	24.9		62		
		10	0	24.9	32.3			
		15	0	24.8				
		20	0	24.8	32.2			
		25	0	24.8				
		30	0	24.6	32.1			
		35	0	24.6				
		40	0	24.5	32.6			
		45	0	24.3				
		50	0	24.3	32.3			
		55	0	24.4				

Positions at the usual hours of observation, from July 23rd, 0h. to July 24th, 12h.

M. Gött. Time				Decl.	Hor. Force.		Vert. Force.	
d.	h.	m.	s.	Sc.-Div	Sc.-Div	Ther.	Sc.-Div	Ther.
23	0	0	0	24.9	47.6	62	48.4	62
	2	0	0	22.8	44.2	63	46.9	63
	3	0	0	25.2	41.5	64	42.0	63
	4	0	0	27.1	39.8	64	38.3	63
	5	0	0	26.1	39.0	64	37.9	63
	6	0	0	25.8	38.9	64	38.3	63
	8	0	0	26.0	40.0	63	39.6	62
	10	0	0	25.9	37.1	63	40.7	62
	11	0	0	26.1	34.8	62	41.4	62
	12	0	0	25.2	35.2	62	41.2	62
	13	0	0	26.2	36.2	62	41.8	61
	14	0	0	25.9	36.9	62	42.0	61
	15	0	0	24.8	40.1	62	41.9	61
	16	0	0	26.0	42.0	62	41.8	61
	18	0	0	25.2	42.2	62	41.7	61
	19	30	0	28.5	42.8	62	42.3	61
	20	0	0	31.1	41.5	62	42.8	61
	20	30	0	32.1	42.7	62	43.5	61
	22	0	0	30.6	41.9	61	44.1	61
	23	0	0	30.1	40.0	62	43.2	61
24	0	0	0	29.0	39.1	62	43.4	62
	2	0	0	28.1	38.2	63	40.1	62
	3	0	0	27.1	36.2	63	39.8	62
	4	0	0	26.0	35.0	63	40.2	63
	5	0	0	24.9	34.9	64	40.2	63
	6	0	0	23.2	34.4	63	40.2	63
	8	0	0	23.5	30.0	62	40.8	62
	10	0	0	25.5	34.1	62	42.9	62
	11	0	0	25.1	34.7	62	42.4	62
	12	0	0[b]	24.9	32.0	62	41.8	62

The Mean Positions during the Month are given in page 60.

Cape of Good Hope $\left\{ \begin{array}{l} \text{Decl. 1 Scale Division} = 0'.75 \\ \text{H. F. } k = .00018; q = .0003 \end{array} \right.$

Positions at the usual hours of observation, from July 23rd, 0h. to July 24th, 10h.

M. Gött. Time				Decl.	Hor. Force.		Vert. Force.	
d.	h.	m.	s.	Sc.-Div	Sc.-Div	Ther.		
23	0	0	0	51.2	72.7	56		
	2	0	0	51.5	65.8	58		
	4	0	0	53.0	67.0	59		
	6	0	0	52.0	67.3	59		
	8	0	0	52.1	67.4	59		
	10	0	0	51.4	65.0	59		
	12	0	0	50.4	64.8	59		
	14	0	0	50.5	65.5	59		

[a] Saturday midnight at Toronto.   [b] Saturday midnight at St. Helena.

July 23 and 24, 1841.

M. Gött. Time.				Decl.	Hor. Force.		Vert. Force.	
d.	h.	m.	s.	Sc.-Divns.	Sc.-Divns.	Ther. °	Sc.-Divns.	Ther. °
23	16	0	0	53·5	70·8	58		
	18	0	0	53·8	69·9	58		
	20	0	0	56·9	73·7	58		
	22	0	0	54·0	69·0	58		
24	0	0	0	53·0	58·7	58		
	2	0	0	52·9	59·6	59		
	4	0	0	51·0	58·9	59		
	6	0	0	50·6	58·1	59		
	8	0	0	49·7	55·0	59		
	10	0	0[a]	51·0	60·0	59		

The Mean Positions during the Month are given in page 60.

Van Diemen Island { Decl. 1 Scale Division = 0′·71; H. F. k = ·0003 ; q = ; V. F. k = ; q =

Positions at the usual hours of observation, from July 23rd, 0h. to July 24th, 2h.

d.	h.	m.	s.	Decl.	Hor. Force	Ther.	Vert. Force	Ther.
23	0	0	0	63·7	20·0	46	18·1	45
	1	0	0		22·5		16·4	
	2	0	0	63·3	22·6	44	16·8	44
	3	0	0		21·8		15·6	
	4	0	0	62·1	23·0	44	12·0	44
	5	0	0		22·9		19·8	
	6	0	0	63·2	28·1	42	21·5	43
	7	0	0		28·0		17·3	
	8	0	0	62·5	24·5	42	16·5	43
	9	0	0		26·0		16·6	
	10	0	0	64·2	25·8	41	18·6	41
	11	0	0		23·8		20·4	
	12	0	0	66·5	24·4	41	19·5	42
	13	0	0		23·8		17·3	
	14	0	0	65·0	22·4	43	17·5	43

July 23 and 24, 1841.

M. Gött. Time.				Decl.	Hor. Force.		Vert. Force.	
h.	d.	m.	s.	Sc.-Divns.	Sc.-Divns.	Ther. °	Sc.-Divns.	Ther. °
23	15	0	0		20·4		20·4	
	16	0	0	68·4	19·8	45	20·6	45
	17	0	0		19·7		18·1	
	18	0	0	73·1	18·7	46	21·1	45
	19	0	0		17·0		19·7	
	20	0	0	68·6	19·1	47	16·6	46
	21	0	0		17·5		17·5	
	22	0	0	61·5	13·9	47	19·6	46
	23	0	0		14·4		17·1	
24	0	0	0	57·9	13·9	47	16·9	46
	1	0	0		17·5		15·7	
	2	0	0[b]	62·8	18·3	48	15·7	47

The Mean Positions during the Month are given in page 61.

Antarctic Expedition at Garden Island, Sydney.

Decl. 1 Scale Division = 0′·73
H. F. k = ·000186; q =
V. F. = ; q =

Positions at the usual hours of observation, from July 23rd, 0h. to July 24th, 2h.

d.	h.	m.	s.	Decl.	Hor. Force	Ther.	Vert. Force	Ther.
23	0	0	0	34·4	48·7	54	50·4	55
	1	0	0	33·8	53·7	53	50·4	55
	2	0	0	33·9	54·9	53	49·4	55
	3	0	0	34·1	54·9	53	48·8	55
	4	0	0	33·3	55·7	53	48·7	55
	5	0	0	34·6	55·3	52	49·5	55
	6	0	0	34·3	56·0	52	51·7	54
	7	0	0	39·9	57·6	51	51·5	55
	8	0	0	34·6	60·6	51	47·5	55
	9	0	0	35·0	61·1	52	47·9	55

July 23 and 24, 1841.

M. Gött. Time.				Decl.	Hor. Force.		Vert. Force.	
d.	h.	m.	s.	Sc.-Divns.	Sc.-Divns.	Ther. °	Sc.-Divns.	Ther. °
23	10	0	0	35·0	61·3	51	50·2	55
	11	0	0	33·9	60·5	50	50·5	55
	12	0	0	35·2	61·2	50	51·6	55
	13	0	0	32·9	60·2	52	48·4	55
	14	0	0	34·1	54·0	54	48·9	55
	15	0	0	35·6	47·5	57	47·7	57
	16	0	0	36·6	43·8	61	47·5	61
	17	0	0	37·7	43·4	61	47·6	61
	18	0	0	39·7	41·4	62	48·6	62
	19	0	0	34·9	39·8	61	47·2	62
	20	0	0	38·2	43·3	61	48·0	61
	21	0	0	37·2	42·3	59	48·8	61
	22	0	0	33·5	41·0	57	48·9	60
	23	0	0	29·1	42·7	57	48·0	58
24	0	0	0	30·0	41·5	56	50·5	57
	1	0	0	29·9	47·1	55	50·6	55
	2	0	0[c]	32·3	49·9	54	52·7	54

The Mean Positions from July 19th to July 31st inclusive, are given in page 61.

August 2 and 3, 1841.

Toronto* { Decl. 1 Scale Division = 0′·72; H. F. k = ·000076 q; = ·0002; V. F. k = ·000093 q; = ·00018

Extra observations.

The V. F. was observed at 1m. 30s. before, and the H. F. 2m. 30s. after the times specified.

d.	h.	m.	s.	Decl.	Hor. Force	Ther.	Vert. Force	Ther.
2	16	20	0	153·2	371·3	73	39·7	73
		25	0	155·1	372·8		42·8	
		30	0	151·8	373·6		43·5	

[a] Saturday midnight at the Cape of Good Hope.
[b] Saturday midnight at Van Diemen Island.
[c] Saturday midnight at Sydney.

* Toronto, August, 1841.—*Times of observation at which the magnetometers were disturbed, but the mean readings were not materially changed.*

d.	h.	
3	2	Decl., H. F., and H. V., moderately disturbed by shocks.
	12	Decl. and H. F. moderately disturbed by shocks.
	14	Decl. and H. F. slightly disturbed by shocks.
5	6	H. F. slightly disturbed and vibrating.
	10	H. F. moderately disturbed, vibrating with shocks.
	18	V. F. vibrating slightly.
	22	Decl. and H. F. considerable vibrations and shocks.
6	0	H. F. slightly disturbed, vibrating.
	6, 8, 10	Decl. and H. F. moderately disturbed with shocks.
	12	H. F. and V. F. considerably disturbed, vibrating.
	14	Decl. and H. F. considerably disturbed, vibrating.
7	0	Decl. much, H. F. moderately disturbed, vibrating.
	2	H. F. moderately disturbed, with shocks.
10	0	Decl. and H. F. much disturbed by shocks; V. F. vibrating very much.
11	2	Decl., slight shocks; H. F. slight vibrations.
	18	H. F. slight vibrations.
	20, 22	Decl. slight vibrations; H. F. moderate vibrations and shocks.
12	0	H. F. slight vibrations.
	2	Decl. slight shocks; H. F. slight vibrations.
13	0	Decl. moderate, vibrations and shocks; V. F. vibrating much.
	2	H. F. vibrating much, with shocks.
	20	V. F. considerable vibrations.
14	2	H. F. disturbed by shocks.
	16	H. F. and V. F. moderate vibrations; H. F. shocks.
15	18, 20	H. F. vibrating slightly.
16	0	Decl. and H. F. slight shocks.
	14	Decl. and H. F. moderate vibrations and shocks.
	16	H. F. moderate vibrations and shocks.
	20	H. F. slight vibrations.
17	8	H. F. moderate vibrations and shocks.
17	10	Decl. and H. F. moderate vibrations.
	16	V. F. slight vibrations.
18	12	H. F. slight vibrations.
	14	V. F. slight vibrations.
19	0	Decl. and H. F. moderate shocks.
	8	H. F. vibrating slightly.
	10	Decl. and H. F. vibrating slightly.
20	0	Decl. and H. F. moderate vibrations and shocks; V. F. moderate vibrations.
	2	H. F. vibrating much.
	14	V. F. vibrating much.
21	0	Decl. and H. F. moderate shocks; V. F. vibrating much.
	8 & 10	H. F. moderate shocks; V. F. slight shocks at 10h.
	16	Decl. slight shocks.
22	18	Decl. and H. F. slight vibrations.
	20	V. F. moderate vibrations.
	22	Decl. moderate shocks.
23	0 & 2	Decl. and H. F. vibrating much, with shocks.
	10	H. F. moderate vibrations and shocks.
	20	H. F. vibrating slightly.
24	0	Decl. moderate vibrations and shocks; H. F. vibrating very much; V. F. vibrating much.
	2	Decl. slight shocks; H. F. vibrating much, with shocks.
	22	Decl., H. F., and V. F. moderate vibrations and shocks.
25	2	Decl. slight shocks, H. V. moderate vibrations and shocks
	4	V. F. vibrating slightly.
	18	V. F. vibrating much.
	20	H. F. vibrating much, with shocks.
	22	Decl. and H. F. vibrations and shocks; V. F. vibrations.
26	4	H. F. moderate vibrations.
	14	V. F. vibrating much.
	16	H. F. vibrations.
	20	H. F. slight vibrations.
27	2	V. F. vibrating much.
28	4	Decl. slight vibrations.
30	6	H. F. moderate vibrations.
	20 & 22	Decl. moderate vibrations and shocks.
31	4	H. F. vibrations and shocks.
	8	H. F. slight vibrations.
	10	V. F. moderate vibrations.

AUGUST 2 and 3, 1841.

M. Gött. Time.				Decl.	Hor. Force.		Vert. Force.	
d.	h.	m.	s.	Sc.-Div^{ns}	Sc.-Div^{ns}	Ther.	Sc.-Div^{ns}	Ther.
2	16	35	0	146·9	374·8		44·5	
		40	0	145·0	376·5		45·	
		45	0	143·7	377·6		46·5	
		50	0	145·0	374·6		47·5	
		55	0	146·5	372·1		46·5	
	17	0	0	149·5	372·8	73	45·6	73
		5	0	150·9	368·7		44·1	
		10	0	151·1	364·2		42·2	
		15	0	153·4	362·8		41·3	
		20	0	153·0	380·9		40·5	
		25	0	153·1	360·1		40·0	
		30	0	154·0	362·3		41·1	
		35	0	153·0	363·0		41·6	
		40	0	152·7	363·4		41·3	
		45	0	151·3	365·0		41·9	
		50	0	149·4	362·8		41·5	

Positions at the usual hours of observation, August 2 and 3.

d.	h.	m.	s.	Decl.	Hor. Force	Ther.	Vert. Force	Ther.
2	0	0	0	150·3	392·7	67	54·7	67
	2	0	0	146·1	387·5	68	54·6	68
	4	0	0	139·0	381·3	70	52·2	69
	6	0	0	123·7	384·3	71	51·5	71
	8	0	0	126·6	390·0	73	51·1	71
	10	0	0	135·0	398·4	74	51·7	72
	12	0	0	137·5	390·3	75	56·3	73
	14	0	0	151·6	352·6	74	61·8	73
	16	0	0	167·4	364·9	73	32·7	73
	18	0	0	146·9	360·1	72	45·9	73
	20	0	0	139·4	385·0	71	53·9	71
	22	0	0	141·4	390·7	70	53·7	70
3	0	0	0	149·6	399·4	69	53·0	70
	2	0	0	147·8	385·6	71	46·4	71
	4	0	0	144·8	364·9	72	48·6	71
	6	0	0	132·6	377·5	73	51·7	72
	8	0	0	129·9	378·0	73	51·9	72
	10	0	0	133·0	383·0	74	55·6	73
	12	0	0	137·1	370·4	74	58·9	73
	14	0	0	138·4	374·0	74	49·4	73
	16	0	0	137·5	377·2	73	52·1	73
	18	0	0	134·9	361·9	73	43·1	72
	20	0	0	141·9	362·0	72	41·3	72
	22	0	0	140·0	364·4	71	45·3	71

Mean Positions at the same hours during the Month.

d.	h.	m.	s.	Decl.	Hor. Force	Ther.	Vert. Force	Ther.
	0	0	0	143·8	391·5	68	53·8	68
	2	0	0	145·5	385·4	69	51·7	69
	4	0	0	137·8	368·9	70	51·6	70
	6	0	0	127·8	374·8	71	51·3	70
	8	0	0	126·9	390·3	72	51·8	71
	10	0	0	131·9	395·2	73	52·1	72
	12	0	0	133·8	387·2	73	53·7	72
	14	0·0	0	139·7	375·6	72	52·0	72
	16	0	0	142·3	375·3	71	49·0	71
	18	0	0	138·4	379·8	70	48·9	70
	20	0	0	138·6	381·1	69	47·5	70
	22	0	0	137·7	386·6	69	50·3	69

St. Helena { Decl. 1 Scale Division = 0'·71
H. F. k = ·00018 q ; = ·00025
V. F. k = ·00017 q ; =

Regular and extra observations.
The V. F. was observed at 2^m. 30^s. before, and the H. F.
2^m. 30^s. after the times specified.

d.	h.	m.	s.	Decl.	Hor. Force	Ther.	Vert. Force	Ther.
2	0	0	0	25·2	48·2	64	39·2	63
	2	0	0	22·9	46·2	64	37·5	64

AUGUST 2 and 3, 1841.

d.	h.	m.	s.	Decl. Sc.-Div^{ns}	Hor. Force Sc.-Div^{ns}	Ther.	Vert. Force Sc.-Div^{ns}	Ther.
2	3	0	0	23·7	44·1	64	38·5	64
	4	0	0	25·9	44·0	65	39·3	64
	5	0	0	24·9	41·9	65	37·1	64
	6	0	0	23·8	41·3	65	36·5	64
	8	0	0	25·0	40·1	64	36·5	64
	10	0	0	25·2	42·0	64	36·1	63
	11	0	0	25·2	41·1	64	36·1	63
	12	0	0	24·3	37·9	64	35·3	63
		10	0	24·2	36·5			
		15	0	24·8				
		20	0	24·9	35·4		35·5	
		25	0	25·0				
		30	0	25·1	34·9		35·9	
		35	0	25·1				
		40	0	25·2	35·1		35·9	
		45	0	25·2				
		50	0	25·0	37·0		35·6	
		55	0	24·9				
	13	0	0	24·7	38·0	64	35·5	63
		10	0	23·9	38·8		35·4	
		20	0	23·9	38·1		35·4	
		25	0	23·9				
		30	0	23·5	37·9		35·6	
		35	0	22·9				
		40	0	23·0	36·8		33·7	
		45	0	23·9				
		50	0	24·0	36·3		35·7	
		55	0	24·1				
	14	0	0	24·1	35·9	64	35·8	
		5	0	24·1				63
		10	0	24·0	35·3		35·8	
		15	0	23·9				
		20	0	23·8	35·0		35·6	
		25	0	23·8				
		30	0	23·8	34·7		35·6	
		35	0	23·6				
		40	0	23·7	34·4		35·0	
		45	0	23·9				
		50	0	23·9	34·0			
		55	0	23·8				
	15	0	0	23·8	33·7	64	35·9	
		5	0	23·9				63
		10	0	23·0	33·7		35·9	
		15	0	23·0				
		20	0	24·0	34·4		35·8	
		25	0	24·2				
		30	0	24·1	36·2		35·8	
		35	0	24·1				
		40	0	24·9	40·0		36·6	
		45	0	24·8				
		50	0	24·8	41·1		36·5	
		55	0	24·8				
	16	0	0	24·9	42·0	64	36·6	
		5	0	25·0				63
		10	0	25·6	41·8		36·6	
		15	0	26·0				
		20	0	26·0	41·2		36·8	
		25	0	26·0				
		30	0	25·9	41·0		36·8	
		35	0	25·9				
		40	0	25·7	41·0		36·8	
		45	0	25·7				
		50	0	25·8	41·0		36·8	
		55	0	25·4				
	17	0	0	25·2	41·0	63	36·8	
		10	0	25·0	41·2		36·8	

AUGUST 2 and 3, 1841.

d.	h.	m.	s.	Decl. Sc.-Div^{ns}	Hor Force Sc.-Div^{ns}	Ther.	Vert. Force Sc.-Div^{ns}	Ther.
2	17	15	0	24·9				
		20	0	24·9	41·1			
		25	0	25·0				
		30	0	25·0	41·0		36·6	
		35	0	24·9				
		40	0	24·9	40·9		36·5	
		45	0	24·9				
		50	0	25·0	40·1		36·5	
		55	0	25·0				
	18	0	0	25·0	40·0	63	36·5	62
		5	0	25·0				
		10	0	25·0	41·0		36·5	
		15	0	25·0				
		20	0	25·0	40·4		36·5	
		25	0	25·1				
		30	0	25·8	40·2		36·5	
		35	0	26·2				
		40	0	26·7	40·3		36·5	
		45	0	26·9				
		50	0	27·0	40·9		36·5	
		55	0	27·0				
	19	0	0	27·1	41·0	63	36·5	62
		5	0	27·3				
		10	0	27·9	40·9		36·5	
		15	0	28·0				
		20	0	28·1	41·0		36·5	
		25	0	28·9				
		30	0	29·1	40·9		36·7	
		35	0	29·2				
		40	0	29·7	41·1			
		45	0	29·9				
		50	0	29·9	42·1		36·9	
		55	0	30·1				
	20	0	0	31·0	42·0	63	37·0	62
		5	0	31·9	42·0		37·6	
		10	0	31·9	42·0		37·6	
		15	0	32·2				
		20	0	32·1	42·0		37·6	
		25	0	32·0				
		30	0	31·9	42·1		37·5	
		35	0	31·9				
		40	0	31·4	42·1		37·7	
		45	0	31·1				
		50	0	31·1	42·8		37·7	
		55	0	30·8				
	21	0	0	30·1	43·1	63	37·9	62
		5	0	30·0				
		10	0	29·9	44·0		37·9	
		15	0	29·5				
		20	0	29·4	44·9		37·9	
		25	0	29·0				
		30	0	29·0	46·0		37·9	
		35	0	29·3				
		40	0	29·4	45·9		38·1	
		45	0	29·0				
		50	0	29·0	46·0		38·1	
		55	0	28·9				
	22	0	0	28·2	47·0	63	38·2	63
		5	0	28·2				
		10	0	28·2	46·9		38·5	
		15	0	28·1				
		20	0	28·0	46·9		38·3	
		25	0	28·0				
		30	0	27·9	46·9		38·3	
		35	0	27·6				
		40	0	27·1	46·5		38·3	
		45	0	27·0				

August 2 and 3, 1841.

M. Gött. Time				Decl.	Hor. Force.		Vert. Force.	
d.	h.	m.	s.	Sc.-Div^ns	Sc.-Div^ns	Ther.	Sc.-Div^ns	Ther.
2	22	50	0	27·0	46·9		38·2	
		55	0	27·4				
	23	0	0	27·9	49·8	63	38·8	63
		5	0	28·0				
		10	0	27·1	50·0		38·9	
		15	0	27·0				
		20	0	27·0	49·1		38·8	
		25	0	27·0				
		30	0	27·3	49·2		38·9	
		35	0	27·1				
		40	0	26·6	50·0		38·9	
		45	0	26·1				
		50	0	26·0	49·1		38·7	
		55	0	25·5				
3	0	0	0	25·2	48·8	64	38·4	63
		5	0	25·1				
		10	0	25·5	49·0		38·4	
		15	0	25·1				
		20	0	25·7	50·0		38·7	
		25	0	25·7				
		30	0	26·9	50·0		38·6	
		35	0	26·3				
		40	0	26·0	48·3		38·9	
		45	0	26·0				
		50	0	26·0	47·8		38·1	
		55	0	26·1				
	1	0	0	26·3	47·1	64	38·3	64
		5	0	26·0				
		10	0	25·0	46·0		37·9	
		15	0	24·9				
		20	0	24·3	45·9		37·7	
		25	0	24·3				
		30	0	24·9	46·0		37·7	
		35	0	25·0				
		40	0	24·8	45·6		37·8	
		45	0	24·1				
		50	0	24·0	44·1		37·8	
		55	0	24·1				
	2	0	0	24·3	43·7	64	37·8	64
		5	0	24·7				
		10	0	24·9	44·0		38·0	
		15	0	24·9				
		20	0	24·2	43·7		38·0	
		25	0	23·9				
		30	0	23·1	41·9		37·7	
		35	0	23·2				
		40	0	23·0	41·0		37·6	
		45	0	23·0				
		50	0	23·3	40·6		37·4	
		55	0	23·6				
	3	0	0	23·7	40·0	65	37·3	65
		5	0	23·5				
		10	0	23·3	40·2		37·3	
		15	0	23·1				
		20	0	22·6	39·9		37·1	
		25	0	22·1				
		30	0	22·2	39·2		37·0	
		35	0	22·4				
		40	0	22·5	39·1		36·3	
		45	0	22·9				
		50	0	23·2	39·9		36·0	
		55	0	23·4				
	4	0	0	23·7	39·1	66	35·7	65
		5	0	23·6				
		10	0	23·5	38·6		34·6	
		15	0	23·5				
		20	0	23·3	38·0		34·3	

August 2 and 3, 1841.

M. Gött. Time				Decl.	Hor. Force.		Vert. Force.	
d.	h.	m.	s.	Sc.-Div^ns	Sc.-Div^ns	Ther.	Sc.-Div^ns	Ther.
3	4	25	0	23·1				
		30	0	23·1	37·8		33·8	
		35	0	23·0				
		40	0	23·0	37·6		33·4	
		45	0	23·0				
		50	0	23·0	37·8		32·9	
		55	0	23·0				
	5	0	0	23·0	37·6	66	32·4	65
		5	0	22·9				
		10	0	22·9	38·0		31·8	
		15	0	23·0				
		20	0	23·0	38·5		31·8	
		25	0	23·1				
		30	0	23·2	38·2		31·8	
		35	0	23·2				
		40	0	23·5	38·0		31·6	
		45	0	23·7				
		50	0	23·8	38·4		31·5	
		55	0	24·0				
	6	0	0	24·1	38·1	66	31·6	65
		5	0	24·3				
		10	0	24·4	38·1			
		15	0	24·2				
		20	0	24·2	38·0			
		25	0	24·2				
		30	0	24·3	37·7		31·8	
		35	0	24·3				
		40	0	24·3	37·1		31·8	
		45	0	24·1				
		50	0	24·1	37·1		31·8	
		55	0	24·2				
	7	0	0	24·2	37·9		32·3	
		5	0	24·5				
		10	0	24·8	37·0		32·3	
		15	0	24·8				
		20	0	24·7	36·7		32·2	
		25	0	24·2				
		30	0	24·2	36·2			
		35	0	24·0				
		40	0	24·0	36·0		32·2	
		45	0	24·1				
		50	0	24·1	35·9		32·2	
		55	0	24·1				
	8	0	0	24·1	35·1	65	32·2	65
		5	0	24·1				
		10	0	24·1	35·1		32·2	
		15	0	24·1				
		20	0	24·5	35·6		32·1	
		25	0	24·7				
		30	0	24·9	35·9		32·1	
		35	0	24·9				
		40	0	24·9	36·0		31·9	
		45	0	24·9				
		50	0	24·9	35·2		31·9	
		55	0	24·9				
	9	0	0	25·1	35·1		31·9	
		5	0	25·0				
		10	0	24·9	35·1		31·9	
		15	0	24·8				
		20	0	24·2	35·0		31·8	
		25	0	24·3				
		30	0	24·6	35·4		31·8	
		35	0	24·7				
		40	0	24·8	36·0		31·7	
		45	0	24·8				
		50	0	24·6	36·3		31·9	
		55	0	24·4				

August 2 and 3, 1841.

M. Gött. Time				Decl.	Hor. Force.		Vert. Force.	
d.	h.	m.	s.	Sc.-Div^ns	Sc.-Div^ns	Ther.	Sc.-Div^ns	Ther.
3	10	0	0	24·6	37·0	65	32·9	65
		5	0	24·8				
		10	0	24·9	37·2		33·2	
		15	0	24·9				
		20	0	24·9	37·2		32·2	
		25	0	25·0				
		30	0	25·0	37·0		32·2	
		35	0	25·4				
		40	0	25·3	37·1		32·5	
		45	0	25·3				
		50	0	25·4	37·3		32·3	
		55	0	25·4				
	11	0	0	25·4	36·9	65	32·3	64
		5	0	25·6				
		10	0	25·0	35·2		32·0	
		15	0	25·0				
		20	0	25·0	35·9		32·2	
		25	0	25·0				
		30	0	25·0	35·4		32·0	
		35	0	24·9				
		40	0	24·8	34·9		32·0	
		45	0	25·0				
		50	0	25·0	35·4		32·0	
		55	0	25·5				
	12	0	0	25·3	36·0	65	31·7	64
		5	0	25·7				
		10	0	25·9	36·2		31·9	
		15	0	26·0				
		20	0	26·0	36·6		33·1	
		25	0	26·2				
		30	0	26·4	37·0		33·0	
		35	0	26·3				
		40	0	26·3	37·1		32·8	
		45	0	26·2				
		50	0	26·2	37·6		33·0	
		55	0	26·2				
	13	0	0	26·1	38·1	65	33·5	64
		5	0	26·0				
		10	0	25·9	38·2		33·2	
		15	0	25·8				
		20	0	25·6	38·0		32·6	
		25	0	25·2				
		30	0	25·2	37·2		32·2	
		35	0	25·1				
		40	0	25·2	37·2		32·2	
		45	0	25·9				
		50	0	26·1	37·4		32·2	
		55	0	26·1				
	14	0	0	27·0	38·9	64	33·0	64
		5	0	27·0				
		10	0	27·0	40·0		33·0	
		15	0	26·9				
		20	0	26·9	40·0		33·2	
		25	0	26·9				
		30	0	26·9	40·8		33·2	
		35	0	26·9				
		40	0	26·9	40·5		33·2	
		45	0	26·9				
		50	0	26·9	40·0		33·2	
		55	0	25·8				
	15	0	0	26·8	39·9	64	33·2	63
		5	0	26·8				
		10	0	26·8	40·0		33·2	
		15	0	26·5				
		20	0	26·2	40·0		33·1	
		25	0	26·1	*			
		30	0	26·1	39·5		33·0	

K

AUGUST 2 and 3, 1841.

d.	h.	m.	s.	Decl. Sc.-Div	Hor. Force Sc.-Div	Ther.	Vert. Force Sc.-Div	Ther.
3	15	35	0	26·1				
		40	0	26·1	39·9		33·0	
		45	0	26·2				
		50	0	26·2	39·8		33·0	
		55	0	26·8				
	16	0	0	26·8	39·8	64	33·0	63
		5	0	26·1				
		10	0	26·1	39·8		33·4	
		15	0	26·1				
		20	0	26·1	39·8		33·2	
		25	0	26·1				
		30	0	26·1	39·9		33·2	
		35	0	26·1				
		40	0	26·1	40·0		33·1	
		45	0	26·1				
		50	0	26·2	40·0		33·5	
		55	0	26·2				
	17	0	0	26·2	40·1	64	33·5	
		5	0	26·1				
		10	0	26·1	40·1		33·4	
		15	0	26·1				
		20	0	26·1	39·9		33·4	
		25	0	26·1				
		30	0	26·1	40·1		33·2	
		35	0	26·1				
		40	0	26·1	41·0		33·2	
		45	0	26·1				
		50	0	26·1	41·0		33·2	
		55	0	26·1				
	18	0	0	26·1	41·0	64	33·1	63
		5	0	26·1				
		10	0	26·5	41·1			
		15	0	26·2				
		20	0	26·7	41·1		33·6	
		25	0	27·0				
		30	0	26·8	41·1		33·4	
		35	0	27·0				
		40	0	27·1	41·7		33·4	
		45	0	27·1				
		50	0	27·6	41·9		33·4	
		55	0	27·9				

Mean Positions at the same hours of observation during the Month.[a]

d.	h.	m.	s.	Decl.	Hor. Force	Ther.
	0	0	0	24·4	52·4	62
	2	0	0	24·7	50·4	63
	3	0	0	25·2	48·4	63
	4	0	0	25·8	46·5	63
	5	0	0	25·7	44·9	64
	6	0	0	25·4	43·8	64
	8	0	0	25·6	42·9	63
	10	0	0	25·9	42·8	62
	11	0	0	26·3	42·7	63
	12	0	0	26·2	43·6	62
	13	0	0	26·1	44·5	62
	14	0	0	26·0	44·0	62
	15	0	0	25·4	45·6	62
	16	0	0	27·1	46·1	62
	18	0	0	30·1	46·4	62
	19	30	0	30·3	46·8	62
	20	0	0			

AUGUST 2 and 3, 1841.

d.	h.	m.	s.	Decl. Sc.-Div	Hor. Force Sc.-Div	Ther.	Vert. Force Sc.-Div	Ther.
	20	30	0	29·6	47·3	62		
	22	0	0	26·3	49·7	62		
	23	0	0	25·3	51·6	62		

CAPE OF GOOD HOPE.[b]

VAN DIEMEN ISLAND { Decl. 1 Scale Division = 0'·71 ; H. F. k = ·0003; q = ; V. F. k = ; q = }

Extra observations.

The V. F. was observed at 2m. 30s. before, and the H. F. 2m. 30s. after the times specified.

d.	h.	m.	s.	Decl. Sc.-Div	Hor. Force Sc.-Div	Ther.	Vert. Force Sc.-Div	Ther.
3	1	20	0	54·2	31·5	50		
		25	0	55·7				
		30	0	53·6	31·5		65·0	50
		35	0	57·2				
		40	0	58·4	31·7		64·6	
		45	0	55·6				
		52	30	51·8	32·2		64·1	
	2	7	30	50·4	31·6	50	62·5	50
		15	0	50·2	30·2		61·9	
		22	30	51·6			63·8	
		30	0	54·4	31·1		65·2	
		37	30	55·7	32·1		65·9	
		45	0	55·5	32·2		64·8	

Positions at the usual hours of observation, August 2 and 3.

d.	h.	m.	s.	Decl. Sc.-Div	Hor. Force Sc.-Div	Ther.	Vert. Force Sc.-Div	Ther.
2	0	0	0	57·3	34·4	49	66·5	48
	1	0	0	61·8	39·3		60·7	
	2	0	0	51·9	35·4	47	65·6	46
	3	0	0	57·2	36·3		60·9	
	4	0	0	57·3	36·6	46	62·2	47
	5	0	0	57·3	37·3		64·5	
	6	0	0	57·8	38·3	46	66·7	45
	7	0	0	56·6	38·6		65·6	
	8	0	0	56·7	39·8	45	65·6	45
	9	0	0	57·2	41·1		66·2	
	10	0	0	57·9	41·1	44	65·0	44
	11	0	0	56·1	40·0		64·5	
	12	0	0	57·6	39·6	44	67·0	44
	13	0	0	55·2	38·5		64·1	
	14	0	0	59·6	35·8	46	67·2	46
	15	0	0	62·5	35·6		64·7	
	16	0	0	64·8	33·1	50	63·9	49
	17	0	0	63·8	35·5		58·7	
	18	0	0	67·1	35·2	50	61·9	50
	19	0	0	67·8	35·8		60·2	
	20	0	0	68·9	37·1	50	60·4	50
	21	0	0	62·7	36·4		57·3	
	22	0	0	62·5	34·8	52	60·1	50
	23	0	0	55·2	35·2		60·0	
3	0	0	0	59·2	34·5	50	64·9	49
	1	0	0	51·0	31·1		64·5	
	2	0	0	51·0	32·0	50	62·9	49
	3	0	0	53·5	33·2		62·0	
	4	0	0	53·4	40·6	50	64·1	49
	5	0	0	54·0	35·6		59·5	
	6	0	0	57·9	35·4	49	63·6	48
	7	0	0	57·6	36·0		63·1	

AUGUST 2 and 3, 1841.

d.	h.	m.	s.	Decl. Sc.-Div	Hor. Force Sc.-Div	Ther.	Vert. Force Sc.-Div	Ther.
3	8	0	0	57·8	36·2	48	63·3	48
	9	0	0	58·4	36·0		63·3	
	10	0	0	60·6	36·9	48	63·6	47
	11	0	0	59·9	38·5		62·4	
	12	0	0	61·5	36·4	47	64·7	47
	13	0	0	59·4	35·7		59·6	
	14	0	0	57·2	36·3	48	64·7	47
	15	0	0	60·4	36·5		59·2	
	16	0	0	62·1	36·8	49	60·5	48
	17	0	0	62·7	36·2		62·3	
	18	0	0	61·3	35·4	50	63·7	48
	19	0	0	65·5	28·8		73·0	
	20	0	0	61·2	32·5	49	64·3	48
	21	0	0	61·3	33·6		66·0	
	22	0	0	59·5	34·0	48	64·3	47
	23	0	0	56·3	38·6		61·9	

Mean Positions at the same hours during the Month.

d.	h.	m.	s.	Decl. Sc.-Div	Hor. Force Sc.-Div	Ther.	Vert. Force Sc.-Div	Ther.
	0	0	0	60·0	35·4	51	58·5	50
	1	0	0	58·8	35·6		59·1	
	2	0	0	58·0	36·2	50	58·3	49
	3	0	0	59·5	36·3		58·1	
	4	0	0	59·1	36·9	49	58·9	49
	5	0	0	59·7	37·4		60·3	
	6	0	0	59·6	37·5	48	60·6	48
	7	0	0	61·1	37·9		61·2	
	8	0	0	61·3	38·1	48	60·8	47
	9	0	0	61·7	38·7		60·5	
	10	0	0	61·6	39·0	47	59·9	47
	11	0	0	61·5	38·7		60·3	
	12	0	0	60·3	38·2	48	58·6	48
	13	0	0	60·0	36·9		57·6	
	14	0	0	60·9	34·9	50	57·6	50
	15	0	0	62·9	34·1		57·2	
	16	0	0	64·6	33·4	52	57·4	51
	17	0	0	66·0	34·1		57·5	
	18	0	0	66·4	34·4	53	57·3	52
	19	0	0	65·9	34·4		57·4	
	20	0	0	64·4	34·5	53	58·3	52
	21	0	0	63·0	34·5		57·9	
	22	0	0	61·8	34·8	52	57·8	51
	23	0	0	60·5	35·9		57·6	

AUGUST 5, 6, and 7, 1841.

TORONTO { Decl. 1 Scale Division = 0'·72 ; H. F. k = ·000076; q = ·0002 ; V. F. k = ·000093; q = ·00018 }

Extra observations.

The V. F. was observed at 1m. 30s. before, and the H. F. 2m. after the times specified.

d.	h.	m.	s.	Decl. Sc.-Div	Hor. Force Sc.-Div	Ther.	Vert. Force Sc.-Div	Ther.
6	2	20	0	125·3	359·7	70	39·7	69
		25	0	122·5	359·2		40·1	
		30	0	120·7	362·8		40·4	
		35	0	121·3	362·7		40·2	
		40	0	122·5	363·0		40·1	
		45	0	123·6	313·7		42·2	
		50	0	126·6	376·3		42·8	
		55	0	128·5	375·1		44·5	
3	0	0	0	127·8	382·0	70	44·5	70

[a] The connexion of the observations with the V. F. magnet at different periods of the month, is too uncertain to allow of mean positions being deduced.

[b] The magnetometers were employed in experimental determinations on the 2nd and 3rd of August.

AUGUST 5, 6, and 7, 1841.

M. Gött. Time			Decl.	Hor. Force		Vert. Force	
d.	h. m. s.		Sc.-Div	Sc.-Div	Ther.	Sc.-Div	Ther.
6	3 5 0		129·5	383·3		46·2	
	10 0		132·0	382·4		45·5	
	15 0		133·1	385·1		45·8	
	20 0		134·8	383·2		48·0	
	25 0		136·5	380·1		47·3	
	30 0		135·8	387·6		47·3	
	35 0		137·4	384·2		47·2	
	40 0		136·9	384·3		47·4	
	45 0		137·9	383·5		48·1	
	50 0		138·4	382·9		48·1	
	14 30 0*		126·3	321·9	73	51·3	72
	35 0		134·0	331·5		69·9	
	40 0		149·0	345·4		78·0	
	45 0			340·9		73·4	
	50 0		156·9	333·2		61·1	
	55 0		161·0	337·2		50·0	
	15 0 0		168·6	346·4	73	40·5	72
	5 0		152·9	370·5		30·2	
	10 0		99·5	327·7		19·0	
	15 0		91·7	328·7		32·6	
	20 0		134·2	370·0		59·3	
	25 0		137·1	379·7		66·0	
	30 0		140·7	361·7		61·5	
	35 0		149·3	373·7		60·7	
	40 0		134·1	364·0		54·3	
	45 0		133·5	360·6		55·7	
	50 0		136·7	359·8		58·4	
	55 0		135·9	359·3		57·2	
	16 5 0		137·3	355·9	72	56·4	72
	10 0		143·3	358·8		56·0	
	15 0		148·1	362·9		56·1	
	20 0		147·0	363·6		53·4	
	25 0		143·7	359·3		49·2	

Positions at the usual hours of observation, August 5, 6, and 7.

5	0 0 0		138·5	380·8	70	51·6	70
	2 0 0		147·9	387·4	70	52·5	70
	4 0 0		142·2	350·5	72	48·0	71
	6 0 0		131·9	373·1	73	49·7	72
	8 0 0		126·1	385·8	74	52·7	73
	10 0 0		129·6	378·1	75	58·7	73
	12 0 0		128·0	368·2	75	52·3	73
	14 0 0		136·8	370·0	73	51·0	73
	16 0 0		145·3	351·0	71	53·1	72
	18 0 0		146·5	364·4	70	45·4	71
	20 0 0		148·6	361·5	69	41·0	69
	22 0 0		148·7	349·0	68	42·5	68
6	0 0 0		147·7	382·7	67	54·0	68
	2 0 0		123·8	349·1	69	37·9	69
	4 0 0		142·1	382·9	71	49·4	70
	6 0 0		117·0	365·7	72	53·6	71

AUGUST 5, 6, and 7, 1841.

M. Gött. Time			Decl.	Hor. Force		Vert. Force	
d.	h. m. s.		Sc.-Div	Sc.-Div	Ther.	Sc.-Div	Ther.
6	8 0 0		127·3	399·2	73	66·0	71
	10 0 0		148·2	415·5	73	66·9	71
	12 0 0		125·5	454·6	74	80·7	72
	14 0 0		173·9	337·5	73	40·7	72
	16 0 0		136·3	356·6	71	57·6	72
	18 0 0		139·0	371·3	70	55·2	71
	20 0 0		138·7	382·8	69	45·7	69
	22 0 0		143·0	394·5	69	43·0	69
7	0 0 0		141·9	360·1	66	51·9	67
	2 0 0		149·0	381·9	68	56·4	68
	4 0 0		141·2	361·9	70	53·1	69
	6 0 0		127·9	373·8	71	54·7	70
	8 0 0		128·3	381·8	72	53·1	71
	10 0 0		132·7	386·4	73	52·3	72
	12 0 0		133·8	376·9	73	52·4	72
	14 0 0		134·2	371·1	73	52·1	73
	16 0 0*		134·5	376·0	70	53·8	71

The Mean Positions at the same hours during the Month are given in page 64.

ST. HELENA { Decl. 1 Scale Division = 0'·71
H. F. k = ·00018; q = ·00025
V. F. k = ·00017; q =

Extra observations.

The V. F. was observed at 2m. 30s. before, and the H. F. 2m. 30s. after the times specified.

			Decl.	Hor. Force		Vert. Force	
6	2 15 0		26·4		64		63
	20 0		26·1	40·6		30·5	
	25 0		26·0			30·5	
	30 0		25·9	39·8		30·5	
	35 0		25·5				
	40 0		25·6	40·1		30·5	
	45 0		25·2				
	50 0		25·1	40·2		30·5	
	55 0		25·1				
3	0 0		25·1	37·9	64	30·0	
	5 0		24·9				
	10 0		24·3	37·7		29·5	
	15 0		24·3				
	20 0		24·1	38·0		29·0	
	25 0		24·2				
	30 0		24·2	38·1		29·3	
	35 0		24·4				
	40 0		24·8	38·5		29·3	
	45 0		24·8				
	50 0		24·8	39·1		29·3	
	55 0		24·8				
4	0 0		24·7	39·8	64	29·3	64
	5 0		24·7				
	10 0		24·7	39·0		29·3	

AUGUST 5, 6, and 7, 1841.

M. Gött. Time			Decl.	Hor. Force		Vert. Force	
d.	h. m. s.		Sc.-Div	Sc.-Div	Ther.	Sc.-Div	Ther.
6	4 15 0		24·8				
	20 0		24·8	38·9		29·3	
	25 0		24·5				
	30 0		24·7	37·8		29·3	
	35 0		24·1				
	40 0		24·0	36·9		29·3	
	45 0		23·9				
	50 0		23·6	35·9		29·1	
	55 0		23·1				
5	0 0		23·0	35·0	65	29·1	
	5 0		22·9				
	10 0		22·9	33·2		29·1	
	15 0		22·9				
	20 0		22·2	32·1		28·6	
	25 0		22·3				
	30 0		22·1	31·7		27·6	
	35 0		21·7				
	40 0		21·9	30·5		27·3	
	45 0		22·2				
	50 0		22·1	29·8		27·3	
	55 0		22·5				
6	0 0		22·3	29·0	65	27·0	64
	5 0		22·1				
	10 0		22·1	28·9		26·8	
	15 0		22·1				
	20 0		22·2	28·4		26·8	
	25 0		22·7				
	30 0		22·9	28·0		26·6	
	35 0		23·0				
	40 0		23·0	27·0		26·5	
	45 0		23·1				
	50 0		23·1	26·0		26·7	
	55 0		23·0				
7	0 0		23·0	25·1	65	26·4	
	5 0		22·9				
	10 0		22·8	24·8		25·8	
	15 0		22·9				
	20 0		23·0	23·4		25·7	
	25 0		22·9				
	30 0		22·9	22·7		25·6	
	35 0		22·9				
	40 0		22·6	22·8		25·7	
	45 0		22·4				
	50 0		22·9	22·1		25·7	
	55 0		22·9				
8	0 0		22·9	22·5	65	25·9	65
	5 0		23·0				
	10 0		23·9	21·0		25·9	
	15 0		24·1				
	20 0		24·0	20·0		25·9	
	25 0		24·2				
	30 0		24·2	20·1		25·8	
	35 0		24·0				

---

h. m.
6 14 0　Wind S.W. very light. Cirri dispersed about; low range of cirro-strati in North; dense haze in South horizon; fair.
　30　Bright streamers in East extending to zenith; became fainter in a few minutes; bright bank of light in North, altitude 45°.
　45　Bank brighter in North; faint streamers in N.W.; streamers in East disappeared.
　50　Streamers still faint in N.W.; a number of faint streamers appearing in North, occasionally becoming bright, and again dying away.
15 5　Faint streamers in the East, several bright streamers North by West.
　10　Several streamers in East shooting across from East by North, to the South of the zenith.

d. h. m.
6 15 15　Magnificent arch of light extending from N.W. to S.S.E. across the sky a little to the South of zenith, light very bright in N.W., and on dying away appearing to move as bright cirri across the sky; in 5 minutes it had entirely disappeared.
　23　No appearance of Aurora except moderately bright light in North; streamers not visible.
　28　Streamers in the North.
　40　Streamers visible at intervals.
　45　No traces of auroral light; cirri generally over the sky; haze in horizon.
16 0　A few light cirri scattered about. No further appearance of Aurora during the night.

b Saturday midnight at Toronto.

August 5, 6, and 7, 1841.

d.	h.	m.	s.	Decl. Sc.-Div	Hor. Force Sc.-Div	Ther.	Vert. Force Sc.-Div	Ther.
6	8	40	0	24·0	20·9		25·7	
		45	0	24·1				
		50	0	23·9	20·9		24·7	
		55	0	23·7				
	9	0	0	23·4	26·2		24·0	
		5	0	23·8				
		10	0	23·9	34·1		24·9	
		15	0	23·2				
		20	0	23·9	36·9		24·8	
		25	0	24·9				
		30	0	25·1	35·0		24·9	
		35	0	25·9				
		40	0	26·0	33·0		25·0	
		45	0	26·0				
		50	0	26·0	32·0		25·0	
		55	0	26·0				
	10	0	0	26·0	30·0	65	24·9	64
		5	0	26·1				
		10	0	26·1	28·0		25·4	
		15	0	26·1				
		20	0	26·0	27·0		25·4	
		25	0	26·0				
		30	0	26·0	26·9		25·4	
		35	0	26·0				
		40	0	26·0	26·0		25·4	
		45	0	26·0				
		50	0	25·9	25·1		25·4	
		55	0	25·2				
	11	0	0	25·1	26·0	64	25·3	
		5	0	25·0				
		10	0	24·9	28·0		25·6	
		15	0	24·9				
		20	0	24·9	31·0		25·7	
		25	0	25·3				
		30	0	25·9	31·8		25·7	
		35	0	26·8				
		40	0	26·9	31·8		25·8	
		45	0	26·2				
		50	0	26·1	31·0		25·7	
		55	0	25·5				
	12	0	0	24·9	30·0	64	25·6	64
		5	0	24·1				
		10	0	24·1	27·9		25·2	
		15	0	24·1				
		20	0	24·1	26·9		25·2	
		25	0	24·8				
		30	0	23·2	25·8		25·0	
		35	0	23·2				
		40	0	23·1	24·9		24·9	
		45	0	22·9				
		50	0	22·9	24·9		24·8	
		55	0	22·9				
	13	0	0	23·1	25·1	64	24·8	
		5	0	23·1				
		10	0	23·2	25·7		25·0	
		15	0	23·4				
		20	0	23·4	27·2		24·6	
		25	0	24·8				
		30	0	25·0	34·9		25·4	
		35	0	25·6				
		40	0	26·0	38·0		25·7	
		45	0	26·7				
		50	0	26·1	36·9		25·8	
		55	0	25·7				
	14	0	0	25·4	36·0	64	25·7	64
		5	0	25·4				
		10	0	25·2	35·2		25·7	

August 5, 6, and 7, 1841.

d.	h.	m.	s.	Decl. Sc.-Div	Hor. Force Sc.-Div	Ther.	Vert. Force Sc.-Div	Ther.
6	14	15	0	25·0				
		20	0	24·8	34·9		25·7	
		25	0	24·6				
		30	0	24·5	33·0		25·3	
		35	0	23·8				
		40	0	23·9	32·8		25·3	
		45	0	24·0				
		50	0	24·0	31·9		25·5	
		55	0	24·1				
	15	0	0	24·2	31·4	64	25·5	
		5	0	24·2				
		10	0	24·2	30·3		25·7	
		15	0	24·3				
		20	0	24·9	30·2		25·8	
		25	0	25·2				
		30	0	25·4	30·7		26·1	
		35	0	25·8				
		40	0	26·0	30·9		26·5	
		45	0	26·4				
		50	0	26·0	31·1		26·4	
		55	0	25·6				
	16	0	0	26·0	31·9	64	26·3	63
		5	0	25·9				
		10	0	26·1	32·1		26·3	
		15	0	26·1				
		20	0	26·0	32·8		26·3	
		25	0	26·2				
		30	0	26·2	32·2			
		35	0	26·2				
		40	0	26·0	34·1		26·3	
		45	0	26·0				
		50	0	25·9	34·1		26·3	
		55	0	25·9				
	17	0	0	25·9	34·8		26·3	
		5	0	25·9				
		10	0	26·1	35·0		26·5	
		15	0	20·1				
		20	0	26·1	35·1		26·5	
		25	0	26·2				
		30	0	26·2	35·1		26·5	
		35	0	26·2				
		40	0	26·2	35·5		26·5	
		45	0	26·2				
		50	0	26·2	35·6		26·3	
		55	0	26·3				
	18	0	0	26·7	35·9	63	26·4	63
		5	0	26·7				
		10	0	26·7	36·0		26·4	
		15	0	26·6				
		20	0	26·6	36·0		26·5	
		25	0	26·6				
		30	0	26·5	36·6		26·5	
		35	0	26·8				
		40	0	26·9	37·1		26·5	
		45	0	27·1				
		50	0	27·2	37·8		26·5	
		55	0	27·6				
	19	0	0	27·9	38·0	63	26·6	
		5	0	28·0				
		10	0	28·1	38·2		26·6	
		15	0	28·2				
		20	0	28·5	38·9		26·6	
		25	0	29·5				
		30	0	29·4	39·2		26·9	
		35	0	29·8				
		40	0	30·5	40·1		26·9	
		45	0	30·9				

August 5, 6, and 7, 1841.

d.	h.	m.	s.	Decl. Sc.-Div	Hor. Force Sc.-Div	Ther.	Vert. Force Sc.-Div	Ther.
6	19	50	0	31·0	40·9		27·2	
		55	0	30·8				
	20	0	0	31·0	41·3	63	27·2	63
		5	0	31·2				
		10	0	31·7	41·9		27·4	
		15	0	30·9				
		20	0	31·5	41·3		27·2	
		25	0	31·6				
		30	0	31·4	42·1		27·2	
		35	0	31·0				
		40	0	31·3	42·2		27·3	
		45	0	31·2				
		50	0	31·0	42·7		27·3	
		55	0	30·9				
	22	0	0	30·0	41·8	63	27·7	62
	23	0	0	29·9	43·0	63	26·8	63

Positions at the usual hours of observation, August 5, 6, and 7.

d.	h.	m.	s.	Decl. Sc.-Div	Hor. Force Sc.-Div	Ther.	Vert. Force Sc.-Div	Ther.
5	0	0	0	26·9	46·8	64	30·0	63
	2	0	0	27·7	46·9	64	30·7	64
	3	0	0	28·1	47·9	65	30·1	64
	4	0	0	27·8	42·9	65	28·1	65
	5	0	0	27·5	43·0	65	27·7	65
	6	0	0	25·9	39·1	65	27·6	65
	8	0	0	26·0	36·0	64	27·1	64
	10	0	0	27·1	36·0	64	27·7	64
	11	0	0	26·9	38·1	64	27·9	64
	12	0	0	27·2	37·8	64	27·8	64
	13	0	0	27·0	39·3	64	29·1	63
	14	0	0	26·4	38·6	64	28·3	63
	15	0	0	26·4	39·7	64	28·4	63
	16	0	0	26·8	41·8	64	28·5	63
	18	0	0	27·1	42·3	64	28·7	63
	19	30	0	31·0	41·0	64	29·2	63
	20	0	0	33·0	41·2	63	30·0	63
	20	30	0	32·9	41·5	63	29·6	63
	22	0	0	31·3	40·8	63	30·0	62
	23	0	0	30·5	42·2	63	29·9	63
6	0	0	0	27·8	43·9	64	30·2	63
	2	0	0	27·0	37·8	64	30·5	63
	3	0	0	25·1	37·9	63	30·0	64
	4	0	0	24·7	39·8	65	29·3	64
	5	0	0	23·0	35·0	65	29·1	64
	6	0	0	22·3	29·0	65	27·0	64
	8	0	0	23·0	25·1	65	25·6	65
	10	0	0	26·0	30·0	65	24·9	64
	11	0	0	25·1	26·0	64	25·3	64
	12	0	0	24·9	30·0	64	25·6	64
	13	0	0	23·1	25·1	64	23·0	64
	14	0	0	25·7	36·0	64	25·7	64
	15	0	0	24·2	31·4	64	23·5	64
	16	0	0	26·0	31·9	64	26·3	63
	18	0	0	26·7	35·9	63	26·4	63
	19	30	0	29·4	39·2	63	26·6	63
	20	0	0	31·0	41·3	63	27·2	63
	20	30	0	31·4	42·1	63	27·3	62
	22	0	0	30·0	41·8	63	27·7	62
	23	0	0	29·9	43·0	63	26·8	63
7	0	0	0	28·1	40·0	63	26·1	63
	2	0	0	27·5	39·9	64	26·8	64
	3	0	0	28·1	40·0	64	26·9	64
	4	0	0	27·6	38·0	64	26·3	64
	5	0	0	27·0	35·1	65	24·7	64
	6	0	0	26·5	36·8	64	24·4	64
	8	0	0	25·9	36·9	64	24·5	64

AUGUST 5, 6, and 7, 1841.

M. Gött. Time.			Decl.	Hor. Force.		Vert. Force.	
d. h.	m.	s.	Sc.-Div^ms	Sc.-Div^ms.	Ther.	Sc.-Div^ms.	Ther.
7 10	5	0	26·1	37·1	64	23·0	63
11	0	0	26·3	37·2	64	23·0	63
12	0	0*	26·3	37·3	64	21·5	63

The Mean Positions at the same hours during the Month are given in page 66.

CAPE OF GOOD HOPE
{ Decl. 1 Scale Division = 0'·75
H. F. k = ·000180 ; q = ·0003
V. F. k = ·000037 ; q =

Positions at the usual hours of observation, August 5, 6, and 7.

d. h.	m.	s.	Decl.	Hor. Force.		Vert. Force.	
			Sc.-Div^ms	Sc.-Div^ms.	Ther.	Sc.-Div^ms.	Ther.
5 0	0	0	48·7	52·5	59	54·1	60
2	0	0	52·0	52·3	59	41·8	60
4	0	0	50·9	48·3	59	66·5	60
6	0	0	51·2	47·0	59	60·0	60
8	0	0	50·5	44·6	59	63·8	60
10	0	0	49·5	46·0	59	63·1	60
12	0	0	49·9	48·2	59	50·7	60
14	0	0	50·4	48·0	59	59·0	60
16	0	0	51·1	51·0	59	45·0	60
18	0	0	53·0	51·2	59	39·2	60
20	0	0	57·9	53·8	59	39·9	60
22	0	0	53·8	53·0	59	66·5	59
6 0	0	0	51·6	54·0	59	62·3	59
2	0	0	55·0	40·1	59	70·5	60
4	0	0	49·2	47·5	60	64·3	60
6	0	0	46·8	35·3	60	85·1	60
8	0	0	44·1	32·0	60	89·6	60
10	0	0	47·8	40·1	60	46·5	60
12	0	0	44·5	43·6	60	66·1	60
14	0	0	48·2	48·9	60	20·9	61
16	0	0	48·0	41·3	59	52·7	60
18	0	0	49·0	44·0	59	47·1	59
20	0	0	52·1	48·9	59	28·4	59
22	0	0	51·9	49·4	59	45·4	59
7 0	0	0	50·9	47·7	60	55·3	59
2	0	0	50·5	46·1	60	57·7	60
4	0	0	50·0	43·5	60	67·3	60
6	0	0	49·8	43·7	60	58·6	60
8	0	0	49·8	44·5	60	54·0	60
10	0	0^b	50·0	44·8	60	54·0	60

Mean Positions at the same hours during the Month.

d. h.	m.	s.	Decl.	Hor. Force.		Vert. Force.	
0	0	0	50·3	57·5	57	54·3	58
2	0	0	52·3	54·1	58	54·1	58
4	0	0	52·6	53·5	58	60·2	58
6	0	0	51·8	52·8	58	62·2	58
8	0	0	51·8	51·9	58	63·6	58
10	0	0	51·8	52·3	58	62·7	58
12	0	0	51·6	53·5	58	61·3	58
14	0	0	52·8	55·5	57	57·0	58
16	0	0	53·3	56·2	57	55·5	58
18	0	0	54·6	56·5	57	50·9	57
20	0	0	54·1	59·5	57	60·7	57
22	0	0	49·7	59·9	57	67·2	57

AUGUST 5, 6, and 7, 1841.

VAN DIEMEN ISLAND
{ Decl. 1 Scale Division = 0'·71
H. F. k = ·0003 ; q =
V. F. k = ; q =

Extra observations.
The V. F. was observed at 2^m. 30^s. before, and the H. F. 2^m. 30^s. after the times specified.

M. Gött. Time.			Decl.	Hor. Force.		Vert. Force.	
d. h.	m.	s.	Sc.-Div^ms	Sc.-Div^ms.	Ther.	Sc.-Div^ms.	Ther.
6 2	10	0	41·8				
	22	30^c	44·2	36·1	46	56·4	45
	30	0	46·9	34·6		58·2	
	37	30	48·2			60·9	
	45	0	46·8	32·1		63·0	
3	5	0	48·2				
	10	0	50·4	32·9	46	65·0	46
	15	0	50·7				
	20	0	51·4	33·9		66·2	
	25	0	52·5				
	30	0	53·2	34·9		64·7	
	35	0	54·7				
	40	0	56·1	35·6		64·8	
	45	0	56·5				
	50	0	56·4	36·2		64·2	
6 15		0	49·5		45		
	20	0	49·6	37·0		61·1	45
	25	0	50·3				
	30	0	51·8	38·6		59·5	
	35	0^d	54·5				
	40	0	56·4	39·7		58·8	
	45	0	57·9				
	50	0	60·1	40·3		57·8	
7 10		0	65·7	39·8	45		
	15	0	62·7				
	20	0	63·9	39·5		56·6	45
	25	0	60·9				
	30	0	60·6	39·8		55·0	
	35	0	61·0				
	40	0	62·6	40·1		54·3	
	45	0	62·4				
	50	0	61·1	38·7		52·4	
	55	0	64·3				
8 5		0	64·3		46		
	10	0	63·1	37·3		53·6	48
	15	0	64·6				
	20	0	63·8	36·2		53·6	
	25	0	63·8				
	30	0	62·8	35·9		53·6	
	35	0	67·4				
	40	0	71·7	38·6		56·3	
	45	0	73·4				
	50	0	72·4	37·5		53·9	
9 10		0	69·3	30·9	46		
	15	0	73·0				
	20	0	74·1	31·7		60·8	49
	25	0	75·1				
	30	0	72·9			60·8	
	35	0	72·9				
	40	0	72·2	42·9		50·1	
	45	0	72·6				
10 10		0	68·6	38·0	46		
	15	0	69·2				
	20	0	69·8	36·9		45·8	50
	25	0	70·4				

AUGUST 5, 6, and 7, 1841.

M. Gött. Time.			Decl.	Hor. Force.		Vert. Force.	
d. h.	m.	s.	Sc.-Div^ms.	Sc.-Div^ms.	Ther.	Sc.-Div^ms.	Ther.
6 10	30	0	69·7	36·1		48·8	
	35	0	70·3				
	40	0	69·5	40·4		50·1	
	45	0	69·0				
	50	0	67·7	35·2		50·1	
11 10		0	66·5	31·8	46		49
	15	0	64·9				
	20	0	63·7	30·9		55·4	
	25	0	62·4				
	30	0	62·1	30·4		56·3	
	35	0	62·5				
	40	0	62·9	28·8		58·5	
	45	0	64·5				
	50	0	65·3	27·8		60·9	
	55	0	66·1				
12 5		0	67·9				
	10	0	68·6	29·5	47	62·9	49
	15	0	67·3				
	20	0	68·0	29·2		62·2	
	25	0	66·8				
	30	0	66·8	32·0		62·1	
	35	0	66·3				
	40	0	66·4	32·0		60·3	
	45	0	65·4				
	50	0	65·1			60·1	

Positions at the usual hours of observation, August 5, 6, and 7.

d. h.	m.	s.	Decl.	Hor. Force.		Vert. Force.	
5 0	0	0	58·9	39·1	44	67·5	43
1	0	0	58·6	39·1		69·6	
2	0	0	58·3	39·9	44	67·2	44
3	0	0	58·9	39·1		66·7	
4	0	0	61·5	39·7	44	66·3	43
5	0	0	57·4	39·8		66·9	
6	0	0	55·6	39·9	44	66·8	43
7	0	0	56·1	39·8		70·5	
8	0	0	56·4	39·9	43	66·7	42
9	0	0	65·3	46·2		63·2	
10	0	0	60·7	44·4	44	56·1	44
11	0	0	60·7	42·7		61·7	
12	0	0	58·0	42·2	44	60·9	43
13	0	0	56·4	39·6		62·7	
14	0	0	58·1	37·0	44	66·0	44
15	0	0	60·5	37·0		66·6	
16	0	0	62·1	35·1	45	66·2	45
17	0	0	64·4	36·5		66·4	
18	0	0	63·8	35·5	46	67·4	46
19	0	0	63·7	35·3		68·2	
20	0	0	62·7	33·6	47	70·0	46
21	0	0	63·2	32·0		72·2	
22	0	0	59·2	32·0	46	70·2	45
23	0	0	57·7	34·3		65·6	
6 0	0	0	56·8	35·8	46	66·2	45
1	0	0	55·2	35·9		63·2	
2	0	0	38·2	35·4	46	62·4	45
3	0	0	45·9	31·3		64·9	
4	0	0	55·7	35·9	47	63·1	47
5	0	0	57·2	37·7		64·5	
6	0	0	48·8	34·4	45	60·8	45
7	0	0	62·7	39·3		56·9	
8	0	0	65·0	37·2	46	54·6	48
9	0	0	67·6	34·4		53·0	
10	0	0	68·7	38·8	46	41·4	50

* Saturday midnight at St. Helena.
b Saturday midnight at the Cape of Good Hope.
c The Decl. magnet oscillating vertically.
d From 6^h. 35^m. to 6^h. 45^m. the Decl. magnet oscillating vertically.

August 5, 6, and 7, 1841.

M. Gött. Time. (d. h. m. s.)	Decl. (Sc.-Div.)	Hor. Force. (Sc.-Div.)	Ther.	Vert. Force. (Sc.-Div.)	Ther.
6 11 0 0	66·9	34·2		50·4	
12 0 0	67·3	29·5	47	63·1	49
13 0 0	65·5	30·0		59·5	
14 0 0	64·4	24·5	48	66·5	48
15 0 0	66·7	29·0		63·0	
16 0 0	61·4	34·6	48	52·6	48
17 0 0	63·3	35·1		56·8	
18 0 0	63·6	34·6	48	59·9	47
19 0 0	66·1	33·2		64·9	
20 0 0	63·6	35·5	48	63·4	47
21 0 0	65·5	33·4		66·3	
22 0 0	59·5	31·7	47	67·6	46
23 0 0	59·4	32·4		69·6	
7 0 0 0	52·1	35·0	46	63·8	46
1 0 0	52·1	34·5		66·0	
2 0 0a	55·8	35·7	46	66·7	45

The Mean Positions at the same hours during the Month are given in page 66.

August 23 and 24, 1841.

Toronto { Decl. 1 Scale Division = 0'·72 / H. F. k = ·000076; q = ·0002 / V. F. k = ·000093; q = ·00018 }

Extra observations.

The V. F. was observed at 1m. 30s. before, and the H. F. 2m. after the times specified.

M. Gött. Time. (d. h. m. s.)	Decl. (Sc.-Div.)	Hor. Force. (Sc.-Div.)	Ther.	Vert. Force. (Sc.-Div.)	Ther.
23 14 15 0	150·1	385·8	70	48·0	70
20 0	145·2	388·7		55·8	
25 0	141·3	385·5		51·3	
30 0	138·3	386·9		51·6	
35 0	138·0	388·2		51·5	
40 0	137·6	392·2		51·6	
45 0	137·0	392·1		51·7	
50 0	136·7	389·5		51·7	
55 0	135·3	386·2		52·3	
15 0 0b	131·8	385·8	70	52·1	70
5 0	131·6			53·2	
10 0	131·1	386·2		52·8	
15 0	132·0	385·3		52·2	

Positions at the usual hours of observation, August 23 and 24.

M. Gött. Time. (d. h. m. s.)	Decl. (Sc.-Div.)	Hor. Force. (Sc.-Div.)	Ther.	Vert. Force. (Sc.-Div.)	Ther.
23 0 0 0	148·7	398·0	66	53·4	67
2 0 0	148·1	398·0	68	52·1	68
4 0 0	140·7	380·2	69	51·3	68
6 0 0	129·3	388·4	70	52·2	69
8 0 0	126·2	377·3	71	54·2	70
10 0 0	123·5	410·0	72	56·4	70
12 0 0	130·4	384·7	72	52·7	70
14 0 0	158·6	369·6	70	54·4	70
16 0 0	137·3	386·0	70	53·8	70
18 0 0	136·9	391·7	68	55·0	68
20 0 0	123·8	379·3	67	28·3	67
22 0 0	141·4	394·7	65	52·7	66
24 0 0 0	138·5	393·7	64	52·4	65
2 0 0	145·3	386·9	66	54·9	66
4 0 0	132·2	373·5	68	54·2	67

a Saturday midnight at Van Diemen Island.

August 23 and 24, 1841.

M. Gött. Time. (d. h. m. s.)	Decl. (Sc.-Div.)	Hor. Force. (Sc.-Div.)	Ther.	Vert. Force. (Sc.-Div.)	Ther.
24 6 0 0	126·0	388·9	70	52·1	69
8 0 0	126·1	399·1	71	51·5	69
10 0 0	131·1	414·4	72	50·5	70
12 0 0	132·4	394·8	72	49·4	71
14 0 0	132·9	392·1	71	50·9	70
16 0 0	134·0	398·0	69	52·2	69
18 0 0	135·8	401·7	67	54·7	67
20 0 0	136·4	398·3	66	55·2	66
22 0 0	138·0	415·0	65	58·0	66

The Mean Positions at the same hours during the Month are given in page 64.

St. Helena { Decl. 1 Scale Division = 0'·71 / H. F. k = ·00018; q = ·00025 / V. F. k = ·00017; q = }

Extra observations.

The V. F. was observed at 2m. 30s. before, and the H. F. 2m. 30s. after the times specified.

M. Gött. Time. (d. h. m. s.)	Decl. (Sc.-Div.)	Hor. Force. (Sc.-Div.)	Ther.	Vert. Force. (Sc.-Div.)	Ther.
23 10 0 0	22·8	39·8	61		61
5 0	22·9				
10 0	23·0	41·0		49·4	
15 0	23·3				
20 0	22·9	42·8		49·3	
25 0	22·9				
30 0	23·0	45·2		49·3	
35 0	23·1				
40 0	23·7	48·3		49·3	
45 0	23·8				
50 0	23·8	49·8		49·3	
55 0	23·7				
11 0 0	22·9	48·8	61	49·4	61
5 0	24·0				
10 0	24·0	47·4		49·6	
15 0	24·0				
20 0	24·0	45·1		49·5	
25 0	24·0				
30 0	24·0	43·9		49·5	
35 0	24·0				
40 0	24·0	43·2		49·5	
45 0	24·0				
50 0	24·0	43·1		49·7	
55 0	24·1				
12 0 0	23·9	43·9	61	49·7	61
5 0	23·9				
10 0	23·9	43·8		49·7	
15 0	23·9				
20 0	23·9	43·4		49·7	
25 0	23·7				
30 0	23·8	43·9		50·3	
35 0	23·9				
40 0	23·9	44·2		50·3	
45 0	23·8				
50 0	24·0	44·8		50·4	
55 0	24·0				
13 0 0	24·0	44·9	61	50·5	61
5 0	23·9				
10 0	23·8	45·0		50·5	
15 0					
20 0	23·5	45·0			
25 0	23·6				

August 23 and 24, 1841.

M. Gött. Time. (d. h. m. s.)	Decl. (Sc.-Div.)	Hor. Force. (Sc.-Div.)	Ther.	Vert. Force. (Sc.-Div.)	Ther.
23 13 30 0	23·9	46·0		50·6	
35 0	24·0				
40 0	24·5	47·1		50·8	
45 0	25·3				
50 0	25·0	50·3		51·1	
55 0	25·5				
14 0 0	25·4	51·1	61	51·1	61
10 0	25·0	51·1		51·5	
15 0	25·4				
20 0	25·1	50·9		51·2	
25 0	25·4				
30 0	25·1	49·9		51·0	
35 0	25·3				
40 0	25·4	49·0		51·0	
45 0	25·3				
50 0	25·0	48·9		51·3	
55 0	24·6				
15 0 0	24·4	47·9	61	51·3	61
10 0	25·0	48·0		51·4	
15 0	24·6				
20 0	24·9	47·9		51·6	
25 0	24·6				
30 0	24·8	48·0		51·2	
35 0	24·8				
40 0	24·7	47·1		51·2	
45 0	24·7				
50 0	24·8	47·3		51·2	
55 0	24·9				
24 1 0 0	19·1	52·0	62	51·1	61
5 0					
10 0	19·0	51·2			
15 0	18·9				
20 0	18·9	51·0		56·5	
25 0	18·9				
30 0	19·0	50·1		56·5	
35 0	19·1				
40 0	19·1	50·0		56·6	
45 0	19·0				
50 0	19·0			56·2	
55 0	19·0				
2 0 0	19·0	49·0	62	55·9	61
5 0	19·0				
10 0	19·2	49·2		55·5	
15 0	19·6				
20 0	19·6	49·9		55·6	
25 0	19·8				
30 0	19·9	49·5		55·5	
35 0	19·9				
40 0	20·0	48·9		55·3	
45 0	20·0				
50 0	20·0	48·0		55·1	
55 0	20·1				
3 0 0	20·9	47·1	62	54·8	61
5 0	20·7				
10 0	20·8	46·2		54·5	
15 0	20·9				
20 0	21·3	45·9		54·2	
25 0	21·5				
30 0	21·6	45·6		54·0	
35 0	21·7				
40 0	21·7	45·2		53·5	

b 23 15 0   Calm; clear and unclouded; very faint auroral light in the North.
    16 0   Ditto.

August 23 and 24, 1841.

M. Gött. Time.				Decl.	Hor. Force.		Vert. Force.	
d.	h.	m.	s.	Sc.-Div^m	Sc.-Div^m	Ther.	Sc.-Div^m	Ther.
24	3	45	0	21·7				
		50	0	21·9	45·1		53·6	
		55	0	21·9				
	4	0	0	22·1	45·0	62	53·3	62
		5	0	22·2				
		10	0	22·4	44·8		53·3	
		15	0	22·6				
		20	0	22·6	44·6		52·9	
		25	0	22·5				
		30	0	22·5	44·4		52·5	
		35	0	22·6				
		40	0	22·8	44·8		52·5	
		45	0	22·8				
		50	0	23·0	45·3		52·3	
		55	0	23·1				

Positions at the usual hours of observation, August 23 and 24.

M. Gött. Time.				Decl.	Hor. Force.		Vert. Force.	
d.	h.	m.	s.	Sc.-Div^m	Sc.-Div^m	Ther.	Sc.-Div^m	Ther.
23	0	0	0	22·9	56·4	61	52·0	61
	2	0	0	24·0	54·5	62	50·1	62
	3	0	0	24·6	52·8	63	49·1	62
	4	0	0	24·7	50·1	62	48·1	62
	5	0	0	22·0	45·0	63	47·1	62
	6	0	0	22·5	46·9	62	47·7	62
	8	0	0	22·8	42·0	62	48·6	62
	10	0	0	22·8	39·8	61	48·8	61
	11	0	0	22·9	48·8	61	49·4	61
	12	0	0	23·9	43·9	61	49·7	61
	13	0	0	24·0	44·9	61	50·5	61
	14	0	0	25·4	51·1	61	51·1	61
	15	0	0	24·4	47·9	61	51·3	61
	16	0	0	24·9	47·1	61	51·5	61
	18	0	0	25·9	47·9	60	51·9	60
	19	30	0	29·8	48·5	61	52·3	60
	20	0	0	30·8	47·8	60	53·5	60
	20	30	0	30·0	49·9	60	52·7	60
	22	0	0	25·0	52·0	60	53·7	60
	23	0	0	21·7	51·2	60	54·1	60
24	0	0	0	20·3	53·1	60	54·9	60
	2	0	0	19·0	49·0	61	55·9	61
	3	0	0	20·9	47·1	62	54·8	62
	4	0	0	22·1	45·0	62	53·3	62
	5	0	0	23·3	45·4	62	52·2	62
	6	0	0	23·7	46·5	62	51·4	62
	8	0	0	24·5	47·1	61	51·5	61
	10	0	0	24·8	48·0	61	50·9	61
	11	0	0	25·0	48·0	61	51·3	61
	12	0	0	25·0	47·9	61	51·4	60
	13	0	0	25·3	49·8	61	51·6	60
	14	0	0	25·9	49·2	60	52·1	60
	15	0	0	25·9	48·8	61	58·1	60
	16	0	0	26·1	49·0	60	60·0	60
	18	0	0	26·5	50·8	60	59·9	60
	19	30	0	28·5	51·3	60	59·7	60
	20	0	0	29·2	51·5	60	58·2	60
	20	30	0	28·4	52·5	60	59·6	60
	22	0	0	24·9	55·6	60	59·6	60
	23	0	0	24·0	57·0	60	55·6	60

The Mean Positions at the same hours during the Month are given in page 66.

August 23 and 24, 1841.

CAPE OF GOOD HOPE

Decl. 1 Scale Division = $0'·75$
H.F. $k = ·000180$; $q = ·0003$
V.F. $k = ·00004$; $q =$

Positions at the usual hours of observation, August 23 and 24.

M. Gött. Time.				Decl.	Hor. Force.		Vert. Force.	
d.	h.	m.	s.	Sc.-Div^m	Sc.-Div^m	Ther.	Sc.-Div^m	Ther.
23	0	0	0	52·4	61·1	56	55·5	57
	2	0	0	54·0	57·0	57	61·1	57
	4	0	0	53·0	56·5	57	65·1	57
	6	0	0	52·0	54·9	58	69·7	58
	8	0	0	49·5	52·1	57	87·1	58
	10	0	0	49·8	50·1	57	81·0	57
	12	0	0	51·6	55·0	56	61·1	57
	14	0	0	55·5	60·1	57	27·8	58
	16	0	0	54·2	57·5	57	60·0	57
	18	0	0	57·2	56·0	57	57·3	57
	20	0	0	58·0	55·7	57	73·4	57
	22	0	0	47·9	63·2	57	69·0	57
24	0	0	0	49·1	57·0	57	62·9	57
	2	0	0	50·4	53·3	56	68·6	57
	4	0	0	53·0	55·0	56	64·0	57
	6	0	0	52·2	57·5	56	66·7	57
	8	0	0	52·7	57·1	56	66·7	56
	10	0	0	53·2	59·0	55	65·0	56
	12	0	0	53·8	59·4	55	67·4	55
	14	0	0	55·6	61·8	55	59·8	55
	16	0	0	54·3	61·8	54	67·4	55
	18	0	0	54·6	63·7	54	61·5	55
	20	0	0	52·1	68·6	53	70·9	54
	22	0	0	47·5	70·0	54	69·7	54

The Mean Positions at the same hours during the Month are given in page 69.

VAN DIEMEN ISLAND

Decl. 1 Scale Division = $0'·71$
H.F. $k = ·0003$; $q =$
V.F. $k =$; $q =$

Extra observations.

The V.F. was observed at 2^m. 30^s. before, and the H.F. 2^m. 30^s. after the times specified.

M. Gött. Time.				Decl.	Hor. Force.		Vert. Force.	
d.	h.	m.	s.	Sc.-Div^m	Sc.-Div^m	Ther.	Sc.-Div^m	Ther.
23	8	15	0	69·7				
		20	0	69·0	35·7	52	54·3	51
		25	0	69·1				
		30	0	67·0	35·1		53·6	
		35	0	64·0				
		40	0	63·0	36·5		51·2	
		45	0	63·6				
		50	0	63·3	36·9		50·9	
	11	15	0	68·1				
		20	0	66·4	36·1	52	52·7	52
		25	0	65·5				
		30	0	64·7	36·1		48·6	
		35	0*	63·7				
		40	0				49·2	

Positions at the usual hours of observation, August 23 and 24.

M. Gött. Time.				Decl.	Hor. Force.		Vert. Force.	
d.	h.	m.	s.	Sc.-Div^m	Sc.-Div^m	Ther.	Sc.-Div^m	Ther.
23	0	0	0	59·4	32·1	55	54·2	54
	1	0	0	56·4	32·5		53·0	
	2	0	0	59·2	33·9	54	53·4	53
	3	0	0	60·9	33·9		49·2	

August 23 and 24, 1841.

M. Gött. Time.				Decl.	Hor. Force.		Vert. Force.	
d.	h.	m.	s.	Sc.-Div^m	Sc.-Div^m	Ther.	Sc.-Div^m	Ther.
23	4	0	0	60·3	34·6	54	51·0	53
	5	0	0	60·5	35·1		52·2	
	6	0	0	60·2	36·8	52	52·9	51
	7	0	0	60·2	36·2		52·9	
	8	0	0	71·2	36·8	52	56·1	51
	9	0	0	64·0	36·9		48·8	
	10	0	0	64·5	37·7	51	49·7	51
	11	0	0	70·1	33·0		59·7	
	12	0	0	62·7	34·3	54	44·8	53
	13	0	0	62·0	33·6		44·9	
	14	0	0	64·1	28·9	57	49·9	56
	15	0	0	68·0	30·0		47·2	
	16	0	0	66·2	29·7	57	47·7	57
	17	0	0	69·0	30·7		49·5	
	18	0	0	69·1	28·7	59	52·3	57
	19	0	0	68·8	29·4		52·7	
	20	0	0	60·0	27·9	56	55·6	55
	21	0	0	66·0	31·0		54·5	
	22	0	0	64·1	33·4	55	51·3	54
	23	0	0	56·6	37·5		44·3	
24	0	0	0	60·3	32·2	53	58·3	53
	1	0	0	58·8	33·0		57·0	
	2	0	0	64·8	37·0	54	54·0	52
	3	0	0	61·5	35·3		53·0	
	4	0	0	62·0	35·1	51	57·8	50
	6	0	0	62·9	37·5	49	58·9	49
	7	0	0	63·2	37·3		59·0	
	8	0	0	63·9	38·0	48	59·1	48
	9	0	0	62·8	38·7		58·9	
	10	0	0	62·1	38·8	48	58·5	48
	11	0	0	60·6	39·2		56·5	
	12	0	0	59·9	39·0	49	54·5	49
	13	0	0	59·7	37·0		54·8	
	14	0	0	61·3	35·1	52	53·4	52
	15	0	0	63·0	33·5		53·1	
	17	0	0	63·5	33·5	54	54·0	54
	18	0	0	68·0	34·2		54·1	
	19	0	0	68·4	35·4	55	53·2	54
	20	0	0	67·0	36·6		52·1	
	21	0	0	64·5	35·7	55	51·6	54
	22	0	0	63·3	36·9		52·4	
	23	0	0	62·1	36·8	53	52·9	52
	23	0	0	63·0	37·3		53·2	

The Mean Positions at the same hours during the Month are given in page 66.

August 26 and 27, 1841.

TORONTO

Decl. 1 Scale Division = $0'·72$
H.F. $k = ·000076$; $q = ·0002$
V.F. $k = ·000093$; $q = ·00018$

Extra observations.

The V.F. was observed at 1^m. 30^s. before, and the H.F. 2^m. after the times specified.

M. Gött. Time.				Decl.	Hor. Force.		Vert. Force.	
d.	h.	m.	s.	Sc.-Div^m	Sc.-Div^m	Ther.	Sc.-Div^m	Ther.
27	1	25	0	136·2	391·2	66	52·1	66
		30	0	136·1	393·6		48·2	
		35	0	137·0	394·2		48·1	
		40	0	136·6	395·5		52·3	
		45	0	133·1			45·9	
	2	40	0	133·0	399·9	66	49·4	66
		45	0	134·1	400·6		50·3	

* The Decl. magnet had a considerable vertical motion.

August 26 and 27, 1841.

M. Gött. Time.				Decl.	Hor. Force.			Vert. Force.	
d.	h.	m.	s.	Sc.-Div⁹ˢ.	Sc.-Div⁹ˢ.	Ther.		Sc.-Div⁹ˢ.	Ther.
27	2	50	0	134·1	401·4			49·6	
		55	0	136·0	403·2			50·8	
	3	0	0	135·2	402·4	66		51·2	66
		5	0	134·9	399·7			51·2	
		10	0	134·9	397·9			51·2	
		15	0	136·0	396·1			51·5	
		20	0	136·1	393·7			51·5	
		25	0	136·7	392·2			51·5	
		30	0	137·1	391·4			52·1	
		35	0	137·6	390·2			52·4	
		40	0	138·0	388·5			53·0	
		45	0	137·7	390·8			56·7	
		50	0	136·5	389·0			51·6	
		55	0	136·1	387·0			55·0	
	4	5	0	135·7	385·8	67		52·5	66
		10	0	135·7	390·2			54·5	
		15	0	133·8	385·1			58·2	
		20	0	134·4	384·5			53·0	
		25	0		383·7			52·5	
		30	0	133·1	381·7			56·1	
		35	0	133·0	380·5			56·6	
		40	0	132·0	379·1			54·4	
		45	0	131·4	376·5			51·8	
		50	0	132·0	374·4			51·8	
		55	0	132·8	376·6			52·8	

Positions at the usual hours of observation,
August 26 and 27.

M. Gött. Time.				Decl.	Hor. Force.			Vert. Force.	
d.	h.	m.	s.	Sc.-Div⁹ˢ.	Sc.-Div⁹ˢ.	Ther.		Sc.-Div⁹ˢ.	Ther.
26	0	0	0	133·9	393·2	66		52·0	66
	2	0	0	144·9	404·1	68		49·5	67
	4	0	0	136·2	376·7	69		49·3	68
	6	0	0	120·9	364·8	69		50·6	68
	8	0	0	124·5	397·7	70		57·5	69
	10	0	0	125·0	398·5	70		60·7	69
	12	0	0	117·7	380·5	69		67·6	69
	14	0	0	146·8	351·0	68		65·5	68
	16	0	0	162·0	340·0	67		40·3	67
	18	0	0	145·0	338·3	67		47·9	67
	20	0	0	154·8	367·9	66		44·3	67
	22	0	0	115·0	390·2	66		48·1	66
27	0	0	0	148·6	403·7	66		61·5	66
	2	0	0	130·6	383·8	66		44·5	66
	4	0	0	135·7	385·5	67		56·4	66
	6	0	0	128·2	382·0	68		55·3	67
	8	0	0	130·5	398·1	69		54·7	68
	10	0	0	133·8	412·0	69		51·4	69
	12	0	0	135·0	400·2	70		57·4	69
	14	0	0	138·1	402·6	70		52·6	71
	16	0	0	134·9	390·1	69		52·5	70
	18	0	0	131·4	377·5	69		34·7	70
	20	0	0	137·0	385·9	69		43·6	70
	22	0	0	139·8	390·1	68		49·3	70

The Mean Positions at the same hours during the Month
are given in page 64.

St. Helena { Decl. 1 Scale Division = 0′·71
H. F. k = ·00018; q = ·00025
V. F. k = ·00017; q =

Extra observations.
The V. F. was observed at 2ᵐ. 30ˢ. before, and the H. F.
2ᵐ. 20ˢ. after the times specified.

M. Gött. Time.				Decl.	Hor. Force.			Vert. Force.	
26	10	15	0	24·0		62			61
		20	0	24·0	33·4			52·6	

August 26 and 27, 1841.

M. Gött. Time.				Decl.	Hor. Force.		Vert. Force.	
d.	h.	m.	s.	Sc.-Div⁹ˢ.	Sc.-Div⁹ˢ.	Ther.	Sc.-Div⁹ˢ.	Ther.
26	10	25	0	23·9				
		30	0	23·5	33·9		51·4	
		35	0	23·4				
		40	0	23·4	34·0		51·2	
		45	0	23·5				
		50	0	23·8	33·9		51·5	
		55	0	23·7				
	11	0	0	23·6	33·9	62	51·4	61
		5	0	23·6				
		10	0	23·5	33·5		51·3	
		15	0	23·6				
		20	0	23·8	34·0		51·3	
		25	0	24·0				
		30	0	24·0	33·8		Vibrⁿ.	
		35	0	24·0				
		40	0	24·0	33·5		51·4	
		45	0	23·8				
		50	0	23·4	33·1		51·2	
		55	0	23·2				
	12	0	0	23·1	33·1	62	51·0	61
		5	0	23·1				
		10	0	23·1	33·2		51·1	
		15	0	23·1				
		20	0	23·1	33·0		51·2	
		25	0	22·9				
		30	0	22·8	32·9		51·1	
		35	0	22·8				
		40	0	22·7	33·0		51·2	
		45	0	22·6				
		50	0	22·6	33·5		51·2	
		55	0	22·7				
	13	0	0	22·6	34·1	62	51·1	61
		5	0	22·4				
		10	0	22·3	34·0		51·2	
		15	0	22·4				
		20	0		34·0			
		25	0	22·2				
		30	0	22·5	34·5		51·7	
		35	0	22·1				
		40	0	22·1	35·0		51·5	
		45	0	22·2				
		50	0	22·3	35·5		51·6	
		55	0	22·5				
	14	0	0	22·5	36·5	61	51·6	61
		5	0	22·5				
		10	0	22·1	37·0		51·6	
		15	0	22·1				
		20	0	22·0	37·0		51·6	
		25	0	22·0				
		30	0	22·0	37·1		51·6	
		35	0	22·0				
		40	0	22·7	37·7		51·6	
		45	0	22·5				
		50	0	22·8	38·1		51·8	
		55	0	22·6				
	15	0	0	22·7	39·0	61	51·8	61
		5	0	22·2				
		10	0	22·2	39·0		51·8	
		15	0	22·2				
		20	0	22·2	39·0		51·8	
		25	0	22·2				
		30	0	22·5	39·9		51·8	
		35	0	22·5				
		40	0	22·5	41·0		52·0	
		45	0	22·3				
		50	0	22·3	42·9		52·0	
		55	0	22·5				

August 26 and 27, 1841.

M. Gött. Time.				Decl.	Hor. Force.		Vert. Force.	
d.	h.	m.	s.	Sc.-Div⁹ˢ.	Sc.-Div⁹ˢ.	Ther.	Sc.-Div⁹ˢ.	Ther.
26	16	0	0	22·7	45·8	61	52·0	61
		5	0	23·0				
		10	0	23·2	49·0		52·4	
		15	0	23·6				
		20	0	23·1	49·2		52·4	
		30	0	23·9	48·9			
		35	0	23·6				
		40	0	23·6	47·9		52·3	
		45	0	23·2				
		50	0	23·1	46·9		52·2	
		55	0	23·0				
	17	0	0	22·5	46·0	61	52·1	
		5	0	22·4				
		10	0	22·1	45·1		51·8	
		15	0	22·0				
		20	0	22·0	45·0		51·7	
		25	0	22·0				
		30	0	22·1	45·0		51·9	
		35	0	22·8				
		40	0	22·9	45·0		52·0	
		45	0	23·2				
		50	0	23·1	45·6		52·1	
		55	0	23·1				
	18	0	0	23·3	45·9	61	51·9	61
		5	0	23·9				
		10	0	24·1	45·1		52·9	
		15	0	24·1				
		20	0	24·1	44·9		52·5	
		25	0	24·2				
		30	0	24·1	44·9		52·2	
		35	0	24·1				
		40	0	24·1	44·2		52·2	
		45	0	24·1				
		50	0	24·1	44·1		52·1	
		55	0	24·1				
	19	0	0	24·1	44·0	61	52·1	61
		5	0	24·3				
		10	0	24·9	43·8		52·8	
		15	0	25·0				
		20	0	25·1	44·1		53·1	
		25	0	25·9				
		30	0	26·0	44·7		52·6	
		35	0	26·0				
		40	0	26·1	45·1		52 7	
		45	0	26·1				
		50	0	26·1	46·1		52·4	
		55	0	26·7				
	20	0	0	26·9	47·1	61	52·0	61
		5	0	26·8				
		10	0	26·1	48·0		52·0	
		15	0	26·1				
		30	0	26·0	48·1		52·0	
		25	0	26·0				
		30	0	26·0	48·5		52·1	
		35	0	26·0				
		40	0	25·8	48·3		52·1	
		45	0	25·4				
		50	0	25·1	48 5		52·1	
		55	0	25·0				

Positions at the usual hours of observation
August 26 and 27.

M. Gött. Time.				Decl.	Hor. Force.		Vert. Force.	
26	0	0		24·7	57·0	61	53·9	60
	2	0		23·9	51·3	62	52·8	61
	3	0		25·2	48·0	62	51·7	61

August 26 and 27, 1841.

M. Gött. Time.				Decl.	Hor. Force.		Vert. Force.	
d.	h.	m.	s.	Sc.-Div.	Sc.-Div.	Ther.	Sc.-Div.	Ther.
26	4	0	0	25·9	46·8	62	51·3	62
	5	0	0	24·3	43·2	62	50·9	62
	6	0	0	23·2	40·2	62	51·1	62
	8	0	0	23·1	40·0	61	50·3	61
	10	0	0	24·3	33·6	62	51·9	61
	11	0	0	23·6	33·9	62	51·4	61
	12	0	0	23·1	33·1	62	51·0	61
	13	0	0	22·6	34·1	62	51·1	61
	14	0	0	22·5	36·5	61	51·6	61
	15	0	0	22·7	39·0	61	51·8	61
	16	0	0	22·7	45·8	61	52·0	61
	18	0	0	23·3	45·9	61	51·9	61
	19	30	0	26·0	44·7	61	52·1	61
	20	0	0	26·9	47·1	61	52·0	61
	20	30	0	26·0	48·5	61	52·2	60
	22	0	0	22·1	48·1	61	52·9	60
	23	0	0	23·6	55·1	61	52·8	60
27	0	0	0	25·0	54·0	61	53·5	61
	2	0	0	23·7	51·0	62	52·5	62
	3	0	0	22·5	48·9	63	51·4	62
	4	0	0	23·6	47·1	63	49·9	63
	5	0	0	25·0	49·9	66	49·3	63
	6	0	0	24·9	46·9	64	48·0	63
	8	0	0	24·9	45·0	62	48·5	62
	10	0	0	24·4	42·1	62	48·8	62
	11	0	0	23·9	41·6	62	49·0	62
	12	0	0	24·9	42·9	62	49·8	62
	13	0	0	26·1	51·8	61	49·9	61
	14	0	0	26·9	52·4	62	50·3	61
	15	0	0	26·0	51·0	62	50·5	61
	16	0	0	25·1	49·1	62	50·7	61
	18	0	0	23·5	48·0	61	51·3	61
	19	30	0	30·0	51·2	61	51·3	61
	20	0	0	29·3	49·0	61	51·3	61
	20	30	0	27·5	48·4	61	51·8	60
	22	0	0	24·4	49·8	61	52·3	60
	23	0	0	22·9	48·0	61	52·0	61

The Mean Positions at the same hours are given in page 66.

CAPE OF GOOD HOPE { Decl. 1 Scale Division = 0'·75
H. F. k = ·000180; q = ·0003
V. F. k = ·000037; q =

Positions at the usual hours of observation, August 26 and 27.

M. Gött. Time				Decl.	Hor. Force.		Vert. Force.	
26	0	0	0	51·8	38·2	55	61·7	56
	2	0	0	52·1	43·0	56	63·7	56
	4	0	0	53·4	44·4	57	67·4	57
	6	0	0	51·7	51·0	57	84·8	57
	8	0	0	51·4	53·7	57	85·5	57
	10	0	0	51·1	55·9	57	89·3	57
	12	0	0	48·0	51·8	56	88·4	57
	14	0	0	48·0	48·2	56	75·4	56
	16	0	0	48·8	41·8	55	50·2	57
	18	0	0	54·0	42·0	56	52·2	56
	20	0	0	53·2	42·0	56	59·0	57
	22	0	0	49·0	43·0	56	72·2	57
27	0	0	0	53·1	40·9	57	29·2	57
	2	0	0	55·0	51·9	58	60·2	58
	4	0	0	53·2	46·2	58	60·6	58
	6	0	0	53·6	45·8	59	60·0	59
	8	0	0	53·0	46·5	58	63·9	59
	10	0	0	52·1	49·0	59	70·5	59
	12	0	0	51·6	47·9	59	62·9	59
	14	0	0	55·7	38·8	59	76·2	59

August 26 and 27, 1841.

M. Gött. Time.				Decl.	Hor. Force.		Vert. Force.	
d.	h.	m.	s.	Sc.-Div.	Sc.-Div.	Ther.	Sc.-Div.	Ther.
27	16	0	0	53·0	41·6	59	54·4	59
	18	0	0	56·1	46·0	59	53·6	59
	20	0	0	55·0	38·8	58	59·9	58
	22	0	0	51·2	37·0	58	66·8	58

The Mean Positions at the same hours during the Month are given in page 69.

VAN DIEMEN ISLAND { Decl. 1 Scale Division = 0'·71
H. F. k = 0003; q =
V. F. k = ; q =

Extra observations.

The V. F. was observed at 2m. 30s. before, and the H. F. 2m. 30s. after the times specified.

M. Gött. Time				Decl.	Hor. Force.		Vert. Force.	
27	2	15	0	46·5	32·5	54		53
		22	30	46·6	32·2		46·4	
		30		48·7	32·5		46·4	
		37	30	50·5	32·6		46·5	
		45		51·2	33·1		46·2	
	3	15	0	57·9		54		53
		20		58·0	31·8		53·2	
		25		57·3				
		30		58·1	32·2		52·9	
		35		60·4				
		40		61·2			53·2	

Positions at the usual hours of observation, August 26 and 27.

M. Gött. Time				Decl.	Hor. Force.		Vert. Force.	
26	0	0	0	57·6	35·7	51	55·7	50
	1	0	0	59·3	33·8		56·9	
	2	0	0	58·6	33·9	51	58·8	51
	3	0	0	58·4	34·8		52·6	
	4	0	0	58·6	35·4	51	52·5	51
	5	0	0	56·6	36·3		52·8	
	6	0	0	54·1	37·1	49	53·0	49
	7	0	0	62·1	36·5		59·5	
	8	0	0	59·4	37·1	48	54·5	48
	9	0	0	62·6	38·0		54·8	
	10	0	0	63·7	40·4	47	52·4	47
	11	0	0	62·3	36·4		55·6	
	12	0	0	63·6	34·5	49	55·7	49
	13	0	0	63·6	33·5		52·0	
	14	0	0	65·4	32·0	53	50·7	53
	15	0	0	66·6	32·6		49·7	
	16	0	0	69·1	37·1	57	52·9	56
	17	0	0	69·4	28·8		55·3	
	18	0	0	70·6	29·1	58	52·9	56
	19	0	0	70·4	29·1		54·6	
	20	0	0	73·0	28·8	58	58·1	56
	21	0	0	68·2	30·3		55·4	
	22	0	0	61·3	32·6	57	51·0	55
	23	0	0	63·6	34·0		51·3	
27	0	0	0	61·4	31·5	55	54·4	55
	1	0	0	60·0	31·4		56·3	
	2	0	0	60·2	32·2	54	52·4	53
	3	0	0	57·7	32·6		53·3	
	4	0	0	60·2	32·8	54	51·2	54
	5	0	0	60·6	36·7		52·5	
	6	0	0	59·8	35·5	51	56·2	52
	7	0	0	61·3	35·8		59·3	
	8	0	0	61·9	36·6	49	58·8	48
	9	0	0	61·5	37·9		60·3	
	10	0	0	60·1	37·0	48	56·0	48
	11	0	0	67·1	37·3		57·7	

August 26 and 27, 1841.

M. Gött. Time.				Decl.	Hor. Force.		Vert. Force.	
d.	h.	m.	s.	Sc.-Div.	Sc.-Div.	Ther.	Sc.-Div.	Ther.
27	12	0	0	60·9	38·0	48	51·3	50
	13	0	0	62·5	34·9		55·1	
	14	0	0	65·5	37·2	51	47·4	52
	15	0	0	67·4	37·5		46·1	
	16	0	0	69·1	37·0	54	48·5	53
	17	0	0	70·6	36·0		54·3	
	18	0	0	72·0	33·9	53	61·6	53
	19	0	0	68·2	35·7		57·4	
	20	0	0	68·6	34·2	53	56·2	53
	21	0	0	68·4	34·0		57·3	
	22	0	5	66·6	36·0	52	54·2	52
	23	0	0	53·7	39·1		46·9	

The Mean Positions at the same hours are given in page 66.

ANTARCTIC EXPEDITION AT THE BAY OF ISLANDS, NEW ZEALAND

Decl. 1 Scale Division = 0'·73
H. F. k = ·00018; q =
V. F. k = ; q =

Positions at the usual hours of observation, August 26 and 27.

M. Gött. Time				Decl.	Hor. Force.		Vert. Force.	
26	0	0	0	36·4	89·1	52	58·9	55
	1	0	0	37·9	90·3		55·9	
	2	0	0	37·8	85·9	52	55·9	55
	3	0	0	38·5	85·7		54·9	
	4	0	0	39·9	85·6	52	57·4	55
	5	0	0	43·0	92·5		57·9	
	6	0	0	39·7	92·9	51	58·8	55
	7	0	0	42·5	91·2		57·1	
	8	0	0	41·4	93·3	50	56·6	56
	9	0	0	40·7	92·7		57·3	
	10	0	0	40·4	91·0	53	56·7	57
	11	0	0	38·8	83·6		54·7	
	12	0	0	39·3	80·0	56	53·7	60
	13	0	0	40·9	76·3		53·5	
	14	0	0	42·5	72·1	59	51·7	62
	15	0	0	44·2	71·3		53·3	
	16	0	0	46·7	64·5	63	52·2	68
	17	0	0	43·5	62·5		48·1	
	18	0	0	44·9	61·7	65	51·0	63
	19	0	0	44·2	62·7		52·9	
	20	0	0	45·1	63·1	61	54·3	62
	21	0	0	40·4	70·4		58·9	
	22	0	0	37·3	79·6	58	58·8	57
	23	0	0	39·8	79·9		58·2	
27	0	0	0	38·9	79·1	57	57·1	55
	1	0	0	37·7	80·0		57·8	
	2	0	0	33·9	86·4	55	59·6	55
	3	0	0	36·9	83·5		56·3	
	4	0	0	38·4	83·3	53	57·8	54
	5	0	0	38·9	90·4		59·3	
	6	0	0	39·5	89·8	51	58·9	53
	7	0	0	40·1	91·2		59·4	
	8	0	0	35·6	93·6	50	59·4	56
	9	0	0	37·5	90·1		58·3	
	10	0	0	37·9	93·2	56	59·4	64
	11	0	0	38·5	87·2		51·8	
	12	0	0	38·3	81·4	62	51·0	74
	13	0	0	39·9	69·6		45·7	
	14	0	0	42·3	68·4	71	46·7	77
	15	0	0	43·6	63·7		45·7	

L

August 26 and 27, 1841.

M. Gött. Time. (d. h. m. s.)	Decl. (Sc.-Div.)	Hor. Force (Sc.-Div.)	Hor. Force (Ther.)	Vert. Force (Sc.-Div.)	Vert. Force (Ther.)
27 16 0 0	44·0	58·8	75	45·9	77
17 0 0	44·1	53·5		46·4	
18 0 0	42·2	52·3	75	45·7	77
19 0 0	41·9	56·5		48·8	
20 0 0	42·4	58·8	69	49·7	71
21 0 0	42·9	61·5		49·4	
22 0 0	41·6	66·1	63	53·1	
23 0 0	37·1	75·6		56·6	68

Mean Positions at the usual hours of observation, during the Month of September.

M. Gött. Time. (d. h. m. s.)	Decl. (Sc.-Div.)	Hor. Force (Sc.-Div.)	Hor. Force (Ther.)	Vert. Force (Sc.-Div.)	Vert. Force (Ther.)
0 0 0	42·2	78·8	59	58·1	61
1 0 0	42·8	81·0		58·1	
2 0 0	42·6	82·1	58	57·8	59
3 0 0	42·9	82·8		56·8	
4 0 0	43·7	83·6	57	57·9	60
5 0 0	43·8	84·1		58·0	
6 0 0	43·8	84·8	56	58·3	58
7 0 0	43·7	85·9		58·7	
8 0 0	42·8	87·4		59·0	58
9 0 0	41·3	87·3		58·7	
10 0 0	39·0	84·4	58	57·6	59
11 0 0	39·7	80·1		55·8	
12 0 0	40·8	75·7	63	53·5	63
13 0 0	43·2	71·7		52·1	
14 0 0	45·2	69·7	66	52·1	67
15 0 0	46·4	68·5		52·2	
16 0 0	46·9	68·6	67	52·7	68
17 0 0	46·2	68·4		53·1	
18 0 0	43·4	68·8	65	53·9	68
19 0 0	44·8	70·1		55·1	
20 0 0	44·2	72·5	63	56·3	65
21 0 0	43·6	74·0		56·1	
22 0 0	43·0	76·1	61	57·8	63
23 0 0	42·2	77·5		57·9	

August 31 and September 1, 1841.

TORONTO [a] { Decl. 1 Scale Division = 0'·72
H. F. k = ·000076 ; q = ·0002
V. F. k = ·000093 ; q = ·00018

Extra observations.

The V. F. was observed at 1m. 30s. before, and the H. F. 2m. after the times specified.

M. Gött. Time. (d. h. m. s.)	Decl. (Sc.-Div.)	Hor. Force (Sc.-Div.)	Hor. Force (Ther.)	Vert. Force (Sc.-Div.)	Vert. Force (Ther.)
31 12 15 0	151·8	379·2	73	102·1	72
20 0	149·9	379·0		101·8	
25 0	146·2	378·2		101·8	
30 0	145·6	380·6		102·1	
35 0	144·5	375·1		102·1	
40 0	144·2	373·3		102·2	
45 0	142·2	371·4		102·1	
50 0	140·1	362·7		101·7	
55 0	141·1	361·8		101·6	
13 0 0	141·8	366·2	74	101·0	73
5 0	142·0	366·9		101·0	
10 0	141·9	366·4		100·2	

Positions at the usual hours of observation, August 31 and September 1.

M. Gött. Time. (d. h. m. s.)	Decl. (Sc.-Div.)	Hor. Force (Sc.-Div.)	Hor. Force (Ther.)	Vert. Force (Sc.-Div.)	Vert. Force (Ther.)
31 0 0 0	155·1	395·5	68	100·0	68
2 0 0	153·7	383·6	69	99·2	69
4 0 0	141·1	366·0	69	100·5	69
6 0 0	132·6	374·0	70	100·3	69
8 0 0	134·2	400·8	71	100·5	70
10 0 0	140·9	404·6	73	98·4	71
12 0 0	163·4	376·5	73	96·3	72
14 0 0	150·8	362·7	73	91·9	72
16 0 0	148·1	389·2	71	90·5	71
18 0 0	147·3	356·2	71	85·9	71
20 0 0	148·4	359·8	70	86·0	70
22 0 0	149·8	365·7	69	85·1	69

August 31 and September 1, 1841.

M. Gött. Time. (d. h. m. s.)	Decl. (Sc.-Div.)	Hor. Force (Sc.-Div.)	Hor. Force (Ther.)	Vert. Force (Sc.-Div.)	Vert. Force (Ther.)
1 0 0 0	157·4	392·2	68	99·6	68
2 0 0	157·8	368·4	69	97·4	68
4 0 0	139·1	342·1	70	96·0	69
6 0 0	133·0	372·0	71	96·7	70
8 0 0	133·0	390·8	72	99·9	71
10 0 0	142·4	389·7	73	91·9	72
12 0 0	143·4	374·9	74	102·4	73
14 0 0	142·4	382·1	72	95·3	72
16 0 0	145·1	379·4	72	98·1	72
18 0 0	142·9	381·7	72	92·5	72
20 0 0	144·0	389·1	71	95·0	71
22 0 0	144·1	385·9	71	95·0	71

Mean Positions at the same hours during the Month of September. [b]

M. Gött. Time. (d. h. m. s.)	Decl. (Sc.-Div.)	Hor. Force (Sc.-Div.)	Hor. Force (Ther.)	Vert. Force (Sc.-Div.)	Vert. Force (Ther.)
0 0 0	142·1	394·5	66	98·9	66
2 0 0	142·9	386·4	66	98·2	66
4 0 0	136·5	368·0	67	98·0	66
6 0 0	130·7	381·6	68	97·7	67
8 0 0	129·3	392·7	69	98·4	67
10 0 0	134·0	396·1	70	95·9	68
12 0 0	136·9	396·9	69	96·8	68
14 0 0	139·6	393·3	69	94·8	68
16 0 0	138·4	388·8	68	95·5	67
18 0 0	140·9	390·7	68	94·2	68
20 0 0	140·1	391·2	67	95·2	67
22 0 0	140·4	390·9	66	95·5	68

ST. HELENA { Decl. 1 Scale Division = 0'·71
H. F. k = ·00018; q = ·00025
V. F. k = ·00017; q =

Positions at the usual hours of observation, August 31 and September 1.

M. Gött. Time. (d. h. m. s.)	Decl. (Sc.-Div.)	Hor. Force (Sc.-Div.)	Hor. Force (Ther.)	Vert. Force (Sc.-Div.)	Vert. Force (Ther.)
31 0 0 0	23·2	53·9	60	53·9	61
2 0 0	21·9	56·5	62	53·5	62
3 0 0	22·6	54·8	63	52·6	62

---

a Toronto, September, 1841.—*Times of observation at which the Magnets were disturbed, but the mean readings were not materially changed.*

d. h.	
2 16	Decl. slight vibrations.
3 4	Decl. moderate vibrations, with shocks.
6	Decl. and H. F. moderate vibrations, with shocks.
6 0	V. F. vibrating much.
8	V. F. vibrating much.
7 12	V. F. vibrating much.
20	H. F. slight vibrations.
23	H. F. slight vibrations.
8 0	V. F. very considerable vibrations.
	V. F. very considerable vibrations.
9 22	H. F. moderate shocks.
11 2	Decl. slight vibrations.
4	Decl. slight vibrations.
16	H. F. moderate shocks.
12 22	Decl. slight vibrations; H. F. vibrating much, with shocks.
13 0	Decl. slight vibrations; H. F. moderate vibrations, with shocks.
2	H. F. vibrating much, with shocks.
6	H. F. moderate vibrations, with shocks.
20	Decl. and H. F. vibrating much, with shocks.
22	Decl. and H. F. vibrating much, with shocks.
14 0	Decl. much disturbed by shocks ; H. F. vibrating very much.
2	Decl. slight shocks ; H. F. vibrating much, with shocks.
4	Decl. and H. F. moderate vibrations, with shocks.
10	H. F. slight vibrations.
18	Decl. much disturbed by shocks ; H. F. moderately.
20	Decl. much disturbed by shocks, vibrating slightly.
15 0	H. F. vibrating much.
16 20	H. F. moderate vibrations, with shocks.

d. h.	
17 22	Decl. vibrating considerably.
18 0	Decl. vibrating much; H. F. slight vibrations, with shocks.
4	H. F. and V. F. vibrating much.
6	Decl. and H. F. disturbed moderately by shocks.
8	H. F. much disturbed by shocks and vibrations.
18 10	V. F. vibrating slightly.
12	Decl. moderate vibrations; H. F. slight shocks.
21 0	V. F. vibrating very considerably.
2	Decl. vibrating slightly ; H. F. moderately disturbed by shocks.
10	Decl. moderately disturbed by shocks.
22 8	H. F. vibrating moderately, with shocks.
23 22	Decl. and H. F. vibrating slightly.
24 0	H. F. moderately disturbed by shocks.
16	H. F. vibrating considerably.
18	H. F. moderate shocks and vibrations.
22	H. F. vibrating much, with shocks.
26 18	H. F. vibrating much.
20	H. F. vibrating much.
27 0	H. F. vibrating much ; Decl. moderate vibrations and shocks.
2	H. F. vibrating much ; Decl. slight shocks.
12	H. F. slight vibrations.
28 0	Decl. much disturbed by shocks ; H. F. vibrating much, with shocks.
6	Decl. slight shocks ; H. F. moderate vibrations, with shocks.
29 6	H. F. vibrating moderately.
30 2	H. F. vibrating much.
8	H. F. vibrating considerably.

b The mean positions of the V. F. magnet are from the 1st to the 8th September, inclusive.

August 31 and September 1, 1841.

M. Gött. Time			Decl.	Hor. Force		Vert. Force	
d. h. m.	s.		Sc.-Div^ns	Sc.-Div^ns	Ther.	Sc.-Div^ns	Ther.
31 4 0	0		25·0	52·0	63	53·0	62
5 0	0		26·6	51·0	63	53·3	62
6 0	0		26·2	49·3	63	51·5	62
8 0	0		26·2	49·1	62	51·9	62
10 0	0		26·1	48·7	62	51·7	61
11 0	0		26·2	45·1	62	51·8	61
12 0	0		27·0	50·0	61	51·9	61
13 0	0		25·9	47·8	61	52·2	61
14 0	0		25·9	48·2	61	a	
15 0	0		26·3	50·0	61		
16 0	0		25·6	49·9	61		
18 0	0		26·0	48·9	60		
19 30	0		27·3	46·0	60		
20 0	0		27·6	45·0	60		
20 30	0		28·0	48·0	60		
22 0	0		24·0	50·0	60		
23 0	0		21·1	49·6	60		
1 0 0	0		15·3	50·0	61		
2 0	0		19·0	49·7	62		
3 0	0		19·0	45·5	61		
4 0	0		19·6	42·0	62		
5 0	0		20·0	43·2	62		
6 0	0		22·0	44·6	62		
8 0	0		23·7	44·0	62		
10 0	0		22·1	43·8	61		
11 0	0		22·4	41·8	61		
12 0	0		22·9	43·1	61		
13 0	0		22·4	45·0	61		
14 0	0		20·5	b			
15 0	0		21·4			53·0	61
16 0	0		22·2			53·8	60
18 0	0		23·5			54·2	60
19 30	0		27·0			54·4	60
20 0	0		26·9			55·7	60
20 30	0		25·9			57·1	60
22 0	0		22·4			58·0	60
23 0	0		18·8			58·0	60

The Mean Positions at the same hours during the Month of August are given in page 66.

CAPE OF GOOD HOPE { Decl. 1 Scale Division = 0'·75 ; H.F. k = ·00018; q = ·0003 ; V.F.k = ·00004; q =

Positions at the usual hours of observation, August 31 and September 1.

d. h. m.	s.		Decl.	Hor. Force		Vert. Force
31 0 0	0		51·9	63·8	56	109·1
2 0	0		54·0	61·3	56	99·2
4 0	0		53·3	58·1	57	105·3
6 0	0		52·9	59·0	57	111·9
8 0	0		53·9	59·2	57	109·6
10 0	0		53·3	59·4	57	112·9
12 0	0		55·5			c
14 0	0		53·0			
16 0	0		52·7			
18 0	0		56·0			
20 0	0		53·8			
22 0	0		49·8			
1 0 0	0		50·6			
2 0	0		52·4			

August 31 and September 1, 1841.

M. Gött. Time			Decl.	Hor. Force	Ther.	Vert. Force	Ther.
d. h. m.	s.		Sc.-Div^ns	Sc.-Div^ns		Sc.-Div^ns	
1 4 0	0		50·7				
6 0	0		51·6				
8 0	0		52·0				
10 0	0		50·6				
12 0	0		51·8				
14 0	0		52·9				
16 0	0		53·7				
18 0	0		55·4				

The Mean Positions at the usual hours of observation during the Month of August are given in page 69.

VAN DIEMEN ISLAND { Decl. 1 Scale Division = 0'·71 ; H.F. k = ·0003; q = ; V.F. k = ; q =

Extra observations.

The V. F. was observed at 2m. 30s. before, and the H. F. 2m. 30s. after the times specified.

d. h. m.	s.		Decl.	Hor. Force	Ther.	Vert. Force	Ther.
1 1 10	0		95·6	60·3	56		
15	0		98·7				
20	0		100·6	59·2		46·0	55
25	0		101·3				
30	0		100·5	57·4		46·0	
35	0		101·4				
40	0		100·7	56·9		47·9	
45	0		101·5				

Positions at the usual hours of observation, August 31 and September 1.

d. h. m.	s.		Decl.	Hor. Force	Ther.	Vert. Force	Ther.
31 0 0	0		113·4	33·7	56	49·9	55
1 0	0		111·7	34·0		50·9	
2 0	0		109·7	34·4	55	48·2	54
3 0	0		112·4	34·6		48·0	
4 0	0		111·3	35·5	54	51·9	53
5 0	0		114·1	36·5		53·4	
6 0	0		114·4	36·7	53	53·8	52
7 0	0		113·4	37·8		55·0	
8 0	0		112·8	38·2	52	55·0	50
9 0	0		112·3	38·6		56·5	
10 0	0		111·8	39·1	51	57·5	49
11 0	0			37·0		d	
12 0	0		107·1	33·8	52		
13 0	0			36·3			
14 0	0		d	34·2	53	53·2	53
15 0	0						
16 0	0			55·3			
17 0	0			55·9			
18 0	0			55·9			
19 0	0		115·8	54·5		55·7	
20 0	0		111·2	53·2	59	49·2	56
21 0	0		106·5	50·1		58·2	
22 0	0		105·8	50·3	57	57·4	55
23 0	0		106·9	51·6		58·4	
1 0 0	0		104·5	55·0	56	53·9	54
1 0	0		100·0	56·9		50·3	
2 0	0		97·6	55·5	55	48·0	55
3 0	0		100·6	57·1		48·4	
4 0	0		103·2	57·9	53	49·9	52
5 0	0		105·5	58·4		50·1	

August 31 and September 1, 1841.

M. Gött. Time			Decl.	Hor. Force	Ther.	Vert. Force	Ther.
d. h. m.	s.		Sc.-Div^ns	Sc.-Div^ns		Sc.-Div^ns	
1 6 0	0		106·1	58·3	51	52·0	51
7 0	0		105·8	58·8		52·8	
8 0	0		107·0	59·5	50	54·9	50
9 0	0		104·6	60·2		55·5	
10 0	0		106·0	60·2	49	58·2	49
11 0	0		111·0	59·4		60·6	
12 0	0		105·5	59·4	50	53·8	50
13 0	0		104·4	57·9		57·1	
14 0	0		104·8	56·5	53	49·9	53
15 0	0		107·2	54·1		51·0	
16 0	0		110·5	55·6	56	50·5	55
17 0	0		110·7	54·8		50·0	
18 0	0		112·3	55·2	57	50·2	56
19 0	0		113·1	53·0		48·6	
20 0	0		113·8	56·5	58	48·6	56
21 0	0		109·7	56·4		49·2	
22 0	0		106·2	57·6	57	45·6	55
23 0	0		107·2	57·1		50·6	

Mean Positions at the same hours during the Month of September.

d. h. m.	s.		Decl.	Hor. Force	Ther.	Vert. Force	Ther.
0 0	0		105·1	57·4	55	51·4	53
1 0	0		105·0	57·6		52·1	
2 0	0		105·0	57·5	53	52·2	53
3 0	0		106·1	59·4		51·1	
4 0	0		106·5	59·8	52	53·2	51
5 0	0		106·6	59·7		53·2	
6 0	0		107·3	60·2	51	55·2	50
7 0	0		108·3	60·4		56·0	
8 0	0		108·4	60·8	48	56·1	49
9 0	0		107·8	60·8		56·5	
10 0	0		106·8	61·4	49	56·3	49
11 0	0		104·9	60·4		55·9	
12 0	0		103·2	59·4	50	54·0	50
13 0	0		104·1	57·7		53·7	
14 0	0		106·6	56·5	52	53·5	52
15 0	0		109·6	55·8		54·0	
16 0	0		112·2	56·3	55	52·4	54
17 0	0		114·0	57·2		51·4	
18 0	0		113·7	56·4	57	51·3	55
19 0	0		112·7	56·1		51·1	
20 0	0		111·3	56·4	58	50·7	55
21 0	0		108·5	56·4		50·9	
22 0	0		106·9	56·8	56	50·3	54
23 0	0		105·7	57·1		51·7	

ANTARCTIC EXPEDITION AT THE BAY OF ISLANDS, NEW ZEALAND.

Decl. 1 Scale Division = 0'·73 ; H.F. k = ·00018 ; q = ; V.F. k = ; q =

Positions at the usual hours of observation, August 31 and September 1.

d. h. m.	s.		Decl.	Hor. Force	Ther.	Vert. Force	Ther.
31 0 0	0		40·4	81·1	54	59·4	55
1 0	0		40·0	83·5		56·4	
2 0	0		40·9	85·1	52	57·6	54
3 0	0		41·7	86·0		54·2	
4 0	0		42·7	88·7	50	59·5	52
5 0	0		43·6	90·4		58·9	

---

a V. F. magnet under adjustment.
b H. F. magnet under adjustment.
c H. F. and V. F. magnets out of adjustment.
d The connexion of the series of readings with each of the three magnetometers was broken about 31d. 12h., for the purpose of determining the magnetic moments of the bars; the readings, prior to the connexion being broken, are comparable with the mean monthly positions of August; the subsequent ones with those of September.

AUGUST 31 and SEPTEMBER 1, 1841.

M. Gött. Time. d. h. m. s.	Decl. Sc.-Div.	Hor. Force. Sc.-Div.	Ther.	Vert. Force. Sc.-Div.	Ther.
31 6 0 0	42·9	90·0	49	60·9	53
7 0 0	41·5	94·1		59·7	
8 0 0	40·0	95·9	47	59·3	52
9 0 0	38·9	96·2		60·8	
10 0 0	37·0	93·4	51	60·0	52
11 0 0	36·9	87·2		57·2	
12 0 0	38·0	86·2	61	50·9	60
13 0 0	42·3	72·3		51·8	
14 0 0	43·5	72·0	62	52·5	62
15 0 0	44·5	72·1		52·8	
16 0 0	45·2	71·6	65	53·0	64
17 0 0	45·8	70·0		52·9	
18 0 0	45·4	69·1	64	53·5	65
19 0 0	43·5	67·9		54·1	
20 0 0	37·9	75·3	62	58·7	62
21 0 0	40·0	66·7		55·2	
22 0 0	39·2	68·9	59	57·1	60
23 0 0	39·6	69·8		57·9	
1 0 0 0	40·5	77·5	54	60·8	55
1 0 0	40·6	84·6		59·5	
2 0 0	40·4	84·2	52	58·4	55
3 0 0	41·4	86·3		59·7	
4 0 0	42·3	88·0	51	60·7	56
5 0 0	43·6	90·6		61·5	
6 0 0	41·7	89·5	47	61·0	55
7 0 0	41·9	93·1		62·3	
8 0 0	40·1	94·3	45	63·2	55
9 0 0	39·8	95·8		63·4	
10 0 0	39·5	89·7	51	59·7	56
11 0 0	40·2	79·5		55·5	
12 0 0	39·6	74·8	60	54·7	58
13 0 0	41·1	72·5		54·4	
14 0 0	44·3	70·3	64	53·3	64
15 0 0	45·3	67·6		52·5	
16 0 0	46·1	66·1	68	51·9	66
17 0 0	45·8	63·8		52·7	
18 0 0	45·5	66·1	66	54·9	67
19 0 0	43·9	69·0		58·1	
20 0 0	44·3	72·2	60	57·7	62
21 0 0	44·2	74·9		58·1	
22 0 0	43·3	80·0	56	60·6	56
23 0 0	41·7			58·3	

The Mean Positions at the same hours during the Month of September are given in page 74.

SEPTEMBER 12, 13, and 14, 1841.

TORONTO { Decl. 1 Scale Division = 0'·72
H. F. k = ·000076 ; q = ·00021
V. F. k = ·000093 ; q = ·00018 }

Extra observations.
The V. F. was observed at 1m. 30s. before, and the H. F. 2m. after the times specified.

M. Gött. Time. d. h. m. s.	Decl. Sc.-Div.	Hor. Force. Sc.-Div.	Ther.	Vert. Force. Sc.-Div.	Ther.
13 20 25 0a	148·8	317·5	65	78·7	65
30 0	148·5	318·4		75·4	
35 0	146·5	323·7		78·2	
40 0	143·2	337·3		80·9	
45 0	142·3	356·3		81·5	
50 0	139·5	353.7		84·7	

SEPTEMBER 12, 13, and 14, 1841.

M. Gött. Time. d. h. m. s.	Decl. Sc.-Div.	Hor. Force. Sc.-Div.	Ther.	Vert. Force. Sc.-Div.	Ther.
13 20 55 0	138·4	368·3		84·3	
21 0 0	136·6	375·9	64	85·5	66
5 0	133·6	379·9		84·7	
10 0	134·7	392·4		84·8	
15 0	137·7	400·9		85·9	
20 0	141·8	406·0		85·2	
25 0	143·2	306·5		85·2	
30 0	142·8	392·1		84·0	
35 0	144·9	385·9		84·0	
40 0	145·5	375·3		83·9	
45. 0	146·8	355·7		82·7	
50 0	148·3	351·6		79·5	
22 0 0	133·4	318·2		79·0	
30 0	139·6	382·5	64	86·8	65
35 0	140·9	391·1		87·8	
40 0	142·3	398·7		88·8	
45 0	142·8	397·0		90·4	
50 0	145·5	400·9		91·3	
55 0	144·4	394·9		91·3	

Positions at the usual hours of observation, September 12, 13, and 14.

M. Gött. Time. d. h. m. s.	Decl. Sc.-Div.	Hor. Force. Sc.-Div.	Ther.	Vert. Force. Sc.-Div.	Ther.
12 18 0 0b	151·9	390·3	68	90·2	68
20 0 0	145·3	382·0	66	91·3	67
22 0 0	147·8	388·5	65	91·2	66
13 0 0 0	141·0	407·4	64	89·0	65
2 0 0	145·9	300·4	66	92·1	66
4 0 0	135·5	351·6	68	92·6	67
6 0 0	130·5	375·0	69	103·6	70
8 0 0	135·1	397·0	69	109·2	69
10 0 0	136·5	399·4	70	107·1	69
12 0 0	137·3	390·7	70	104·5	69
14 0 0	136·5	303·9	69	104·3	68
16 0 0	140·7	394·5	67	104·1	67
18 0 0	151·5	360·2	65	104·3	67
20 0 0	146·8	345·8	65	83·6	65
22 0 0	133·4	318·2	64	79·0	65
14 0 0 0	143·2	405·6	62	95·3	64
2 0 0	142·8	395·9	64	103·2	64
4 0 0	138·6	375·4	66	106·6	65
6 0 0	127·1	371·6	67	108·3	66
8 0 0	130·9	399·0	68	107·4	67
10 0 0	136·4	399·9	68	108·1	67
12 0 0	138·0	399·5	68	106·1	67
14 0 0	137·8	396·6	67	106·7	67
16 0 0	140·2	418·2	66	102·5	66
20 0 0	137·9	401·1	64	106·6	64
22 0 0	138·5	403·5	63	109·6	63
	138·5	407·5	62	110·1	62

The Mean Positions at the same hours during the Month are given in page 74.

ST. HELENA { Decl. 1 Scale Division = 0'·71
H. F. k = ·00018 ; q = ·00025
V. F. k = ·0002 ; q = }

Positions at the usual hours of observation, September 12, 13, and 14.

M. Gött. Time. d. h. m. s.	Decl. Sc.-Div.	Hor. Force. Sc.-Div.	Ther.	Vert. Force. Sc.-Div.	Ther.
12 14 0 0	23·8c	157·1	60	50·6	60
15 0 0	24·0	157·1	60	50·2	60

SEPTEMBER 12, 13, and 14, 1841.

M. Gött. Time. d. h. m. s.	Decl. Sc.-Div.	Hor. Force. Sc.-Div.	Ther.	Vert. Force. Sc.-Div.	Ther.
12 16 0 0	24·1	157·5	60	50·0	60
18 0 0	23·9	159·0	60	50·5	59
19 30 0	30·1	160·0	60	50·2	59
20 0 0	29·1	159·8	60	50·2	59
20 30 0	28·0	159·8	61	50·6	59
22 0 0	26·9	160·2	60	51·2	60
23 0 0	25·6	163·2	61	50·2	60
13 0 0 0	24·8	159·2	61	50·1	61
2 0 0	24·1	153·7	62	50·3	61
3 0 0	24·0	152·0	62	51·1	62
4 0 0	22·0	149·0	62	51·2	62
5 0 0	22·9	151·0	62	50·9	62
6 0 0	24·0	150·0	62	50·7	62
8 0 0	26·2	152·2	62	50·8	61
10 0 0	26·2	151·1	61	50·7	61
11 0 0	26·8	155·4	61	50·7	61
12 0 0	26·1	153·4	61	50·7	60
13 0 0	26·0	153·0	61	50·9	60
14 0 0	25·8	153·1	61	51·1	60
15 0 0	25·5	153·9	60	51·1	60
16 0 0	25·9	154·7	60	51·3	60
18 0 0	26·1	154·9	60	51·2	60
19 30 0	27·9	155·1	60	51·3	59
20 0 0	26·2	155·8	60	50·8	59
20 30 0	25·9	154·0	60	50·2	59
22 0 0	23·4	150·0	60	49·6	59
23 0 0	22·9	154·2	60	47·8	60
14 0 0 0	23·0	152·0	60	47·2	61
2 0 0	22·3	152·1	61	44·3	61
3 0 0	22·9	152·6	61	44·1	61
4 0 0	24·2	151·8	62	43·3	61
5 0 0	24·5	151·2	62	43·3	61
6 0 0	25·0	148·2	62	43·5	61
8 0 0	25·8	150·2	61	44·6	61
10 0 0	25·9	150·9	61	46·6	60
11 0 0	25·8	151·3	61	46·5	60
12 0 0	26·0	152·9	61	47·1	60
13 0 0	26·1	154·9	61	47·6	60
14 0 0	26·1	154·0	60	48·3	60
15 0 0	26·0	153·8	60	48·6	60
16 0 0	26·5	155·1	60	48·8	60
18 0 0	26·7	154·9	60	49·0	59
19 30 0	28·5	155·0	60	49·0	59
20 0 0	27·0	155·0	60	49·1	59
20 30 0	26·8	155·3	59	49·1	59
22 0 0	25·0	159·5	60	48·6	59
23 0 0	25·9	160·0	60	49·1	60

Mean Positions at the same hours during the Month.

d. h. m. s.	Decl.	Hor. Force.	Ther.
0 0 0	23·0	161·9	61
2 0 0	23·9	159·6	62
3 0 0	23·5	136·8	62
4 0 0	23·2	153·9	62
5 0 0	23·1	152·3	62
6 0 0	23·3	151·1	62
8 0 0	24·1	150·0	62
10 0 0	24·0	151·0	61
11 0 0	24·2	152·0	61
12 0 0	24·2	152·6	61
13 0 0	24·3	153·1	61
14 0 0	24·2	154·2	61

d. h. m.
a 13 20 0 Calm and clear. Auroral light, and a few faint streamers in the North.
21 30
and } Bank of Auroral light in the North, with faint streamers and pulsations.
22 0

b Commencing after midnight of Sunday at Toronto.
c Commencing after midnight of Sunday at St. Helena.

SEPTEMBER 12, 13, and 14, 1841.

M. Gött. Time	Decl.	Hor. Force		Vert. Force	
d. h. m. s.	Sc.-Div.	Sc.-Div.	Ther.	Sc.-Div.	Ther.
15 0 0	24·2	154·5	61		
16 0 0	24·3	155·3	61		
18 0 0	24·8	156·1	60		
19 30 0	26·6	155·7	60		
20 0 0	25·7	155·9	60		
20 30 0	25·0	156·2	60		
22 0 0	23·4	160·8	60		
23 0 0	23·0	156·2	60		

CAPE OF GOOD HOPE { Decl. 1 Scale Division = 0'·75
H. F. k = ·00018; q = ·0003
V. F. k = ; q =

Extra observations.
The H. F. was observed at 2m. 30s. after the times specified.

M. Gött. Time	Decl.	Hor. Force		Vert. Force	
13 2 35 0	52·2				
40 0	52·5	63·2	56		
45 0	52·4				
50 0	52·5	61·8			
55 0	52·2				
3 0 0	52·5	61·3			
5 0	52·2				
10 0	51·5	60·6			
15 0	51·0				
20 0	50·2	61·1			
25 0	50·0				
30 0	50·0	60·0			
35 0	47·3				
40 0	49·8	58·2			
45 0	50·0				
50 0	49·8	57·7			
55 0	49·4				
4 0 0	49·2	58·9	57		
5 0	49·1				
10 0	49·7	60·8			
15 0	50·0				
20 0	50·3	60·6			
25 0	50·5				
30 0	50·5	59·4			
35 0	50·3				
40 0	50·0	59·4			
45 0	49·0				
50 0	49·0	62·2			
55 0	49·3				
5 0 0	49·2	63·2	57		
5 0	49·3				
10 0	49·3	63·8			
15 0	49·3				
20 0	49·4	63·4			
25 0	49·5				
30 0	49·5	62·8			
35 0	50·0				
40 0	49·7	62·8			
45 0	49·4				
50 0	49·8	62·6			
55 0	49·8				
6 0 0	50·0	62·3	58		
5 0	50·1				
10 0	50·0	62·3			
7 0 0	50·0	61·8	59		
11 0 0	53·0	70·1	57		

SEPTEMBER 12, 13, and 14, 1841.

Positions at the usual hours of observation, September 12, 13, and 14.

M. Gött. Time	Decl.	Hor. Force		Vert. Force	
d. h. m. s.	Sc.-Div.	Sc.-Div.	Ther.	Sc.-Div.	Ther.
12 12 0 0	52·5*	72·2	55		
14 0 0	50·4	75·5	55		
16 0 0	50·8	76·7	55		
18 0 0	52·0	75·7	54		
20 0 0	53·4	81·0	54		
22 0 0	53·4	79·0	55		
13 0 0 0	53·2	66·0	56		
2 0 0	51·0	65·0	56		
4 0 0	49·2	58·9	58		
6 0 0	50·0	62·3	58		
8 0 0	52·2	66·4	58		
10 0 0	51·2	67·1	58		
12 0 0	52·0	69·6	57		
14 0 0	51·8	68·5	56		
16 0 0	52·6	70·1	56		
18 0 0	54·5	70·5	56		
20 0 0	54·4	72·0	56		
22 0 0	49·0	65·0	57		
14 0 0 0	50·5	63·0	57		
2 0 0	50·0	60·8	58		
4 0 0	51·2	63·7	59		
6 0 0	50·9	62·9	59		
8 0 0	51·0	65·1	58		
10 0 0	50·8	66·3	58		
12 0 0	52·0	68·6	58		
14 0 0	52·0	69·8	58		
16 0 0	53·1	69·4	58		
18 0 0	54·0	70·9	57		
20 0 0	50·1	74·2	56		
22 0 0	46·8	74·9	57		

Mean Positions at the same hours during the Month.[b]

M. Gött. Time	Decl.	Hor. Force		Vert. Force	
0 0 0	50·4	73·3	57		
2 0 0	53·3	70·7	58		
4 0 0	52·1	66·9	58		
6 0 0	51·8	66·3	58		
8 0 0	51·7	65·7	58		
10 0 0	52·0	68·5	58		
12 0 0	52·5	72·0	57		
14 0 0	52·7	73·7	57		
16 0 0	53·1	74·7	57		
18 0 0	54·3	75·3	57		
20 0 0	51·2	78·8	56		
22 0 0	48·8	76·9	56		

VAN DIEMEN ISLAND { Decl. 1 Scale Division = 0'·71
H. F. k = ·0003; q =
V. F. k = ; q =

Extra observations.
The V. F. was observed at 2m. 30s. before, and the H. F. 2m. 30s. after the times specified.

M. Gött. Time	Decl.	Hor. Force		Vert. Force	
12 23 15 0	98·1				
20 0	95·3	54·4	45	70·2	44
25 0	93·3				
30 0	90·9	50·4		70·0	
35 0	88·0				
40 0	89·8	53·4		71·7	

SEPTEMBER 12, 13, and 14, 1841.

M. Gött. Time	Decl.	Hor. Force		Vert. Force	
d. h. m. s.	Sc.-Div.	Sc.-Div.	Ther.	Sc.-Div.	Ther.
12 23 45 0	92·5				
50 0	95·8	52·9		71·8	
55 0	98·9				
13 0 10 0	100·2	55·0	45		
15 0	100·2				
20 0	101·7	53·3		72·8	44
25 0	103·2				
30 0	103·8	50·4		75·4	
35 0	103·5				
40 0	104·3	55·8		74·7	
45 0	105·6				
50 0	105·8	56·2		73·4	
2 15 0	103·7	61·2	45		
22 30	101·4	61·7			
30 0	99·5	61·5		62·6	44
37 30	97·0	62·5		61·6	
45 0	95·1	62·8		59·2	
3 10 0	96·9	59·5	45		
15 0	97·4				
20 0	98·3	59·4		61·9	44
25 0	98·3				
30 0	98·8	64·8		62·5	
35 0	101·6				
40 0	105·8	68·2		59·3	
45 0	106·5				
50 0	106·6	67·9		54·3	
55 0	106·5				
4 5 0	106·4		45		45

Positions at the usual hours of observation, September 12, 13, and 14.

M. Gött. Time	Decl.	Hor. Force		Vert. Force	
12 3 0 0	107·6	67·1		61·8	
4 0 0	106·6	66·3	45	64·1	44
5 0 0	105·7	65·9		63·5	
6 0 0	104·3	65·8	44	63·5	43
7 0 0	105·0	64·6		66·5	
8 0 0	105·2	64·8	43	66·6	41
9 0 0	106·2	66·1		66·2	
10 0 0	106·8	66·2	42	65·7	42
11 0 0	105·1	66·3		64·2	
12 0 0	103·4	65·7	43	63·8	42
13 0 0	103·0	63·9		63·6	
14 0 0	106·4	62·9	43	66·0	43
15 0 0	111·0	62·0		67·9	
16 0 0	115·3	64·3	44	65·8	44
17 0 0	120·8	66·7		66·3	
18 0 0	119·9	64·8	45	67·0	44
19 0 0	117·0	63·7		64·6	
20 0 0	114·6	63·6	45	66·3	44
21 0 0	110·4	61·8		66·7	
22 0 0	112·1	60·6	45	73·1	43
23 0 0	105·4	56·3		71·4	
13 0 0 0	102·4	52·8	45	75·3	44
1 0 0	106·3	58·0		72·2	
2 0 0	106·7	61·4	45	66·9	44
3 0 0	97·3	61·5		57·8	
4 0 0	106·6	66·0	45	54·2	45
5 0 0	106·8	63·2		56·1	
6 0 0	107·1	62·8	43	64·3	43
7 0 0	107·3	61·8		66·4	
8 0 0	109·0	62·1	43	67·5	42
9 0 0	108·2	63·4		65·1	
10 0 0	105·1	62·0	43	66·0	42

* Commencing after midnight of Sunday at the Cape of Good Hope.

[b] The mean positions of the H. F. magnet are from the 13th to 30th September, inclusive.

, 12, 13, and 14, 1841.					SEPTEMBER 12, 13 and 14, 1841.					SEPTEMBER 12, 13, and 14, 1841.						
,me.	Decl.	Hor. Force.		Vert. Force.	M. Gött. Time.	Decl.	Hor. Force.		Vert. Force.	M. Gött. Time.	Decl.	Hor. Force.		Vert. Force.	
h. m. s.	Sc.-Div^{ns}.	Sc.-Div^{ns}.	Ther.	Sc.-Div^{ns}.	Ther.	d. h. m. s.	Sc.-Div^{ns}.	Sc.-Div^{ns}.	Ther.	Sc.-Div^{ns}.	Ther.	d. h. m. s.	Sc.-Div^{ns}.	Sc.-Div^{ns}.	Ther.	Sc.-Div^{ns}.	Ther.

h. m. s.	Decl.	Hor. F.	Th.	Vert. F.	Th.	d. h. m. s.	Decl.	Hor.F	Th	Vert	Th	d. h. m. s.	Decl.	Hor.F	Th	Vert	Th
13 11 0 0	105·6	61·7		68·4		12 7 0 0	41·7	93·9		59·8		14 15 0 0	46·8	73·8		55·3	
12 0 0	104·1	63·0	43	63·4	43	8 0 0	39·7	95·0	53	57·6	58	16 0 0	46·9	72·6	63	54·6	65
13 0 0	103·9	62·8		59·7		9 0 0	38·3	93·1		57·8		17 0 0	46·0	74·4		55·5	
14 0 0	107·3	63·0	46	58·4	46	10 0 0	36·9	90·4	57	57·2	56	18 0 0	45·3	72·0	65	54·4	66
15 0 0	109·9	62·3		58·8		11 0 0	36·7	84·2		54·5		19 0 0	44·0	72·5		56·0	
16 0 0	111·9	62·6	49	58·0	48	12 0 0	38·5	74·8	67	46·9	68	20 0 0	44·5	74·3	63	56·7	63
17 0 0	112·0	59·8		60·9		13 0 0	41·9	68·8		45·8		21 0 0	43·8	77·5		57·9	
18 0 0	113·7	58·6	50	63·8	49	14 0 0	44·0	63·4	75	45·5	76	22 0 0	43·8	79·1	60	58·5	62
19 0 0	106·7	56·3		65·3		15 0 0	46·1	60·2		46·9		23 0 0	43·5	80·0		59·2	
20 0 0	110·8	53·4	51	69·5	49	16 0 0	46·1	61·7	75	48·6	75						
21 0 0	107·5	54·1		67·4		17 0 0	47·4	62·1		51·4		The Mean Positions at the same hours during the Month are given in page 74.					
22 0 0	97·2	56·5	50	58·0	49	18 0 0	45·6	63·1	71	51·9	73						
23 0 0	106·6	54·0		64·9		19 0 0	44·9	65·7		53·1							
14 0 0 0	102·1	56·1	49	61·9	48	20 0 0	42·2	67·7	67	54·4	69	SEPTEMBER 24, 25, 26, 27, 28, and 29, 1841.					
1 0 0	102·1	55·4		64·6		21 0 0	42·6	66·2		55·3							
2 0 0	105·3	58·0	48	63·4	47	22 0 0	42·0	65·1	65	56·3	66	TORONTO { Decl. 1 Scale Division = 0'·72					
3 0 0	107·7	59·2		61·3		23 0 0	38·9	64·0		55·7		H. F. k = ·000076 ; q = ·0002					
4 0 0	108·8	59·4	49	59·2	47	13 0 0 0	35·2	63·2	64	56·7	65	V. F. k = ·000093 ; q = ·00018					
5 0 0	107·5	60·8		58·1		1 0 0	39·8	67·3		56·1							
6 0 0	107·3.	61·3	46	62·0	45	2 0 0	41·3	71·0	63	56·7	65	Regular and extra observations.					
7 0 0	108·9	61·2		64·5		3 0 0	38·7	75·7		56·9							
8 0 0	108·6	61·6	43	65·4	44	4 0 0	42·6	76·9	63	57·5	66	The V. F. was observed at 2^m. 30^s. before, and the H. F. 2^m. after the times specified.					
9 0 0	108·5	61·3		65·5		5 0 0	42·8	73·4		55·8							
10 0 0	107·1	62·7	44	65·1	44	6 0 0	41·6	75·7	61	57·0	61						
11 0 0	105·4	62·7		64·1		7 0 0	40·7	75·6		57·1		24 0 0 0	142·0	418·0		83·6	
12 0 0	102·7	61·5	44	62·0	44	8 0 0	39·8	76·7	60	56·6	62	2 0 0	141·3	416·1		83·0	
13 0 0	102·6	60·0		61·7		9 0 0	38·7	77·8		57·2		4 0 0	140·9	413·2		81·1	
14 0 0	105·8	58·7	48	60·6	47	10 0 0	37·4	74·2	65	54·7	64	6 0 0	137·7	412·7		79·7	
15 0 0	110·0	58·6		59·8		11 0 0	37·9	69·2		52·5		8 0 0	126·2	388·2		81·9	
16 0 0	111·2	56·2	51	60·0	50	12 0 0	38·7	64·4	71	49·9	72	10 0 0	121·6	395·2		89·0	
17 0 0	112·9	59·0		55·3		13 0 0	40·7	60·7		49·1		12 0 0	120·4	408·2		98·6	
18 0 0	112·5	58·3	53	55·7	51	14 0 0	43·1	57·8	76	48·5	76	14 0 0	125·0	384·9		102·2	
19 0 0	110·5	58·4		54·1		15 0 0	43·5	55·9		48·6		16 0 0	146·3	326·3		79·1	
20 0 0	109·2	58·5	54	54·0	52	16 0 0	44·3	56·1	76	49·3	77	18 0 0	142·8	360·6		75·3	
21 0 0	107·3	58·6		53·7		17 0 0	44·4	55·3		49·3		20 0 0	156·9	334·6		72·4	
22 0 0	107·2	59·5	53	53·2	51	18 0 0	44·1	56·1	71	50·6	72	22 0 0	169·4	306·2	66	37·5	66
23 0 0	106·5	59·4		53·5		19 0 0	39·4	58·8		53·2		20 0	176·0	297·5		31·1	
						20 0 0	40·0	55·6	68	52·8	70	25 0	176·0	303·7		32·6	
The Mean Positions at the same hours during the Month are given in page 75.						21 0 0	44·0	57·0		55·4		30 0	171·7	292·6		30·9	
						22 0 0	39·7	66·8	67	57·0	68	35 0	166·0	280·5		30·9	
						23 0 0	42·4	59·5		54·4		40 0	167·9	280·9		31·8	
ANTARCTIC EXPEDITION AT THE BAY OF ISLANDS, NEW ZEALAND.						14 0 0 0	41·4	61·2	67	53·8	68	45 0	168·9	275·4		30·5	
						1 0 0	41·3	63·1		56·7		50 0	169·8	265·1		27·4	
Decl. 1 Scale Division = 0'·73						2 0 0	44·3	66·5	65	57·2	66	55 0	156·4	284·2		41·3	
H. F. k = .00018 ; q =						3 0 0	44·9	67·1		56·0		23 0 0	156·2	285·1	66	48·5	67
V. F. k = ; q =						4 0 0	45·2	69·7	64	56·5	65	5 0	129·1	298·9		42·5	
						5 0 0	44·5	71·1		56·5		10 0	131·0	290·1		40·8	
Positions at the usual hours of observation, September 12, 13, and 14.						6 0 0	45·1	71·5	63	56·5	64	15 0	132·9	287·0		40·8	
						7 0 0	44·7	71·5		57·3		20 0	135·5	296·1		43·5	
12 1 0 0	41·6	91·7		58·6		8 0 0	44·4	72·4	63	57·5	64	25 0	138·0	316·5		44·1	
2 0 0	41·4	94·1	54	57·8	55	9 0 0	43·1	71·0		55·9		30 0	145·9	299·7		40·5	
3 0 0	40·4	94·2		57·2		10 0 0	42·4	72·1	63	55·9	64	35 0	144·5	285·1		35·4	
4 0 0	41·0	91·8	54	57·6	56	11 0 0	41·7	73·2		56·3		40 0	142·1	282·0		35·4	
5 0 0	41·4	92·9		59·4		12 0 0	43·2	72·8	62	54·1	63	45 0	140·2	244·6		22·5	
6 0 0	41·0	94·2	53	59·6	56	13 0 0	45·2	71·9		54·6		50 0	149·4	180·5		26·9	
						14 0 0	45·8	72·4	62	55·1	64	55 0	146·9	215·6		31·5	

d. h. m.
* 24 23 0 Very densely clouded.
  25 9 0 Clearing.
    13 12 Sky almost entirely clear; bright bank of Auroral light moving across the sky like light cirri.
    13 17 Several bright streamers and patches of light appearing and disappearing rapidly.
    13 32 Light brighter. Clouds rising rapidly.
    13 35 Sky overcast. The sky remaining overcast, nothing further was seen of the Aurora until 15^h. 27^m., when several splendid streamers and bright pulsations were observed in the N.E., which continued visible until 15^h. 52^m., growing gradually fainter.

d. h. m.
  25 15 52 Partially clouded round N. horizon; low bank of light in the N.
    16 37 Clearing slightly in N. Light larger and brighter.
    17 37 Densely clouded ; nothing of Aurora visible.
    17 57 Densely clouded.
    18 0 No further observations were made after this hour, being midnight of Saturday at Toronto. In the course of the night the wind, from being light, commenced to blow in heavy squalls, accompanied by violent rain; it continued to blow a gale the whole of Sunday, without rain, but lulled towards evening. At night, the sky being nearly clear for a short time, the Aurora was again seen.

September 24, 25, 26, 27, 28, and 29, 1841.

M. Gött. Time				Decl.	Hor. Force.		Vert. Force.	
d.	h.	m.	s.	Sc.-Div^m.	Sc.-Div^m.	Ther.	Sc.-Div^m.	Ther.
25	0	0	0	143·1	223·4	66	20·5	67
		5	0	139·4	212·9		15·7	
		10	0	137·9	177·8		10·6	
		15	0	131·1	196·0		28·7	
		20	0	122·5	200·2		32·6	
		25	0	127·9	236·0		29·9	
		30	0	127·4	245·2		35·1	
		35	0	128·0	238·1		35·5	
		40	0	125·7	258·0		32·2	
		45	0	125·4	266·1		31·7	
		50	0	126·8	253·1		32·7	
		55	0	121·9	247·6		37·8	
	1	0	0	111·8	251·5	66	33·5	66
		5	0	110·9	250·1		33·6	
		10	0	117·6	234·3		36·4	
		15	0	126·6	221·7		42·3	
		20	0	127·7	224·7		49·2	
		25	0	122·5	250·0		54·2	
		30	0	110·8	258·2		58·9	
		35	0	109·6	266·4		58·4	
		40	0	103·0	287·9		61·9	
		45	0	102·0	278·1		60·9	
		50	0	98·1	238·7		48·7	
		55	0	103·5	220·1		48·0	
	2	0	0	115·6	217·6	66	55·9	66
		5	0	121·4	231·8		58·6	
		10	0	118·4	262·6		60·5	
		15	0	116·9	260·9		60·1	
		20	0	117·4	242·9		66·6	
		25	0	115·5	225·9		66·2	
		30	0	111·9	225·6		64·1	
		35	0	116·9	230·2		67·9	
		40	0	117·2	240·0		72·3	
		45	0	117·4	238·7		69·7	
		50	0	121·7	245·7		76·3	
		55	0	120·1	254·3		74·9	
	3	0	0	120·1	236·9	66	73·7	66
		5	0	98·5	226·3		70·0	
		10	0	100·1	222·2		72·4	
		15	0	99·2	214·2		67·6	
		20	0	92·2	206·7		67·0	
		25	0	96·6	223·7		71·2	
		30	0	110·3	227·3		70·7	
		35	0	111·9	225·1		73·8	
		40	0	115·2	220·5		77·5	
		45	0	103·4	223·5		82·2	
		50	0	107·2	207·2		80·4	
		55	0	105·5	224·0		83·3	
	4	0	0	110·1	226·8	66	83·8	66
		5	0	112·4	235·6		87·2	
		10	0	112·3	252·4		90·3	
		15	0	111·5	230·7		90·1	
		20	0	109·5	234·3		92·5	
		25	0	108·0	230·6		90·0	
		30	0	105·2	241·5		93·2	
		35	0	102·8	257·6		97·0	
		40	0	108·2	284·7		99·8	
		45	0	109·3	284·9		99·0	
		50	0	110·6	275·3		97·4	
		55	0	109·0	294·7		99·0	
	5	0	0	115·1	292·9	66	102·4	65
		5	0	109·7	303·4		99·3	
		10	0	122·8	313·4		102·0	
		15	0	128·7	319·7		100·9	
		20	0	129·0	298·4		98·8	
		25	0	123·8	306·6		98·7	
		30	0	125·7	316·3		99·5	
	5	35	0	131·0	325·3		101·0	
		40	0	137·1	328·6		100·1	
		45	0	141·0	334·4		102·9	
		50	0	143·0	337·6		102·8	
		55	0	144·9	338·4		100·3	
	6	0	0	140·2	348·9	66	105·0	66
		5	0	139·6	356·4		105·2	
		10	0	141·1	352·9		103·9	
		15	0	141·0	332·8		103·3	
		20	0	141·7	326·5		104·7	
		25	0	142·4	319·3		105·0	
		30	0	133·0	336·6		108·3	
		35	0	140·2	361·9		108·4	
		40	0	139·0	373·3		109·1	
		45	0	142·3	365·2		107·9	
		50	0	128·2	347·1		107·9	
		55	0	130·5	357·7		111·2	
	7	0	0	131·0	363·0	66	112·1	66
		5	0	130·5	369·2		112·0	
		10	0	127·6	364·3		112·6	
		15	0	130·2	372·5		113·2	
		20	0	126·7	375·2		113·4	
		25	0	124·6	375·6		114·0	
		30	0	122·6	380·1		114·0	
		35	0	118·7	367·2		113·6	
		40	0	113·9	366·5		113·9	
		45	0	116·4	400·9		114·0	
		50	0	111·7	390·6		112·7	
		55	0	113·5	388·9		110·8	
	8	0	0	109·5	365·2	66	109·7	66
		5	0	111·0	391·5		109·1	
		10	0	106·6	387·6		109·1	
		15	0	108·0	380·9		107·4	
		20	0	110·6	385·4		106·6	
		25	0	110·9	388·0		105·6	
		30	0	114·8	372·7		102·4	
		35	0	115·6	371·2		101·5	
		40	0	116·4	372·7		100·3	
		45	0	116·0	369·6		98·9	
		50	0	117·2	375·4		97·3	
		55	0	118·3	382·9		98·3	
	9	0	0	118·0	378·2	67	97·7	65
		5	0	118·1	365·9		97·4	
		10	0	119·2	369·2		97·4	
		15	0	119·7	369·7		98·3	
		20	0	121·8	360·7		99·5	
		25	0	124·7	365·8		101·1	
		30	0	125·0	373·2		101·0	
		35	0	125·9	374·7		104·2	
		40	0	126·3	373·2		104·2	
		45	0	127·8	370·4		103·3	
		50	0	128·6	364·1		101·8	
		55	0	130·0	360·7		101·1	
	10	0	0	131·0	361·7	67	101·6	66
		5	0	128·5	368·1		103·1	
		10	0	123·5	372·5		113·7	
		15	0	123·5	388·9		113·7	
		20	0	125·6	394·5		114·6	
		25	0	125·6	399·2		115·7	
		30	0	122·3	406·7		119·0	
		35	0	123·9	414·2		121·3	
		40	0	125·0	402·6		122·0	
		45	0	129·8	415·2		122·5	
		50	0	129·0	406·1		122·9	
		55	0	127·1	407·6		126·0	
	11	0	0	128·3	443·9	68	126·0	66
		5	0	117·1	449·0		131·8	
	11	10	0	126·5	423·0		125·8	
		25	0	128·7	411·1		130·8	
		30	0	127·5	403·5		125·2	
		35	0	135·2	397·1		125·2	
		40	0	126·4	397·0		134·9	
		45	0	134·1	424·9		135·7	
		50	0	131·1	430·6		142·4	
		55	0	126·4	433·6		140·1	
	12	0	0	123·1	442·1	67	146·6	67
		5	0	123·6	470·8		145·0	
		10	0	126·1	499·7		158·5	
		15	0	126·4	491·2		146·2	
		20	0	126·4	459·3		139·3	
		25	0	135·5	462·4		138·7	
		30	0	128·7	483·7		141·3	
		35	0	129·1	455·1		137·5	
		40	0	135·6	438·4		135·6	
		45	0	136·4	423·4		137·1	
		50	0	138·1	420·0		137·5	
		55	0	143·5	415·2		133·9	
	13	0	0	149·5	394·4	67	133·1	67
		5	0	154·5	398·7		130·4	
		10	0	156·7	398·0		130·4	
		15	0	154·1	407·0		126·1	
		20	0	156·5	391·6		114·9	
		25	0	157·9	400·5		115·2	
		30	0	153·6	418·9		107·2	
		35	0	131·3	409·2		113·8	
		40	0	135·7	416·5		110·8	
		45	0	137·2	486·5		118·1	
		50	0	141·9	435·2		113·2	
		55	0	150·0	396·3		133·9	
	14	0	0	158·1		67	82·1	67
		5	0	163·3	325·6		82·0	
		10	0	151·1	319·7		86·9	
		15	0	141·0	335·2		100·9	
		20	0	142·4	348·5		107·9	
		25	0	151·9	344·7		100·6	
		30	0	157·5	325·4		96·3	
		35	0	156·2	335·1		96·8	
		40	0	154·3	365·8		113·3	
		45	0	153·2	359·3		105·0	
		50	0	149·9	341·1		99·2	
		55	0	148·3	339·1		95·5	
	15	0	0	148·7	318·4	66	89·3	67
		5	0	150·3	302·0		81·0	
		10	0	148·5	289·8		83·2	
		15	0	149·0	305·4		85·2	
		20	0	149·2	317·6		91·2	
		25	0	148·3	317·8		88·0	
		30	0	154·0	325·0		81·3	
		35	0	166·4	307·9		66·1	
		40	0	139·2	299·4		77·7	
		45	0	135·5	320·2		64·5	
		50	0	158·4	324·4		25·2	
		55	0	128·7	308·9		28·1	
	16	0	0	101·6	294·7	66	34·9	67
		5	0	85·7	283·6		41·6	
		10	0	91·2	267·6		37·6	
		15	0	99·0	262·6		38·8	
		20	0	104·7	290·1		65·4	
		25	0	154·3	318·7		59·4	
		30	0	173·0	352·9		42·6	
		35	0	150·7	332·8		42·6	
		40	0	148·7	359·0		40·8	
		45	0	126·6	328·9		51·7	
		50	0	134·0	284·0		52·7	

SEPTEMBER 25, 26, 27, 28, and 29, 1841.

d.	h.	m.	s.	Decl. Sc.-Div^ns	Hor. Force Sc.-Div^ns	Ther.	Vert. Force Sc.-Div^ns	Ther.
25	16	55	0	160·0	283·3		54·7	
	17	0	0	172·2		66	54·7	67
		5	0	163·5	335·9		63·1	
		10	0	162·8	351·6		68·0	
		15	0	160·3	348·5		70·3	
		20	0	156·3	355·1		74·4	
		25	0	154·4	355·8		73·2	
		30	0	148·1	355·6		75·3	
		35	0	146·0	346·8		75·1	
		40	0	146·5	342·0		78·2	
		45	0	149·6	342·6		78·4	
		50	0	152·0	353·3		80·1	
		55	0	152·3	358·3			
26	18	0	0*	140·8	383·2	60	88·7	61
	20	0	0	135·2	368·5	60	77·0	60
	22	0	0	131·2	364·5	59	55·5	60
27	0	0	0	142·6	410·9	58	83·2	59
	2	0	0	128·0	386·6	58	91·7	59
		15	0	129·6	384·8		90·5	
		20	0	132·0	391·3		90·2	
		25	0	133·0	388·8		90·5	
		30	0	134·0	387·4		90·3	
		35	0	135·7	368·5		91·0	
		40	0	135·2	386·5		91·4	
		45	0	137·8	392·8		92·5	
		50	0	139·8	402·1		92·5	
		55	0	137·9	391·8		93·7	
	3	0	0	138·4	382·9	59	92·8	59
		5	0	140·7	381·5		91·9	
		10	0	140·8	389·0		91·9	
		15	0	139·5	395·0		93·3	
		20	0	135·8	392·2		94·6	
		25	0	138·0	387·7		94·3	
		30	0	139·1	379·8		93·3	
		35	0	139·2	374·0		93·1	
		40	0	137·6	367·7		93·1	
		45	0	136·4	362·6		93·2	
		50	0	134·1	361·4		94·6	
		55	0	134·9	349·4		94·4	
	4	0	0	133·4	335·3	59	94·5	59
		5	0	133·4	325·3		94·0	
		10	0	135·4	317·7		94·0	
		15	0	138·7	325·2		94·8	
		20	0	139·7	336·3		98·9	
		25	0	137·3	339·4		102·3	
		30	0	135·4	332·7		103·9	
		35	0	132·1	337·1		102·7	
		40	0	133·8	336·9		102·9	
		45	0	134·9	348·5		106·6	
		50	0	132·5	346·7		105·6	
		55	0	128·9	347·9		104·3	
	5	0	0	126·6	351·2	59	103·7	59
		5	0	123·0	363·9		103·1	
		10	0	120·4	365·2		102·2	
		15	0	118·6	371·6		101·1	
		20	0	121·7	373·8		99·9	
		25	0	123·1	378·4		100·5	
		30	0	121·9	376·1		99·5	
		35	0	120·7	370·7		98·6	
		40	0	118·7	372·9		97·2	
		45	0	117·5	372·7		96·8	

SEPTEMBER 25, 26, 27, 28, and 29, 1841.

d.	h.	m.	s.	Decl. Sc.-Div^ns	Hor. Force Sc.-Div^ns	Ther.	Vert. Force Sc.-Div^ns	Ther.
27	5	50	0	119·6	380·1		96·7	
		55	0	119·8	385·4		96·6	
	6	0	0	121·7	391·1	59	97·0	59
		5	0	122·4	394·7		96·6	
		10	0	121·0	398·1		97·2	
		15	0	123·9	399·0		96·9	
		20	0	124·4	403·7		96·8	
		25	0	126·8	409·5		97·1	
		30	0	128·8	413·1		98·6	
		35	0	129·0	408·9		98·6	
		40	0	128·9	412·9		98·6	
		45	0	129·4	412·7		99·3	
		50	0	129·1	408·1		99·6	
		55	0	128·8	415·9		99·6	
	7	0	0	129·1	406·2	60	99·8	59
		5	0	128·9	400·7		99·5	
		10	0	129·0	402·8		99·6	
		15	0	130·0	398·6		101·5	
		20	0	130·0	397·4		102·3	
		25	0	130·1	390·9		102·6	
		30	0	128·9	389·5		102·5	
		35	0	128·7	380·7		103·1	
		40	0	131·1	385·4		101·2	
		45	0	132·8	392·2		102·2	
		50	0	133·4	397·0		101·3	
		55	0	133·1	394·6		101·3	
	8	0	0	134·2	401·9	60	101·2	60
		5	0	135·7	417·9		104·0	
		10	0	137·8	421·6		105·0	
		15	0	137·5	414·5		106·6	
		20	0	138·5	420·5		106·1	
		25	0	138·2	418·5		104·8	
		30	0	137·0	422·1		104·8	
		35	0	136·6	422·8		104·1	
		40	0	136·9	421·9		105·7	
		45	0	136·1	420·8		105·7	
		50	0	135·3	406·5		104·8	
		55	0	134·4	403·3		103·3	
	9	0	0	132·9	395·5	60	102·3	60
		5	0	132·3	390·4		101·0	
		10	0	132·2	390·6		100·2	
		15	0	132·0	394·3		99·5	
		20	0	131·0	394·3		99·3	
		25	0	131·0	394·0		98·8	
	10	0	0	135·4	410·3	60	100·0	60
	12	0	0	137·1	398·9	60	99·9	59
	14	0	0	136·6	410·1	60	97·1	59
	16	0	0	134·9	408·7	59	97·3	59
	18	0	0	141·5	422·0	58	83·9	59
	20	0	0	138·4	422·8	58	88·6	59
	22	0	0	132·5	412·8	58	95·5	59
28	0	0	0	124·8	366·5	58	84·9	59
	2	0	0	142·9	421·0	58	93·6	59
	4	0	0	135·4	382·1	57	96·6	57
	6	0	0	127·5	391·3	57	97·6	57
	8	0	0	125·6	396·7	58	99·3	58
	10	0	0	133·5	419·6	58	102·5	58
	12	0	0	133·6	416·6	58	100·5	58
	14	0	0	140·4	409·1	58	99·6	58
	16	0	0	134·9	420·8	58	98·5	58
	18	0	0	142·9	383·6	58	92·5	58

SEPTEMBER 25, 26, 27, 28, and 29, 1841.

d.	h.	m.	s.	Decl. Sc.-Div^ns	Hor. Force Sc.-Div^ns	Ther.	Vert. Force Sc.-Div^ns	Ther.
28	20	0	0	149·9	398·1	57	84·7	58
	22	0	0	142·6	403·0	57	83·8	58
29	0	0	0	142·3	410·2	57	94·0	58
	2	0	0	145·0	412·1	57	96·5	57
	4	0	0	127·3	361·0	57	95·2	58
	6	0	0	127·1	408·9	58	97·0	58
	8	0	0	127·1	412·9	58	100·3	58
	10	0	0	133·9	423·5	58	100·1	58
	12	0	0	135·9	424·0	58	99·0	58
	14	0	0	135·6	423·4	58	98·6	58
	16	0	0	149·8	374·0	58	89·7	58
	20	0	0	135·3	419·7	57	96·1	58
	22	0	0	137·3	426·7	56	98·6	57

Mean Positions at the usual hours of observation during the Month.[b]

d.	h.	m.	s.	Decl. Sc.-Div^ns	Hor. Force Sc.-Div^ns	Ther.	Vert. Force Sc.-Div^ns	Ther.
	0	0	0	142·1	394·5	66	81·2	63
	2	0	0	142·9	386·4	66	86·8	63
	4	0	0	136·5	368·0	67	89·1	63
	6	0	0	130·7	381·6	68	90·4	63
	8	0	0	129·3	392·8	69	92·5	64
	10	0	0	134·0	396·1	70	92·4	64
	12	0	0	136·0	396·9	69	96·1	64
	14	0	0	139·6	393·3	69	92·8	64
	16	0	0	138·4	388·8	68	84·0	64
	18	0	0	140·9	390·7	68	85·8	64
	20	0	0	140·1	391·2	67	84·8	64
	22	0	0	140·4	390·9	66	81·6	63

ST. HELENA $\left\{\begin{array}{l}\text{Decl. 1 Scale Division} = 0'·71 \\ \text{H. F. } k = ·00018; q = ·00025 \\ \text{V. F. } k = ·00022; q =\end{array}\right.$

Regular and extra observations.

The V. F. was observed at 2m. 30s. before, and the H. F. 2m. 30s. after the times specified.

d.	h.	m.	s.	Decl. Sc.-Div^ns	Hor. Force Sc.-Div^ns	Ther.	Vert. Force Sc.-Div^ns	Ther.
24	0	0	0	29·0	172·0	60	56·3	60
		5	0	29·1				
		10	0	29·1	172·1		56·0	
		15	0	29·7				
		20	0	29·9	173·0		55·8	
		25	0	30·0				
		30	0	30·1	173·0		55·7	
		35	0	30·1				
		40	0	30·8	173·0		55·6	
		45	0	30·8				
		50	0	30·9	173·0		55·5	
		55	0	30·9				
	1	0	0	30·9	173·0	61	55·4	61
		5	0	30·9				
		10	0	30·9	172·8		55·1	
		15	0	30·3				
		20	0	30·2	172·8		55·0	
		25	0	30·5				
		30	0	30·4	172·6		54·7	
		35	0	30·1				
		40	0	30·0	172·1		54·9	
		45	0	30·1				

* The Magnetometers showed some disturbance at an earlier hour; the sky being clear from 26d. 18h. to 20h., no Auroral light was perceptible. Additional observations were made at 27d. 2h.; the readings, &c., of the Decl. magnet being considerably below the average at that hour. (25d. 18h. to 26d. 18h. fell on Sunday at Toronto.)

b The mean positions of the V. F. magnet are from the 17th to the 30th September, inclusive.

September 24, 25, 26, 27, 28, and 29, 1841.

d.	h.	m.	s.	Decl. Sc.-Div⁰ˢ.	Hor. Force Sc.-Div⁰ˢ.	Ther.	Vert. Force Sc.-Div⁰ˢ.	Ther.
24	1	50	0	29·9	171·1		54·6	
		55	0	29·9				
	2	0	0	29·3	170·2	62	54·0	61
	3	0	0	27·9	167·0	62	54·7	61
	4	0	0	26·0	163·2	61	54·7	61
	5	0	0	25·2	158·9	62	54·0	61
	6	0	0	23·4	153·0	62	53·5	62
	8	0	0	22·2	143·9	61	55·9	61
		5	0	22·3				
		10	0	22·3	143·9		55·9	
		15	0	22·4				
		20	0	22·4	144·0		55·9	
		25	0	22·1				
		30	0	22·0	144·9		55·9	
		35	0	22·0				
		40	0	21·8	145·9		55·9	
		45	0	21·9				
		50	0	22·1	144·2		56·3	
		55	0	22·0				
	9	0	0	21·6	144·9	61	Vibrⁿ.	61
		5	0	21·5				
		10	0	21·5	145·0		57·9	
		15	0	21·5				
		20	0	21·3	145·4		58·2	
		25	0	21·2				
		30	0	21·1	144·8		58·1	
		35	0	21·5				
		40	0	21·4	144·5		58·5	
		45	0	21·1				
		50	0	21·1	143·1		58·4	
		55	0	21·1				
	10	0	0	21·1	142·0	61	58·7	61
		5	0	21·1				
		10	0	21·1	141·5		58·9	
		15	0	21·1				
		20	0	21·1	142·5		59·0	
		25	0	21·0				
		30	0	21·1	143·9		58·8	
		35	0	21·1				
		40	0	21·1	143·9		58·9	
		45	0	21·5				
		50	0	21·4	143·9		59·0	
		55	0	21·1				
	11	0	0	20·9	144·0	60	59·0	60
		5	0	20·9				
		10	0	19·6	143·9			
		15	0	19·7				
		20	0	19·5	143·0		58·9	
		25	0	19·2				
		30	0	19·2	142·5		58·8	
		35	0	20·0				
		40	0	20·1	142·3		58·9	
		45	0	20·1				
		50	0	20·1	142·8		59·1	
		55	0	20·1				
	12	0	0	20·7	144·6	60	59·3	60
		5	0	21·2				
		10	0	21·4	144·9		59·8	
		15	0	21·2				
		20	0	21·1	144·9		59·9	
		25	0	21·0				
		30	0	20·9	145·3		59·9	
		35	0	20·7				
		40	0	20·0	145·9		59·9	
		45	0	19·8				
		50	0	19·8	146·2		59·8	
		55	0	19·8				

September 24, 25, 26, 27, 28, and 29, 1841.

d.	h.	m.	s.	Decl. Sc.-Div⁰ˢ.	Hor. Force Sc.-Div⁰ˢ.	Ther.	Vert. Force Sc.-Div⁰ˢ.	Ther.
24	13	0	0	19·6	144·9	60	59·8	60
		5	0	19·7				
		10	0	19·9	144·1		59·9	
		15	0	20·4				
		20	0	20·7	144·1		60·4	
		25	0	20·9				
		30	0	20·9	144·1		60·4	
		35	0	21·0				
		40	0	21·2	144·9		61·3	
		45	0	21·4				
		50	0	21·3	145·0		61·3	
		55	0	21·1				
	14	0	0	21·0	145·6	60	61·7	60
		5	0	21·1				
		10	0	21·1	146·8		61·9	
		15	0	21·3				
		20	0	21·6	148·0		61·9	
		25	0	20·8				
		30	0	21·0	150·9		61·9	
		35	0	21·5				
		40	0	21·8	151·8		61·9	
		45	0	21·9				
		50	0	21·9	149·0		62·0	
		55	0	22·0				
	15	0	0	22·1	148·2	60	61·9	60
		5	0	22·3				
		10	0	22·8	148·7		62·0	
		15	0	22·7				
		20	0	22·7	150·2		62·0	
		25	0	22·8				
		30	0	22·9	152·0		62·1	
		35	0	22·9				
		40	0	22·9	152·6		62·1	
		45	0	23·0				
		50	0	23·1	153·0		62·1	
		55	0	23·2				
	16	0	0	23·2	152·9	60	62·1	60
		5	0	24·0				
		10	0	24·4	152·4		62·2	
		15	0	24·5				
		25	0	25·1				
		30	0	25·1	152·3		62·2	
		35	0	25·2				
		40	0	25·5	153·0		62·6	
		45	0	25·8				
		50	0	26·0	153·0		62·6	
		55	0	26·2				
	17	0	0	26·8	151·8	60	62·9	60
		5	0	26·9				
		10	0	26·3	151·0		62·7	
		15	0	26·1				
		20	0	26·1	152·0		62·7	
		25	0	27·0				
		30	0	27·1	152·9		63·0	
		35	0	27·1				
		40	0	27·1	153·0		62·8	
		45	0	27·2				
		50	0	27·1	153·8		62·5	
		55	0	27·1				
	18	0	0	27·2	153·9	60	62·6	59
		5	0	27·1				
		10	0	27·0	154·1		63·5	
		15	0	27·0				
		20	0	27·1	154·2		63·5	
		25	0	27·5				
		30	0	27·5	154·0		62·3	
		35	0	27·7				

September 24, 25, 26, 27, 28, and 29, 1841.

d.	h.	m.	s.	Decl. Sc.-Div⁰ˢ.	Hor. Force Sc.-Div⁰ˢ.	Ther.	Vert. Force Sc.-Div⁰ˢ.	Ther.
24	18	40	0	27·8	153·0		62·3	
		45	0	28·0				
		50	0	28·0	152·2		62·0	
		55	0	28·1				
	19	0	0	28·1	152·0	60	62·0	59
		5	0	28·1				
		10	0	28·1	151·7		62·0	
		15	0	28·1				
		20	0	28·0	151·5		61·9	
		25	0	27·9				
		30	0	27·7	151·5		61·9	
		35	0	27·2				
		40	0	27·0	151·5		61·8	
		45	0	27·0				
		50	0	27·1	151·5		61·6	
		55	0	27·1				
	20	0	0	27·1	151·5	60	61·6	59
		5	0	27·0				
		10	0	26·7	151·9		61·6	
		15	0	26·6				
		20	0	26·4	151·5		61·8	
		25	0	26·4				
		30	0	26·3	150·5		61·8	
		35	0	26·3				
		40	0	26·3	150·8		61·8	
		45	0	26·5				
		50	0	26·7	151·1		61·9	
		55	0	26·9				
	21	0	0	27·1	151·5	60	62·1	59
		5	0	27·1				
		10	0	27·5	152·0		62·1	
		15	0	27·0				
		20	0	27·0	152·0		62·0	
		25	0	27·0				
		30	0	27·0	153·1		62·0	
		35	0	27·0				
		40	0	26·9	154·0		62·0	
		45	0	27·0				
		50	0	26·9	155·0		61·9	
		55	0	26·9				
	22	0	0	26·8	155·4	60	61·9	59
		5	0	26·8				
		10	0	26·1	155·0		61·9	
		15	0	26·5				
		20	0	26·9	154·9		62·1	
		25	0	26·7				
		30	0	26·1	152·7		62·1	
		35	0	25·7				
		40	0	25·2	150·0		60·9	
		45	0	25·2				
		50	0	25·2	149·2		60·8	
		55	0	25·1				
	23	0	0	25·1	148·0	60	60·4	60
		5	0	25·2				
		10	0	24·9	147·4		60·4	
		15	0	24·9				
		20	0	24·9	148·0		60·5	
		25	0	24·9				
		30	0	21·9	147·8		60·5	
		35	0	24·6				
		40	0	24·4	148·9		60·7	
		45	0	24·4				
		50	0	24·7	148·7		60·7	
		55	0	24·0				
25	0	0	0	23·6	144·2	60	60·7	60
		5	0	23·1				
		10	0	23·3	142·9		60·7	

M

SEPTEMBER 24, 25, 26, 27, 28, and 29, 1841.

M. Gött. Time (d. h. m. s.)				Decl. (Sc.-Div^ns)	Hor. Force (Sc.-Div^ns)	Ther.	Vert. Force (Sc.-Div^ns)	Ther.
25	0	15	0	23·3	140·9		60·6	
		20	0	23·1			60·6	
		25	0	23·1				
		30	0	23·8	140·0		60·6	
		35	0	23·9				
		40	0	23·4	140·5		60·5	
		45	0	22·9				
		50	0	22·0	138·0		60·5	
		55	0	22·1				
	1	0	0	22·4	136·8	6	60·2	61
		5	0	21·9				
		10	0	21·8	135·2		59·9	
		15	0	21·9				
		20	0	22·0	135·0		59·9	
		25	0	22·1				
		30	0	21·9	134·1		59·9	
		35	0	21·9				
		40	0	22·5	134·4		59·9	
		45	0	22·7				
		50	0	22·6	134·0		60·2	
		55	0	23·1				
	2	0	0	23·7	132·9	62	60·2	61
		5	0	23·2				
		10	0	23·2	130·5		60·6	
		15	0	23·0				
		20	0	21·8	127·9		60·5	
		25	0	21·8				
		30	0	21·9	127·9		60·8	
		35	0	21·2				
		40	0	21·0	125·1		60·8	
		45	0	20·5				
		50	0	20·7	123·9		60·6	
		55	0	20·9				
	3	0	0	21·1	123·1	62	60·5	61
		5	0	21·1				
		10	0	21·1	122·0		60·5	
		15	0	21·0				
		20	0	20·5	119·0		60·3	
		25	0	20·8				
		30	0	21·8	121·0		60·1	
		35	0	22·8				
		40	0	23·5	114·3		60·8	
		45	0	22·5				
		50	0	22·9	113·0		60·9	
		55	0	22·1				
	4	0	0	21·5	110·7	62	60·7	61
		5	0	20·6				
		10	0	20·7	111·1		60·7	
		15	0	20·6				
		20	0	21·0	111·9		60·6	
		25	0	20·1				
		30	0	20·1	109·3		60·6	
		35	0	19·2				
		40	0	18·9	110·3		60·2	
		45	0	18·0				
		50	0	18·0	110·0		60·1	
		55	0	17·9				
	5	0	0	17·6	109·0	62	59·9	62
		5	0	17·5				
		10	0	17·1	108·0		59·8	
		15	0	17·2				
		20	0	17·1	108·1		59·5	
		25	0	17·8				
		30	0	17·9	107·9		59·5	

SEPTEMBER 24, 25, 26, 27, 28, and 29, 1841.

M. Gött. Time (d. h. m. s.)				Decl. (Sc.-Div^ns)	Hor. Force (Sc.-Div^ns)	Ther.	Vert. Force (Sc.-Div^ns)	Ther.
25	5	35	0	17·9				
		40	0	18·0	107·6		59·5	
		45	0	18·8				
		50	0	19·0	104·8		59·7	
		55	0	18·8				
	6	0	0	18·8	103·5	62	59·7	61
		5	0	18·8				
		10	0	18·5	102·5		59·6	
		15	0	18·9				
		20	0	19·7	100·8		60·0	
		25	0	20·0				
		30	0	20·0	97·0		60·1	
		35	0	19·6				
		40	0	19·9	95·1		60·2	
		45	0	20·1				
		50	0	20·3	91·9		60·2	
		55	0	20·8				
	7	0	0	21·1	90·9	62	60·3	61
		5	0	21·2				
		10	0	21·4	90·8		60·0	
		15	0	21·4				
		20	0	21·4	90·2		61·2	
		25	0	21·8				
		30	0	21·6	89·9		63·2	
		35	0	21·9				
		40	0	21·2	88·8		62·5	
		45	0	21·4				
		50	0	21·2	89·7		62·7	
		55	0	21·4				
	8	0	0	21·6	90·0	62	63·1	61
		5	0	22·0				
		10	0	21·7	94·8		63·9	
		15	0	21·0				
		20	0	20·9	98·2		63·9	
		25	0	20·5				
		30	0	20·5	100·8		64·0	
		35	0	20·1				
		40	0	20·0	102·1		63·9	
		45	0	20·0				
		50	0	19·7	106·0		63·5	
		55	0	19·2				
	9	0	0	19·1	107·4	61	63·3	61
		5	0	19·1				
		10	0	19·1	109·1		64·0	
		15	0	19·1				
		20	0	20·0	111·1		64·7	
		25	0	20·0				
		30	0	20·0	114·0		64·8	
		35	0	20·9				
		40	0	21·0	115·1		64·9	
		45	0	21·2				
		50	0	21·4	116·0		65·2	
		55	0	21·2				
	10	0	0	21·0	116·9	61	65·2	61
		5	0	20·8				
		10	0	20·5	117·5		66·0	
		15	0	20·7				
		20	0	20·7	118·5		65·7	
		25	0	20·8				
		30	0	20·8	118·6		65·5	
		35	0	20·8				
		40	0	20·7	119·4		65·8	
		45	0	20·6				
		50	0	20·4	120·0		65·8	

SEPTEMBER 24, 25, 26, 27, 28, and 29, 1841.

M. Gött. Time (d. h. m. s.)				Decl. (Sc.-Div^ns)	Hor. Force (Sc.-Div^ns)	Ther.	Vert. Force (Sc.-Div^ns)	Ther.
25	10	55	0	20·2				
	11	0	0	20·3	121·8	61	66·1	61
		5	0	20·4				
		10	0	20·2	121·8		66·5	
		15	0	20·1				
		20	0	20·1	122·1		66·5	
		25	0	19·8				
		30	0	19·4	122·4		64·3	
		35	0	19·2				
		40	0	19·2	123·0		59·0	
		45	0	19·1				
		50	0	19·1	123·2		60·1	
		55	0	19·2				
	12	0	0	19·5	123·2	61	61·7	60
		5	0	19·8				
		10	0	19·7	123·3		64·3	
		15	0	19·7				
		20	0	19·6	123·3		64·3	
		25	0	19·2				
		30	0	19·1	123·1		65·3	
		35	0	19·1				
		40	0	19·1	123·0		65·5	
		45	0	19·1				
		50	0	19·1	123·0		66·2	
		55	0	19·1				
	21	0	0*	22·8	148·2		66·0	
	23	0	0	22·2	151·6		[b]	
26	3	0	0	22·8	148·2		66·0	
	8	0	0	22·2	143·8		63·2	
	13	30	0	23·9	149·2		64·4	
	14	0	0	23·4	147·9	60	64·1	59
	14	30	0	23·4	147·5		64·1	
	15	0	0	22·9	147·0	60	64·0	59
	15	30	0	22·8	147·2		63·8	
	16	0	0	22·4	147·9	60	63·8	59
	18	0	0	22·0	149·6	60	64·5	59
	18	30	0	22·9	149·6		64·5	
	19	0	0	24·0	149·0		64·6	59
	19	30	0	24·0	148·5		64·4	
	20	0	0	22·0	150·0	60	64·2	59
	20	30	0	21·5	148·5		64·5	
	21	0	0	20·0	146·0		64·1	59
	21	30	0	18·7	144·6		63·6	
	22	0	0	19·5	147·0	59	63·2	59
	22	30	0	19·6	148·1		61·6	
	23	0	0	19·0	145·1		60·4	59
		5	0	19·3				
		10	0	19·6	145·1		59·9	
		15	0	19·1				
		20	0	19·1	146·0		59·8	
		25	0	19·9				
		30	0	20·2	148·0		59·8	
		35	0	20·9				
		40	0	21·2	149·2		59·7	
		45	0	21·2				
		50	0	21·9	150·9		59·7	
		55	0	22·0				
27	0	0	0	22·9	152·0	60	59·7	60
		5	0	23·9				
		10	0	24·1	152·0		59·6	
		15	0	24·7				
		20	0	25·0	151·0		59·2	
		25	0	25·0				
		30	0	25·0	149·8		58·8	

* 25d. 13h., to 26d. 12h., fell on Sunday at St. Helena.

b Vibrating too much for a correct reading.

September 24, 25, 26, 27, 28, and 29, 1841.

M. Gött. Time (d. h. m. s.)	Decl. Sc.-Div	Hor. Force Sc.-Div	Ther.	Vert. Force Sc.-Div	Ther.
27 0 35 0	24·9				
40 0	24·9	149·0		58·3	
45 0	24·9				
50 0	25·0	148·6		58·1	
55 0	25·2				
1 0 0	26·0	149·0	61	58·0	61
5 0	26·5				
10 0	26·4	147·3		57·9	
15 0	26·1				
20 0	26·1	145·7		57·5	
25 0	25·9				
30 0	26·0	144·7		57·3	
35 0	26·0				
40 0	26·0	144·6		57·2	
45 0	26·0				
50 0	25·6	144·9		57·1	
55 0	25·1				
2 0 0	25·1	145·8	62	56·8	61
5 0	25·1				
10 0	25·4	144·0		57·0	
15 0	25·3				
20 0	25·9	144·8		56·9	
25 0	26·1				
30 0	26·1	145·0		57·1	
40 0	25·8	145·9		57·1	
45 0	25·6				
50 .0	25·1	146·0		56·9	
55 0	25·0				
3 0 0	25·0	146·1	62	56·5	62
5 0	25·6				
10 0	25·8	146·0		56·9	
15 0	25·2				
20 0	24·9	145·8		56·4	
25 0	24·7				
30 0	24·8	145·1		57·0	
35 0	24·8				
40 0	24·4	143·8		57·0	
45 0	24·2				
50 0	23·9	142·0		57·0	
55 0	23·8				
4 0 0	23·9	145·4	63	57·0	62
5 0	24·0				
10 0	24·4	142·3		57·3	
15 0	24·7				
20 0	24·1	142·6		57·3	
25 0	23·3				
30 0	23·1	141·1		57·3	
35 0	22·8				
40 0	22·7	140·8		57·5	
45 0	22·1				
50 0	21·8	140·2		57·5	
55 0	21·4				
5 0 0	21·1	139·4	63	57·5	62
5 0	20·8				
10 0	20·6	140·0		57·5	
15 0	20·8				
20 0	20·9	141·0		57·5	
25 0	21·1				
30 0	21·2	140·9		57·7	
35 0	21·4				
40 0	21·6	140·6		57·7	
45 0	21·8				
50 0	21·9	140·9		57·7	
55 0	21·9				
6 0 0	22·0	141·1	63	57·7	62
5 0	21·9				
10 0	21·9	141·2		57·7	

September 24, 25, 26, 27, 28, and 29, 1841.

M. Gött. Time (d. h. m. s.)	Decl. Sc.-Div	Hor. Force Sc.-Div	Ther.	Vert. Force Sc.-Div	Ther.
27 6 15 0	22·0				
20 0	22·0	141·1		57·7	
25 0	22·0				
30 0	22·8	141·1		58·2	
35 0	22·8				
40 0	23·0	140·0		58·2	
45 0	23·0				
50 0	23·1	139·0		58·2	
55 0	23·7				
7 0 0	23·2	139·0	62	58·5	62
5 0	23·1				
10 0	23·1	139·7		58·4	
15 0	23·5				
20 0	23·5	140·9		58·3	
25 0	23·1				
30 0	23·1	142·0		58·3	
35 0	23·2				
40 0	23·9	142·8		58·6	
45 0	23·9				
50 0	23·9	143·0		58·8	
55 0	23·9				
8 0 0	23·9	144·0	62	59·2	62
5 0	23·7				
10 0	23·9	144·0		59·0	
15 0	23·9				
20 0	23·9	143·0		59·1	
25 0	23·9				
30 0	23·9	142·0		59·5	
35 0	23·1				
40 0	23·1	140·1		59·4	
45 0	23·0				
50 0	23·0	139·1		59·5	
55 0	23·0				
9 0 0	22·9	140·0	62	59·5	62
5 0	23·0				
10 0	23·0	140·1		60·2	
15 0	23·2				
20 0	23·2	140·7		60·2	
25 0	23·3				
30 0	23·1	141·2		60·1	
35 0	23·1				
40 0	23·1	142·0		60·3	
45 0	23·2				
50 0	23·3	142·4		60·2	
55 0	23·2				
10 0 0	23·3	142·0	62	60·7	61
5 0	23·3				
10 0	23·2	142·0		60·7	
15 0	23·2				
20 0	23·3	142·5		60·7	
25 0	23·5				
30 0	23·8	142·9		60·9	
35 0	23·9				
40 0	23·9	143·0		61·0	
45 0	23·9				
50 0	23·9	143·8		61·1	
55 0	23·9				
11 0 0	24·0	144·1	61	61·1	61
5 0	24·0				
10 0	24·1	144·8		61·2	
15 0	24·1				
20 0	24·2	145·1		61·2	
25 0	24·2				
30 0	24·2	145·5		61·4	
35 0	24·4				
40 0	24·5	146·0		61·5	
45 0	24·6				

September 24, 25, 26, 27, 28, and 29, 1841.

M. Gött. Time (d. h. m. s.)	Decl. Sc.-Div	Hor. Force Sc.-Div	Ther.	Vert. Force Sc.-Div	Ther.
27 11 50 0	24·6	146·2		61·6	
55 0	24·6				
12 0 0	24·6	147·1	61	61·7	61
5 0	24·3				
10 0	24·9	149·0		61·7	
15 0	24·9				
20 0	24·9	150·0		61·9	
25 0	15·0				
30 0	24·9	150·0		61·9	
35 0	24·9				
40 0	24·9	149·9		62·0	
45 0	24·9				
50 0	24·8	149·2		62·0	
55 0	24·5				
13 0 0	24·3	149·2	61	62·1	61
5 0	14·3				
10 0	24·1				62·2
30 0	24·3	149·9		62·4	
14 0 0	24·7	150·0	61	62·7	
30 0	24·1	150·6		63·1	
15 0 0	24·0	150·0	61	63·4	
30 0	24·0	150·0		63·7	
16 0 0	24·0	150·0	60	63·8	
18 0 0	23·4	151·2	60	63·7	
30 0	24·2	151·9		63·8	
19 0 0	25·9	150·9	60	63·8	
30 0	25·1	150·0		63·9	
20 0 0	23·9	151·2	60	64·0	
30 0	23·8	151·9		64·2	
21 0 0	23·1	153·0	60	64·4	
22 0 0	23·0	157·0	61		
23 0 0	24·1	159·1	60		
28 0 0 0	24·2	158·8	62		
2 0 0	26·1	156·0	62		
3 0 0	25·9	154·4	62		
4 0 0	22·0	151·0	62		
5 0 0	22·0	153·6	62		
6 0 0	22·2	149·4	62		
8 0 0	23·9	146·0	61		
10 0 0	24·2	146·9	61		
11 0 0	25·0	148·0	61		
12 0 0	25·0	149·0	61		
13 0 0	25·0	151·1	60		
14 0 0	24·6	152·8	60		
15 0 0	24·2	152·3	61		
16 0 0	24·0	152·5	60		
19 30 0	23·2	154·9	60		
20 0 0	21·1	154·8	60		
20 30 0	20·9	155·0			
22 0 0	23·2	155·0	61		
23 0 0	26·1	154·1	61		
29 0 0 0	22·3	158·7	61		
2 0 0	24·3	158·2	62		
3 0 0	22·0	150·0	62		
4 0 0	20·8	143·9	63		
5 0 0	18·1	143·3	63		
6 0 0	18·9	146·6	63		
8 0 0	20·0	146·0	62		
10 0 0	20·2	147·1	62		
11 0 0	21·2	153·7	62		
12 0 0	21·0	150·0	62		
13 0 0	20·8	150·1	61		
14 0 0	20·9	150·9	61		
15 0 0	21·8	157·1	61		
16 0 0	21·1	155·1	61		
18 0 0	19·7	152·4	61		
19 30 0	19·1	150·9	61		

M 2

Panel 1 — SEPTEMBER 24, 25, 26, 27, 28, and 29, 1841.

M. Gött. Time (d. h. m. s.)	Decl. (Sc.-Div)	Hor. Force (Sc.-Div)	Ther.	Vert. Force (Sc.-Div)	Ther.
29 20 0 0	18·8	151·2	61		
20 30 0	18·9	152·4			
22 0 0	20·9	158·5	60		
23 0 0	22·1	161·0	61		

The Mean Positions at the same hours during the Month of September are given in page 76.

CAPE OF GOOD HOPE   {   Decl. 1 Scale Division = 0′·75   H. F. $k = ·000180$; $q = ·0003$ V. F.[a] }

Regular and extra observations.
The H. F. was observed at 2ᵐ. 30ˢ. after the times specified.

M. Gött. Time (d. h. m. s.)	Decl. (Sc.-Div)	Hor. Force (Sc.-Div)	Ther.	Vert. Force (Sc.-Div)	Ther.
24 0 0 0	53·0	90·0	56		
2 0 0	55·3	85·0	57		
25 0	55·5				
30 0	55·4	83·6	57		
35 0	55·6				
40 0	55·6	83·4			
50 0	55·5	82·8			
3 55 0	54·2				
4 0 0	54·0	75·0	57		
6 0 0	52·2	70·0	58		
5 0	52·1				
10 0	52·5	69·1			
15 0	52·3				
20 0	51·4	68·0			
25 0	50·7				
30 0	50·3	67·0			
35 0	50·2				
40 0	50·5	65·6			
50 0	50·7	65·2			
55 0	50·1				
7 0 0	50·0	63·9	59		
5 0	50·0				
10 0	49·5	63·0			
15 0	49·5				
20 0	49·6	62·0			
25 0	49·1				
30 0	48·7	62·0			
35 0	48·7				
40 0	48·0	61·3			
45 0	48·0				
50 0	47·9	61·4			
55 0	47·6				
8 0 0	47·5	61·2	59		
5 0	47·2				
10 0	47·5	61·8			
15 0	47·7				
20 0	47·9	62·8			
25 0	47·0				
30 0	47·4	64·0			
35 0	48·2				
40 0	48·2	64·0			
45 0	48·8				
50 0	48·3	62·8			
55 0	47·4				
9 0 0	47·0	64·8	58		
5 0	46·9				
10 0	46·9	66·1			
15 0	46·9				
20 0	46·9	66·2			

Panel 2 — SEPTEMBER 24, 25, 26, 27, 28, and 29, 1841.

M. Gött. Time (m. s.)	Decl. (Sc.-Div)	Hor. Force (Sc.-Div)	Ther.	Vert. Force (Sc.-Div)	Ther.
24 9 25 0	46·9				
30 0	46·9	64·7			
35 0	46·9				
40 0	46·9	64·9			
45 0	46·5				
50 0	40·9	64·2			
55 0	46·9				
10 0 0	46·8	63·2	59		
5 0	45·9				
10 0		63·0			
15 0	46·3				
20 0	46·1	64·9			
25 0	45·9				
30 0	46·1	67·6			
35 0	46·4				
40 0	46·2	66·2			
45 0	47·1				
50 0	46·9	66·6			
55 0	46·0				
11 0 0	45·0	68·8	58		
5 0	43·0				
10 0	42·0	69·3			
15 0	41·9				
20 0	41·5	68·0			
25 0	41·5				
30 0	41·9	67·1			
35 0	43·0				
40 0	43·9	67·2			
45 0	44·0				
50 0	44·5	68·0			
55 0	45·0				
12 0 0	46·1	69·6	58		
5 0	47·1				
10 0	47·7	69·0			
15 0	47·0				
20 0	46·5	69·0			
25 0	46·5				
30 0	46·2	70·0			
35 0	46·0				
40 0	45·2	72·1			
45 0	45·1				
50 0	45·2	73·1			
55 0	45·0				
13 0 0	44·0	71·1	58		
5 0	43·8				
10 0	44·1	70·1			
15 0	44·4				
20 0	44·2	70·0			
25 0	44·5				
30 0	44·9	70·1			
35 0	45·0				
40 0	45·4	71·0			
45 0	45·2				
50 0	45·2	71·1			
55 0	45·0				
14 0 0	45·2	71·9	58		
5 0	45·2				
10 0	45·2	72·9			
15 0	46·3				
20 0	47·0	78·1			
23 0	45·5				
30 0	46·0	79·5			
35 0	47·9				
40 0	49·4	79·0			

Panel 3 — SEPTEMBER 24, 25 26, 27, 28, and 29, 1841.

M. Gött. Time (d. h. m. s.)	Decl. (Sc.-Div)	Hor. Force (Sc.-Div)	Ther.	Vert. Force (Sc.-Div)	Ther.
24 14 45 0	49·4				
55 0	49·1	75·8			
15 0 0	50·0				
5 0	50·9	73·7	57		
10 0	51·7				
15 0	52·8	72·9			
20 0	53·0				
25 0	53·4	75·8			
30 0	53·9				
35 0	53·7	77·9			
40 0	53·6				
45 0	53·1	78·7			
50 0	53·2				
55 0	53·2	76·8			
16 0 0	53·8				
5 0	54·6	74·4	57		
10 0	55·7				
15 0	56·1	73·0			
20 0	56·9				
25 0	56·9	72·2			
30 0	57·0				
35 0	57·0	74·0			
40 0	57·5				
45 0	58·1	74·5			
50 0	58·0				
55 0	59·0	73·0			
17 0 0	59·0				
5 0	58·7	71·0	57		
10 0	57·3				
15 0	57·8	70·8			
20 0	57·4				
25 0	58·0	71·3			
30 0	59·0				
35 0	59·6	71·7			
40 0	59·8				
45 0	58·7	72·5			
50 0	59·0				
55 0	59·8	72·7			
18 0 0	60·1				
5 0	60·1	74·2	57		
10 0	59·8				
15 0	60·1	76·1			
20 0	59·5				
25 0	59·9	77·0			
30 0	59·3				
35 0	59·0	77·0			
40 0	59·4				
45 0	58·4	77·0			
50 0	57·5				
55 0	57·7	77·8			
19 0 0	57·2				
5 0	57·5	78·1	57		
10 0	56·2				
15 0	55·9	78·0			
20 0	55·2				
25 0	55·5	77·8			
30 0	54·7				
35 0	54·7	77·4			
40 0	54·7				
45 0	54·1	77·3			
50 0	54·4				
55 0	54·2	77·2			
20 0 0	54·0				
	53·8	76·8	56		

* The V. F. magnet was not in satisfactory adjustment.

SEPTEMBER 24, 25, 26, 27, 28, and 29, 1841.

M. Gött. Time (d. h. m. s.)	Decl. Sc.-Divst	Hor. Force Sc.-Divst	Ther.	Vert. Force Sc.-Divst	Ther.
24 20 5 0	53·7				
10 0	53·3	76·7			
15 0	52·9				
20 0	52·9	75·5			
25 0	52·7				
30 0	52·5	74·5			
35 0	52·2				
40 0	52·1	73·6			
45 0	52·2				
50 0	52·9	73·4			
55 0	53·0				
21 0 0	53·4	73·0	56		
5 0	54·0				
10 0	54·7	71·5			
15 0	54·7				
20 0	54·8	70·8			
25 0	54·6				
30 0	54·8	71·1			
35 0	54·5				
40 0	54·5	72·2			
45 0	54·9				
50 0	54·6	72·2			
55 0	54·9				
22 0 0	54·6	71·8	56		
5 0	54·4				
10 0	53·9	71·0			
15 0	53·3				
20 0	53·6	70·6			
25 0	53·3				
30 0	52·7	68·7			
35 0	51·8				
40 0	51·3	66·9			
45 0	50·9				
50 0	50·8	65·8			
55 0	50·2				
23 0 0	49·8	64·6	56		
5 0	49·9				
10 0	49·8	62·2			
15 0	49·4				
20 0	49·5	63·3			
25 0	49·2				
30 0	49·0	62·7			
35 0	48·8				
40 0	48·0	62·9			
45 0	48·0				
50 0	48·1	62·6			
55 0	47·8				
25 0 0 0	48·0	57·9	57		
5 0	48·9				
45 0	49·0				
50 0	48·0	50·0			
55 0	48·1				
1 0 0	49·0	48·6	57		
5 0	49·0				
10 0	48·9	47·4			
15 0	49·0				
20 0	49·2	48·0			
25 0	49·5				
30 0	49·2	47·2			
35 0	49·9				
40 0	50·7	48·3			
45 0	50·9				
50 0	50·9	48·0			
55 0	50·2				

SEPTEMBER 24, 25, 26, 27, 28, and 29, 1841.

M. Gött. Time (d. h. m. s.)	Decl. Sc.-Divst	Hor. Force Sc.-Divst	Ther.	Vert. Force Sc.-Divst	Ther.
25 2 0 0	53·5	45·3	58		
5 0	54·2				
10 0	54·9	41·8			
15 0	53·9				
20 0	51·3	41·2			
25 0	49·1				
30 0	49·0	40·1			
35 0	49·1				
40 0	49·1	37·9			
45 0	48·9				
50 0	49·2	36·1			
55 0	48·9				
3 0 0	49·2	36·8	58		
5 0	50·0				
10 0	50·0	35·1			
15 0	50·4				
20 0	50·0	31·3			
25 0	50·0				
30 0	51·1	34·3			
35 0	53·4				
40 0	54·3	27·0			
45 0	51·1				
50 0	48·4	24·1			
55 0	45·2				
4 0 0	43·1	23·7	58		
5 0	43·8				
10 0	46·0	23·7			
15 0	46·2				
20 0	46·3	23·3			
25 0	47·2				
30 0	47·0	20·0			
35 0	45·5	22·1			
40 0	46·0	24·0			
45 0	45·1	25·5			
50 0	47·8	22·7			
55 0	46·1	21·8			
5 0 0	46·5	21·8	58		
5 0	47·0	21·9			
10 0	45·5	20·7			
15 0	44·0	22·6			
20 0	44·0	23·2			
25 0	45·6	22·8			
30 0	45·9	22·1			
35 0	46·5	21·8			
40 0	46·8	21·0			
45 0	47·0	19·0			
50 0	46·7	17·8			
55 0	44·6	18·0			
6 0 0	45·6	17·1	59		
5 0	45·9	17·2			
10 0	45·9	18·0			
15 0	47·2	16·2			
20 0	49·1	13·8			
25 0	49·0	10·3			
30 0	48·2	9·2			
35 0	46·5	10·2			
40 0	46·5	9·8			
45 0	47·0	7·7			
50 0	47·0	5·1			
55 0	47·3	6·0			
7 0 0	47·9	5·2	59		
5 0	47·4	5·1			
10 0	46·1	5·0			
15 0	45·1	4·9			

SEPTEMBER 24, 25, 26, 27, 28, and 29, 1841.

M. Gött. Time (m. s.)	Decl. Sc.-Divst	Hor. Force Sc.-Divst	Ther.	Vert. Force Sc.-Divst	Ther.
25 7 20 0	44·4	4·2			
25 0	43·6	3·8			
30 0	42·6	3·6			
35 0	43·9	1·5			
40 0	41·9	4·1			
45 0	41·1	5·7			
50 0	40·8	6·1			
55 0	40·8	5·5			
8 0 0	41·0	5·6	59		
5 0	41·7	9·2			
10 0	40·5	12·5			
15 0	40·4	15·1			
20 0	40·8	16·8			
25 0	41·0	19·0			
30 0	41·0	20·1			
35 0	41·0	20·4			
40 0	41·0	22·2			
45 0	40·7	24·7			
50 0	40·1	27·5			
55 0	40·0	29·2			
9 0 0	39·8	30·0	59		
5 0	39·8	31·3			
10 0	39·8	32·8			
15 0	40·5	34·5			
20 0	41·0	35·7			
25 0	41·8	37·7			
30 0	42·0	39·8			
35 0	42·3				
40 0	42·5	41·2			
45 0	43·5				
50 0	43·5	40·3			
55 0	42·8				
10 0 0	41·4	41·7	59		
5 0	40·8				
10 0	40·0	42·8			
15 0	39·8	44·0			
20 0	40·1	44·6			
25 0	41·4				
30 0	41·5	44·0			
35 0	42·2				
40 0	42·0	44·1			
45 0	41·6				
50 0	40·9	46·0			
55 0	41·0				
11 0 0	42·0	47·5	58		
5 0	43·0				
10 0	43·1	46·2			
15 0	43·5				
20 0	43·9	45·7			
25 0	43·4				
30 0	42·3	46·1			
35 0	41·9				
40 0	42·0	47·8			
45 0	41·9				
50 0	41·9	48·3			
55 0	41·7				
12 0 0	41·3	49·2	58		
5 0	41·5				
10 0	41·6	50·0			
15 0	41·7				
20 0	41·8	50·3			
25 0	41·9				
30 0	42·0	50·2			
19 25 0*	46·7				

* 25^{d}. 13^{h}., to 26^{d}. 12^{h}., fell on Sunday at the Cape of Good Hope.

SEPTEMBER 24, 25, 26, 27, 28, and 29, 1841.

M. Gött. Time. d. h. m. s.	Decl. Sc.-Div.	Hor. Force. Sc.-Div.	Ther.	Vert. Force. Sc.-Div.	Ther.
25 22 0 0	52·1	65·3	56		
26 0 0 0	52·4	58·9	57		
1 50 0	53·0	66·9	57		
2 0 0	53·2	67·0	57		
2 7 30	53·4	66·9			
4 0 0	49·0	57·0	57		
6 0 0	49·2	61·0	57		
8 0 0	48·5	65·0			
11 0 0	53·0	71·4			
13 0 0	52·8	74·5	56		
14 0 0	51·1	70·3			
5 0	50·6				
10 0	50·9	70·0			
15 0	50·3				
20 0	50·4	70·2			
25 0	50·2				
30 0	50·3	70·8			
35 0	50·2				
40 0	50·1	69·8			
45 0	50·0				
50 0	50·0	70·0			
15 0 0	50·0	70·0	56		
16 0 0	50·2	68·7	56		
17 0 0	51·4	69·7	55		
18 0 0	50·6	72·7	55		
20 0 0	46·0	75·8	55		
22 0 0	44·1	63·7	55		
27 0 0 0	50·0	67·0	56		
2 0 0	53·1	60·9	56		
4 0 0	48·2	60·6	57		
6 0 0	50·4	61·4	57		
8 0 0	49·0	69·1	57		
10 0 0	50·0	65·0	56		
12 0 0	52·5	71·0	56		
14 0 0	52·8	73·5	56		
16 0 0	50·7	73·4	56		
18 0 0	52·9	72·9	55		
20 0 0	46·5	80·0	56		
22 0 0	46·9	80·0	55		
28 0 0 0	46·9	65·2	56		
2 0 0	53·5	71·5	57		
4 0 0	51·0	69·0	57		
6 0 0	50·2	69·0	58		
8 0 0	49·0	68·0	58		
10 0 0	50·5	69·2	58		
12 0 0	52·0	72·2	58		
14 0 0	52·6	75·0	57		
16 0 0	50·5	74·2	57		
18 0 0	50·5	74·6	57		
20 0 0	49·3	77·7	57		
22 0 0	51·6	71·8	57		
29 0 0 0	53·2	71·5	58		
2 0 0	55·0	71·9	58		
4 0 0	51·6	61·5	58		
6 0 0	50·0	67·1	58		
8 0 0	50·0	66·8	58		
10 0 0	50·8	69·0	57		
12 0 0	50·8	73·0	57		
13 0 0	51·4	73·2	57		
14 0 0	51·0	73·8	57		
15 0 0	56·0	78·2	57		
16 0 0	52·2	79·5	57		
17 0 0	51·0	76·2	57		
18 0 0	51·0	76·0	57		
19 0 0	50·0	77·8	57		
20 0 0	48·9	75·6	57		
21 0 0	49·2	73·1	57		

SEPTEMBER 24, 25, 26, 27, 28, and 29, 1841.

M. Gött. Time. d. h. m. s.	Decl. Sc.-Div.	Hor. Force. Sc.-Div.	Ther.	Vert. Force. Sc.-Div.	Ther.
29 22 0 0	50·2	72·3	57		
23 0 0	52·2	71·6	57		

The Mean Positions at the same hours during the Month are given in page 76.

VAN DIEMEN ISLAND { Decl. 1 Scale Division = 0'·71; H. F. k = ·0003; q = ; V. F. k = ; q = }

Regular and extra observations.

The V. F. was observed at 2^m. 30'. before, and the H. F. 2^m. 30'. after the times specified.

M. Gött. Time. d. h. m. s.	Decl. Sc.-Div.	Hor. Force. Sc.-Div.	Ther.	Vert. Force. Sc.-Div.	Ther.
24 0 0 0	108·6	60·4	57	96·6	56
1 0 0	106·6	60·1		97·5	
2 0 0	104·4	60·2	56	98·3	54
3 0 0	104·1	60·1		96·9	
4 0 0	103·9	60·4	55	100·7	53
5 0 0	103·3	59·4		101·3	
6 0 0	110·2	63·7	53	101·6	52
10 0	106·8	61·5			
15 0	106·4				
20 0	108·7	61·3		98·8	
25 0	109·1				
30 0	109·4	60·3		99·2	
35 0	110·1				
40 0	109·5	61·1		102·3	
45 0	108·8				
50 0	109·3	61·1		100·5	
7 0 0	114·7	62·5		101·1	
10 0	111·3	61·0			
15 0	109·4				
20 0	110·2	59·7		96·9	
25 0	112·4				
30 0	114·1	59·4		97·4	
35 0	114·4				
40 0	114·1	59·8		97·8	
45 0	113·3				
50 0	113·9	59·7		98·3	
55 0	114·2				
8 0 0	114·2	59·6	55	98·3	55
5 0	115·4				
10 0	115·2	58·6		98·7	
15 0	116·7				
20 0	118·7	58·0		96·9	
25 0	121·3				
30 0	123·5	58·1		97·1	
35 0	122·3				
40 0	121·8	58·3		96·9	
45 0	123·1				
50 0	121·2	58·5		96·9	
55 0	120·0				
9 0 0	119·8	58·1		96·9	
10 0	120·1				
15 0		57·7			
20 0	119·5				
25 0				96·4	
30 0	117·9				
35 0		58·3			
40 0	118·1				
45 0				95·8	
50 0	115·9				

SEPTEMBER 24, 25, 26, 27, 28, and 29, 1841.

M. Gött. Time. d. h. m. s.	Decl. Sc.-Div.	Hor. Force. Sc.-Div.	Ther.	Vert. Force. Sc.-Div.	Ther.
24 10 0 0	115·4	59·7	54	96·7	54
10 0	113·1				
15 0		58·9			
20 0	112·6				
25 0				93·8	
30 0	112·4				
35 0		57·6			
40 0	112·3				
11 0 0	112·5	56·9	54	97·4	54
12 0 0	109·9	51·9	54	97·5	54
15 0	108·1				
20 0	109·7	50·6		97·0	
25 0	110·8				
30 0	110·2	49·0		97·6	
35 0	110·1				
40 0	110·2	48·2		98·6	
45 0	109·2				
50 0	109·4	52·9		97·6	
55 0	109·2				
13 0 0	110·8	48·9		97·3	
10 0	109·2	48·7			
15 0	108·7				
20 0	109·3	48·2		95·1	
25 0	109·5				
30 0	109·5	47·0		95·2	
35 0	109·3				
40 0	109·2	46·8		105·0	
45 0	109·7				
50 0	110·0	46·6			
55 0	110·3				
14 0 0	110·7	46·4	60	94·9	58
15 0 0	111·5	43·5		97·4	
16 0 0	115·6	45·5	62	98·0	60
17 0 0	117·5	44·8		100·8	
18 0 0	110·1	47·6	64	98·7	61
19 0 0	117·7	50·6		98·1	
20 0 0	119·7	49·5	63	101·6	60
21 0 0	102·1	45·0		102·9	
5 0	104·5				
10 0	105·2	44·2		104·0	
15 0	105·2				
22 0 0	103·2	41·1	60	108·8	58
23 0 0	98·7	42·2		109·2	
25 0 0 0	71·3	46·8	58	80·2	56
10 0	59·6	53·4			
15 0	72·1				
20 0	95·2	53·5		75·9	
25 0	102·3				
30 0	95·9	44·4		82·5	
35 0	89·6				
40 0	85·1	45·1		86·5	
45 0	84·9				
50 0	94·8	43·1		89·5	
1 0 0	88·4	42·0	58	90·2	58
10 0	86·2	40·0			
15 0	83·1				
20 0	78·8	34·7		88·5	
25 0	73·5				
30 0	70·6	31·9		87·1	
35 0	69·2				
40 0	69·0	28·1		87·1	
45 0	69·2				
2 0 0	93·9	24·7	58	88·3	58
15 0	73·5	22·7		65·1	
20 0	77·8				
30 0	75·1	27·4		69·6	
35 0	68·6				

SEPTEMBER 24, 25, 26, 27, 28, and 29, 1841.

M. Gött. Time.				Decl.		Hor. Force.		Vert. Force.	
d.	h.	m.	s.	Sc.-Div.	Ther.	Sc.-Div.	Ther.	Sc.-Div.	Ther.
25	2	45	0	71·1		21·8		68·2	
	3	0	0						
26	4	0	0	106·4		54·7	55	101·4	54
	5	0	0	106·3		52·2		101·9	
	6	0	0	105·0		53·9	54	99·8	53
	7	0	0	110·2		53·1		103·2	
	8	0	0	109·1		53·7	54	101·2	53
	9	0	0	107·5		56·5		98·2	
	10	0	0	107·1		53·5	54	100·7	53
	11	0	0	102·2		51·4		100·2	
	12	0	0	103·3		51·4	55	97·1	55
	13	0	0	106·2		47·6		101·2	
	14	0	0	108·4		48·2	59	97·0	57
	15	0	0	112·5		47·6			
	16	0	0	113·7		50·3	64	94·4	62
	17	0	0	115·6		46·1		94·0	
	18	0	0	116·8		48·5	65	92·7	63
	19	0	0	116·8		46·7		96·2	
	20	0	0	113·8		48·5	65	98·5	62
	21	0	0	105·7		46·4		101·0	
	22	0	0	105·4		46·6	63	94·3	60
	23	0	0ᵇ	91·5		50·7	62	90·6	61
		10	0	87·1		51·6			
		15	0	91·1				114·6	61
		20	0	92·0		52·7			
		25	0	100·1					
		30	0	102·3		49·6		111·7	
		35	0	101·5					
		40	0	100·2		46·0		108·6	
		45	0	97·4					
		50	0	100·3		48·6		106·0	
		55	0	103·8					61
27	0	0	0	104·5		48·2	61	95·5	
		5	0	103·0					
		10	0	102·1		46·7		95·1	
		15	0	101·6					
		20	0	100·3		45·8		95·4	
		25	0	99·2					
		30	0	99·1		45·4		95·3	
		35	0	98·8					
		40	0	99·3		45·5		96·0	
		45	0	101·0					
		50	0	102·6		51·5	61	96·4	
		55	0	103·5					
	1	0	0	102·8		51·1		88·8	
	2	0	0	95·7		54·3	59	93·4	59
		15	0	100·1		52·3		87·7	
		22	30	105·5		51·6		88·2	
		30	0	103·7		51·4		89·2	
		37	30	102·8		53·9		88·1	
		45	0	105·6		52·1		88·2	
	3	0	0	104·8		51·3		90·6	
	4	0	0	110·0		52·8	56	95·1	57
	5	0	0	102·6		52·2		84·0	
	6	0	0	105·1		50·2	57	95·6	55
	7	0	0	105·1		52·8		93·6	
	8	0	0	110·1		53·2	55	100·2	54
	9	0	0	108·2		53·5		98·7	
	10	0	0	105·5		54·4	56	96·3	55
	11	0	0	106·0		55·0		95·3	
	12	0	0	103·0		51·1	58	95·8	57
	13	0	0	104·8		50·5		93·6	
	14	0	0	109·0		49·6	60	91·8	61

SEPTEMBER 24, 25, 26, 27, 28, and 29, 1841.

M. Gött. Time.				Decl.		Hor. Force.		Vert. Force.	
d.	h.	m.	s.	Sc.-Div.	Ther.	Sc.-Div.	Ther.	Sc.-Div.	Ther.
27	15	0	0	112·0		49·2		90·6	
	16	0	0	115·0		50·5	64	88·8	63
	17	0	0	115·4		48·4		88·8	
	18	0	0	116·8		46·9	67	93·1	65
	19	0	0	112·0		46·4		91·6	
	20	0	0	113·0		49·9	68	87·9	65
	21	0	0	108·6		49·2		87·9	
	22	0	0	105·6		51·2	66	86·7	64
	23	0	0	105·8		51·7		88·7	
28	0	0	0	113·9		59·1	64	83·3	62
		10	0	110·9		50·3			
		15	0	98·0					
		20	0	92·7		52·1		78·5	
		25	0	91·7					
		30	0	94·8		54·1		81·1	
		35	0	99·2					
		40	0	101·7		51·7		82·5	
		45	0	102·6					
		50	0	103·1				86·5	
	1	0	0	100·0		49·1		89·3	
	2	0	0	103·2		49·0	63	95·4	62
	3	0	0	104·8		50·6		92·8	
	4	0	0	108·5		54·1	61	91·1	60
	5	0	0	107·8		54·0		88·8	
	6	0	0	109·9		55·1	58	94·1	58
	7	0	0	113·4		55·5		96·2	
	8	0	0	108·4		55·5	58	91·3	57
	9	0	0	105·4		55·2		97·0	
	10	0	0	105·5		55·1	57	96·4	56
	11	0	0	102·2		52·5		97·0	
	12	0	0	100·9		51·5	58	95·6	58
	13	0	0	104·4		50·5		96·2	
	14	0	0	108·1		49·8	61	95·6	59
	15	0	0	111·4		51·8		93·6	
	16	0	0	113·5		52·0	63	92·5	62
	17	0	0	115·7		52·5		92·6	
	18	0	0	116·5		51·5	66	94·5	62
	19	0	0	116·6		46·9		101·3	
	20	0	0	115·9		48·5	64	95·7	63
	21	0	0	111·9		48·2		96·4	
	22	0	0	106·2		51·3	63	80·8	61
	23	0	0	104·2		49·2		91·5	
29	0	0	0	107·7		51·5	62	99·1	61
	1	0	0	107·4		52·3		98·5	
	2	0	0	106·2		53·6	60	94·2	58
	3	0	0	109·9		57·3		88·5	
	4	0	0	92·3		53·9	58	90·3	57
		10	0	90·1		53·1		88·4	
		15	0	89·4					
		20	0	92·4		53·2		87·8	
		25	0	94·9					
		30	0	95·7		53·4		89·0	
		35	0	98·1					
		40	0	98·4		52·1		90·5	
		45	0	99·1					
		50	0	98·3		51·8		91·9	
		55	0	98·7					
	5	0	0	99·7		51·5	58	92·9	58
		10	0	101·8		51·6			
		15	0	104·3					
		20	0	105·3		51·6		95·1	
		25	0	105·8					
		30	0	107·1		51·7		95·8	

SEPTEMBER 24, 25, 26, 27, 28, and 29, 1841.

M. Gött. Time.				Decl.		Hor. Force.		Vert. Force.	
d.	h.	m.	s.	Sc.-Div.	Ther.	Sc.-Div.	Ther.	Sc.-Div.	Ther.
29	5	35	0	107·6					
		40	0	108·6		51·7		96·9	
		45	0	108·4					
		50	0	108·1		52·1		96·8	
	6	0	0	107·4		52·0	58	95·7	58
	7	0	0	110·6		53·3		97·5	
	8	0	0	107·8		54·8	57	96·5	56
	9	0	0	106·1		55·2		97·0	
	10	0	0	104·1		54·3	56	98·5	55
	11	0	0	101·9		50·6		101·9	
	12	0	0	102·5		52·6	56	99·4	55
	13	0	0	105·2		52·2		97·3	
	14	0	0	108·4		53·0	57	99·2	56
	15	0	0	110·7		52·8		108·0	
	16	0	0	114·4		51·0	59	103·2	57
	17	0	0	116·1		53·0		99·2	
	18	0	0	117·2		53·0	61	98·6	59
	19	0	0	117·2		52·2		96·1	
	20	0	0	110·9		52·5	63	93·0	60
	21	0	0	108·5		52·1		93·0	
	22	0	0	108·8		52·2	63	91·8	61
	23	0	0	108·8		52·4		93·0	

The Mean Positions at the same hours during the Month are given in page 75.

ANTARCTIC EXPEDITION AT THE BAY OF ISLANDS, NEW ZEALAND.

Decl. 1 Scale Division = 0'·73
H. F. k = ·00018 ; q =
V. F. k =

Positions at the usual hours of observation,
September 24, 25, 26, 27, 28 and 29.

d.	h.	m.	s.	Decl.	Hor. Force.	Ther.	Vert. Force.	Ther.
24	0	0	0	45·6	82·5	58	56·6	64
	1	0	0	45·6	85·0		58·3	
	2	0	0	45·5	86·0	56	58·1	61
	3	0	0	44·9	86·8		58·0	
	4	0	0	44·6	86·4	57	58·8	61
	5	0	0	44·1	85·9		58·2	
	6	0	0	47·4	85·6	58	57·9	61
	7	0	0	50·0	83·4		57·5	
	8	0	0	49·1	81·1	58	57·3	60
	9	0	0	47·0	80·0		55·6	
	10	0	0	45·1	76·9	59	56·3	61
	11	0	0	44·4	71·5		53·4	
	12	0	0	46·4	65·4	62	51·8	65
	13	0	0	47·0	61·1		52·1	
	14	0	0	49·5	61·1	63	53·7	65
	15	0	0	51·2	57·8		53·8	
	16	0	0	52·7	57·7	63	55·6	65
	17	0	0	50·3	57·6		56·0	
	18	0	0	46·6	60·4	62	58·7	64
	19	0	0	50·3	60·5		57·4	
	20	0	0	49·5	61·3	62	58·7	63
	21	0	0	42·8	63·3		57·8	
	22	0	0	37·9	61·8	60	60·3	63
	23	0	0	36·3	61·0		61·1	
25	0	0	0ᶜ	38·5	72·4	59	61·2	63
26	1	0	0	44·1	71·8	57	59·9	
	2	0	0	45·0	70·0	58	57·4	62

ᵃ 25ᵈ. 3ʰ., to 26ᵈ. 2ʰ., fell on Sunday at Van Diemen Island; the observation at 26ᵈ. 3ʰ. was missed.

ᵇ "Calm. Clear, bright moonlight; no appearance of Aurora from the hill near the Observatory."
ᶜ 25ᵈ. 1ʰ., to 26ᵈ. 0ʰ., fell on Sunday at the Bay of Islands.

SEPTEMBER 24, 25, 26, 27, 28, and 29, 1841.

M. Gött. Time	Decl.	Hor. Force		Vert. Force	
m. s.	Sc.-Div.	Sc.-Div.	Ther.	Sc.-Div.	Ther.
26 3 0 0	44·8	75·0	60	58·6	
4 0 0	47·4	74·7	60	59·4	63
5 0 0	46·5	70·6	59	58·6	
6 0 0	46·0	72 3	59	59·2	62
7 0 0	47·0	71·6	58	59·2	
8 0 0	46·5	74·6	58	59·9	60
9 0 0	45·9	76·7	58	60·7	
10 0 0	43·9	72·1	59	59·2	60
11 0 0	42·9	70·6	60	58·4	
12 0 0	44·5	71·7	61	58·0	63
13 0 0	46·6	62·2	62	54·7	
14 0 0	47·8	61·7	66	52·7	70
15 0 0	50·3	60·5	67	54·9	
16 0 0	51·3	59·5	67	54·6	69
17 0 0	50·9	61·1	66	56·3	
18 0 0	50·9	63·5	65	57·5	66
19 0 0	47·9	62·5	63	57·6	
20 0 0	46·9	65·2	62	60·4	63
21 0 0	43·0	62·4	62	58·7	
22 0 0	44·1	64·6	61	58·7	64
23 0 0	38·6	76·6	61	63·9	
27 0 0 0	44·0	67·5	61	56·3	66
1 0 0	44·2	70·1	60	58·1	
2 0 0	40·3	70·6	60	56·9	63
3 0 0	47·1	69·9	61	54·8	
4 0 0	46·7	69·3	61	56·1	66
5 0 0	46·9	74·1	60	59·6	
6 0 0	46·2	70·4	59	58·7	61
7 0 0	45·5	72·9	59	60·1	
8 0 0	48·4	68·5	59	57·3	60
9 0 0	47·3	71·5	59	58·6	
10 0 0	44·5	69·7	59	57·3	61
11 0 0	45·3	69·2	60	57·1	
12 0 0	47·7	67·3	61	56·3	62
13 0 0	49·1	66·7	63	54·2	
14 0 0	49·8	66·3	64	54·8	67
15 0 0	50·5	65·6	66	53·3	
16 0 0	51·2	65·9	66	54·1	69
17 0 0	50·6	63·3	66	54·3	
18 0 0	50·4	60·5	65	55·2	67
19 0 0	47·0	64·8	64	58·1	
20 0 0	49·0	68·5	62	59·4	64
21 0 0	46·9	71·0	62	60·0	
22 0 0	46·7	74·1	61	61·4	63

SEPTEMBER 24, 25, 26, 27, 28, and 29, 1841.

M. Gött. Time	Decl.	Hor. Force		Vert. Force	
d. h. m. s.	Sc.-Div.	Sc.-Div.	Ther.	Sc.-Div.	Ther.
27 23 0 0	45·5	74·2	60	60·6	
28 0 0 0	47·9	77·0	60	60·3	61
1 0 0	45·8	73·3		58·5	
2 0 0	44·8	73·5	59	57·8	60
3 0 0	45·2	72·9		55·3	
4 0 0	49·0	74·9	60	58·0	62
5 0 0	47·0	73·6		58·5	
6 0 0	47·1	73·2	60	58·4	62
7 0 0	47·1	73·9		58·5	
8 0 0	45·2	74·1	59	57·9	62
9 0 0	42·7	76·1		58·8	
10 0 0	42·8	74·8	61	58·5	63
11 0 0	43·1	73·0		58·0	
12 0 0	45·0	70·4	63	56·7	64
13 0 0	48·5	66·8		54·4	
14 0 0	49·9	63·2	68	53 9	69
15 0 0	49·9	63·1		53·8	
16 0 0	49·8	61·9	71	53·9	71
17 0 0	50·3	59·9		52·8	
18 0 0	49·9	58·8	70	53·4	72
19 0 0	47·2	53·6		54·6	
20 0 0	48·9	59·0	66	56·4	70
21 0 0	47·8	61·6		57·6	
22 0 0	47·1	68·5	63	58·7	64
23 0 0	44·3	70·8		57·4	
29 0 0 0	46·0	71·5	59	55·9	63
1 0 0	48·9	75·3		57·5	
2 0 0	47·4	76·4	57	58·0	59
3 0 0	47·2	84·2		59·0	
4 0 0	44·3	84·7	56	58·7	59
5 0 0	44·6	80·6		57·5	
6 0 0	47·6	78·5	55	58·1	58
7 0 0	48·1	80·0		59·2	
8 0 0	46·7	82·3	54	60·9	58
9 0 0	44·4	82·9		60·7	
10 0 0	43·8	79·0	56	59·7	58
11 0 0	45·0	73·6		56·5	
12 0 0	46·2	74·4	60	55·9	60
13 0 0	48·6	72·3		54·0	
14 0 0	51·2	71·4	64	54·8	64
15 0 0	51·6	61·8		53·4	
16 0 0	51·2	63·0	66	53·9	67
17 0 0	50·8	63·2		53·2	
18 0 0	49·2	62·7	65	55·3	66

SEPTEMBER 24, 25, 26, 27, 28, and 29, 1841.

M. Gött. Time	Decl.	Hor. Force		Vert. Force	
d. h. m. s.	Sc.-Div.	Sc.-Div.	Ther.	Sc.-Div.	Ther.
29 19 0 0	49·5	64·1		57·6	
20 0 0	46·6	69·3	63	59·8	64
21 0 0	45·8	71·1		58·8	
22 0 0	47·7	70·2	62	58·7	64
23 0 0	47·5	70·7		58·1	

The Mean Positions at the same hours during the Month are given in page 74.

OCTOBER 8 and 9, 1841.

TORONTO* $\begin{cases} \text{Decl. 1 Scale Division} = 0'\cdot72 \\ \text{H. F. } k = \cdot000076 \,;\, q = \cdot0002 \\ \text{V. F. } k = \cdot000092\,;\, q = \cdot00018 \end{cases}$

Extra observations.

The V. F. was observed at $1^m. 30^s.$ before, and the H. F. $2^m.$ after the times specified.

d. h. m. s.	Decl.	Hor. Force		Vert. Force	
	Sc.-Div.	Sc.-Div.	Ther.	Sc.-Div.	Ther.
9 13 20 0	141·3	394·8	59	60·3	
25 0	141·4	390·8		59·5	
35 0	159·9	405·5			
40 0	155·6	427·8			
45 0	136·6	420·4			
50 0	128·0	405·9			
55 0	129·6	396·3			
14 0 0	138·6	397·1	59		
5 0	142·4	392·5			
10 0	152·7	403·8		55·5	59
15 0	154·9	424·5		54·5	
20 0	155·8	417·7		54·0	
25 0	145·3	421·8			
30 0	140·4	409·1		47·0	
35 0	145·2	420·7		50·1	
40 0	142·9	418·9		48·7	
45 0	138·7	419·6		47·0	
50 0	132·8	410·8		46·9	
55 0	138·1	401·0		48·1	
15 0 0	141·7	401·5	59	50·3	60
5 0	147·1	408·1		50·8	
10 0	145·6	409·9		50·8	
15 0	147·5	413·4		50·4	
20 0	148·5	420·1		51·2	
25 0	143·6	421·5		49·8	

* TORONTO, October, 1841.—*Times of observation at which the Magnets were disturbed, but the mean readings were not materially changed.*

d. h.
Sept. 30 22 Decl. and H. F. disturbed, much vibration.
Oct. 2 0 Decl. and H. F. disturbed, moderate vibration.
    2 H. F. disturbed, much vibration and shocks.
   4 0 H. F. slight vibration.
    2 Decl. and H. F. slight shocks.
   5 0 H. F. moderate vibrations.
    2 Decl. and H. F. slight shocks.
   9 & 2 H. F. moderate vibrations and shocks; Decl. slight shocks at 2ʰ
    4 H. F. moderate vibrations and shocks; Decl. slight shocks at 2ʰ.
  11 12 H. F. slight vibration.
   14 Decl. and H. F. moderate shocks.
   22 H. F. slight vibrations and shocks.
  12 6 H. F. considerable vibrations and shocks.
  13 0 H. F. moderate vibrations and shocks.
    2 Decl. and H. F. moderate shocks.
  14 0 & 2 Decl. and H. F. moderate shocks.
  15 22 H. F. and V. F. slight vibration.
  16 2 H. F. much vibration.
    4 H. F. much vibration; Decl. moderate shocks.
    8 H. F. moderate vibrations.
  12 & 14 H. F. moderate vibrations and shocks.
   16 H. F. moderate vibrations; Decl. moderate shocks.
  17 18 H. F. slight vibration.
  18 0 & 2 Decl. and H. F. moderate shocks.

d. h.
Oct. 18 8 H. F. moderate shocks.
   18 H. F. moderate shocks.
   22 Decl. and H. F. vibrating much.
  19 0 Decl. and H. F. slight vibrations and shocks.
    2 Decl. moderate shocks; H. F. moderate vibrations and shocks.
   4 & 6 Decl. and H. F. moderate shocks.
    8 H. F. slight shocks.
  25 0 H. F. strong shocks.
  26 2 Decl. and H. F. moderate shocks.
   10 H. F. moderate vibrations.
  27 22 H. F. slight vibrations.
  28 2 Decl. and H. F. strong shocks, and slight vibrations.

d. h. m.
• 9 13 30 Aurora brilliant; very bright streamers shooting up to an altitude of 40°., and appearing to travel with great rapidity from W. to E.; streamers gradually dying away, and at 13ʰ. 50ᵐ., no further appearance of Aurora, except a very faint light in N. Bank of strati rising in W.
14 10 Bank of light growing brighter; two or three very faint streamers.
  15 Auroral light very faint; clouds clearing away.
  25 Three faint streamers rose in N.E., and progressed slowly to Westward, dying away when due North. No light remained visible.
  30 No traces of the Aurora left.
  40 No auroral light.
  45 Two faint streamers (scarcely visible) in N.N.E.
  55 No auroral light.

OCTOBER 8 and 9, 1841.

Positions at the usual hours of observation, October 8 and 9.

M. Gött. Time.			Decl.	Hor. Force.		Vert. Force.	
d.	h.	m.	s.	Sc.=Div⁰⁵.	Sc.=Div⁰⁵.	Ther.	Sc.=Div⁰⁵.	Ther.
8	0	0	0	141·4	443·4	55	53·2	56
	2	0	0	141·3	447·1	55	54·1	56
	4	0	0	140·4	420·4	57	51·8	56
	6	0	0	137·7	400·0	57	55·4	57
	8	0	0	131·0	410·0	59	60·2	58
	10	0	0	127·4	406·2	60	58·0	59
	12	0	0	143·7	399·1	61	61·0	59
	14	0	0	143·4	421·5	59	52·4	59
	16	0	0	135·5	416·9	59	52·2	59
	18	0	0	137·8	417·1	57	54·0	58
	20	0	0	134·8	420·5	56	54·0	58
	22	0	0	139·5	431·0	55	55·8	56
9	0	0	0	140·0	438·7	55	55·8	56
	2	0	0	141·5	432·0	55	58·1	56
	4	0	0	138·8	417·5	57	54·9	57
	6	0	0	131·5	409·1	58	53·6	57
	8	0	0	132·8	433·5	59	56·2	58
	10	0	0	143·0	398·5	60	57·1	59
	12	0	0	135·7	402·2	59	61·0	58
	14	0	0	138·6	397·1	59	55·5	60
	16	0	0ᵃ	139·9	410·0	58	52·5	59

Mean Positions at the same hours during the Month.

	0	0	0	138·2	442·4	52	56·9	53
	2	0	0	140·4	436·3	52	58·3	53
	4	0	0	138·6	427·3	53	57·3	53
	6	0	0	131·2	421·4	54	57·5	54
	8	0	0	130·5	430·4	55	57·9	55
	10	0	0	134·1	434·6	56	57·4	56
	12	0	0	137·6	433·6	56	57·1	56
	14	0	0	139·7	432·4	55	56·9	56
	16	0	0	139·8	442·1	55	56·5	55
	18	0	0	138·7	432·1	53	54·9	54
	20	0	0	139·7	438·8	53	53·9	54
	22	0	0	139·7	441·5	52	55·5	53

St. Helena { Decl. 1 Scale Division = 0'·71
H. F. k = ·00019 ; q = ·00025
V. F.ᵇ }

Extra observations.
The H. F. was observed at 2ᵐ. 30ˢ. after the times specified.

8	8	5	0	20·5	38·0	62
		10	0	20·6	38·0	
		15	0	20·7	38·0	
		20	0	20·7	37·9	
		25	0	20·8	37·9	
		30	0	20·9	37·8	
		35	0	20·8	37·4	
		40	0	21·0	37·9	
		45	0	21·1	38·2	
		50	0	21·6	38·4	
		55	0	21·8	39·0	
	9	0	0	22·4	40·0	63
		5	0	22·7	40·4	
		10	0	22·9	41·1	
		15	0	22·9	41·2	

OCTOBER 8 and 9, 1841.

M. Gött. Time.			Decl.	Hor. Force.		Vert. Force.	
d.	h.	m.	s.	Sc.=Div⁰⁵.	Sc.=Div⁰⁵.	Ther.	Sc.=Div⁰⁵.	Ther.
8	9	20	0	22·9	41·2		
		25	0	22·9	41·2		
		30	0	23·0			
		35	0	23·1	41·1		
		40	0	23·3	40·4		
		45	0	23·2	40·0		
		50	0	23·1	39·0		
		55	0	23·0	38·1		
	10	0	0	23·0	38·0	63	
		5	0	23·0	38·0		
		10	0	23·0	38·9		
		15	0	23·1	39·0		
		20	0	23·2	40·0		
		25	0	23·9	40·9		
		30	0	24·0	41·0		
		35	0	24·0	42·0		
		40	0	24·0	42·0		
		45	0	24·0	42·0		
		50	0	23·9	42·0		
		55	0	23·2	42·1		

Positions at the usual hours of observation, October 8 and 9.

8	0	0	0	25·7	62·9	61
	2	0	0	26·9	56·9	62
	4	0	0	25·2	55·9	64
	4	0	0	24·0	55·0	63
	6	0	0	23·0	51·0	64
	6	0	0	21·6	44·6	63
	8	0	0	20·6	38·0	62
	10	0	0	23·0	38·0	63
	11	0	0	23·0	43·0	63
	12	0	0	22·1	45·5	62
	13	0	0	21·7	44·0	62
	14	0	0	22·9	47·2	61
	15	0	0	22·2	47·2	61
	16	0	0	22·4	48·7	61
	19	30	0	22·9	52·0	60
	20	0	0	22·0	52·7	61
	20	30	0	21·2	53·4	61
	22	0	0	22·0	57·9	61
	23	0	0	23·3	57·4	61
9	0	0	0	24·6	57·6	62
	2	0	0	25·1	55·1	63
	3	0	0	26·3	54·4	63
	4	0	0	24·0	53·9	63
	5	0	0	24·0	51·8	63
	6	0	0	23·0	51·5	65
	8	0	0	24·0	51·1	65
	10	0	0	26·0	52·2	62
	11	0	0	23·4	47·3	62
	12	0	0	25·7	46·7	62
	13	0	0ᶜ	23·1	49·3	62

Mean Positions at the same hours during the Month.

0	0	0	22·3	61·8	61
2	0	0	23·9	59·0	62
3	0	0	22·9	57·2	63

OCTOBER 8 and 9, 1841.

M. Gött. Time.			Decl.	Hor. Force.		Vert. Force.	
d.	h.	m.	s.	Sc.=Div⁰⁵.	Sc.=Div⁰⁵.	Ther.	Sc.=Div⁰⁵.	Ther.
	4	0	0	21·2	56·1	63		
	5	0	0	20·3	53·7	63		
	6	0	0	20·5	52·7	63		
	8	0	0	21·6	51·0	62		
	10	0	0	22·3	52·0	62		
	11	0	0	22·5	52·0	62		
	12	0	0	22·5	53·1	61		
	13	0	0	22·2	54·2	61		
	14	0	0	21·7	54·6	61		
	15	0	0	21·7	55·0	61		
	16	0	0	21·6	55·5	61		
	18	0	0	21·6	55·7	61		
	19	30	0	20·8	55·5	61		
	20	0	0	20·1	56·1	61		
	20	30	0	19·5	55·7	61		
	22	0	0	19·7	60·0	61		
	23	0	0	21·4	57·3	61		

Cape of Good Hope { Decl. 1 Scale Division = 0'·75
H. F. k = ·000180 ; q = ·0003
V. F.ᵈ }

Positions at the usual hours of observation, October 8 and 9.

8	0	0	0	55·5	67·0	62
	1	0	0	56·9	65·7	62
	2	0	0	56·5	65·9	62
	3	0	0	55·1	67·0	63
	4	0	0	54·5	65·4	63
	5	0	0	52·5	60·9	63
	6	0	0	52·2	53·7	63
	7	0	0	48·0	52·0	63
	8	0	0	46·5	52·0	63
	9	0	0	49·0	54·0	63
	10	0	0	50·4	52·3	62
	11	0	0	50·4	60·0	62
	12	0	0	50·5	63·0	62
	13	0	0	49·0	62·2	62
	14	0	0	51·8	61·9	62
	15	0	0	50·0	61·0	62
	16	0	0	50·4	61·2	62
	17	0	0	50·0	63·5	62
	18	0	0	50·4	65·9	62
	19	0	0	49·2	66·8	62
	20	0	0	49·1	66·3	61
	21	0	0	50·8	66·0	61
	22	0	0	52·9	68·0	61
	23	0	0	53·5	66·4	61
9	0	0	0	54·9	67·2	61
	1	0	0	55·1	67·0	61
	2	0	0	55·0	66·7	61
	3	0	0	55·5	66·3	61
	4	0	0	52·8	65·2	61
	5	0	0	53·2	63·5	61
	6	0	0	53·2	63·6	61
	7	0	0	52·0	66·4	61
	8	0	0	53·2	66·0	61
	9	0	0	52·9	69·5	61
	10	0	0	56·0	69·0	61
	11	0	0ᵉ	51·0	65·3	61

ᵃ Saturday midnight at Toronto.
ᵇ V. F. magnet not in adjustment.
ᶜ Saturday midnight at St. Helena.

ᵈ V. F. magnet not in satisfactory adjustment.
ᵉ Saturday midnight at the Cape of Good Hope.

N

OCTOBER 8 and 9, 1841.

Mean Positions at the same hours during the Month.

M. Gött. Time d. h. m. s.	Decl. Sc.-Div.	Hor. Force Sc.-Div.	Ther.	Vert. Force Sc.-Div.	Ther.
0 0 0	54·1	69·5	60		
1 0 0	55·3	68·8	61		
2 0 0	55·6	68·3	61		
3 0 0	54·6	68·0	61		
4 0 0	53·6	67·7	61		
5 0 0	52·7	65·4	61		
6 0 0	52·3	64·7	61		
7 0 0	52·5	64·9	61		
8 0 0	52·2	65·7	61		
9 0 0	52·8	66·9	62		
10 0 0	52·9	67·9	61		
11 0 0	52·7	68·4	61		
12 0 0	53·2	70·2	60		
13 0 0	53·2	70·9	60		
14 0 0	53·2	71·5	60		
15 0 0	52·4	72·0	60		
16 0 0	52·2	72·2	60		
17 0 0	51·7	72·6	60		
18 0 0	51·1	73·5	60		
19 0 0	49·7	74·3	60		
20 0 0	48·8	74·1	60		
21 0 0	48·5	72·3	60		
22 0 0	49·7	71·3	60		
23 0 0	52·0	70·4	60		

VAN DIEMEN ISLAND { Decl. 1 Scale Division = 0'·71
H. F. k = ·0003; q =
V. F. k = ; q =

Extra observations.

The V. F. was observed at 2m. 30s. before, and the H. F. 2m. 30s. after the times specified.

M. Gött. Time d. h. m. s.	Decl. Sc.-Div.	Hor. Force Sc.-Div.	Ther.	Vert. Force Sc.-Div.	Ther.
8 5 10 0	62·4	59·3	50		
15 0	62·1			51·4	50
20 0	62·2	60·0			
25 0	59·3				
30 0	57·2	60·6		51·2	
35 0	56·7				
6 10 0	47·3	60·9	50		
15 0	49·2				
20 0	50·5	57·2		43·5	50
25 0	54·9				
30 0	59·3	55·6		50·2	
35 0	61·4				
40 0	62·1	56·3		54·9	
45 0	64·0				
50 0	64·8	57·2		53·7	
55 0	62·5				
7 10 0	59·9	53·3	51		
15 0	60·0				
20 0	60·4	51·5		54·9	51
25 0	60·0				
30 0	63·0	51·0		56·7	
35 0	65·1				
40 0	65·9	50·0		57·4	
45 0	64·6				
50 0	65·0			56·3	
8 10 0	61·6	50·0	51		
15 0	60·1				

a Saturday midnight at Van Diemen Island.

OCTOBER 8 and 9, 1841.

M. Gött. Time d. h. m. s.	Decl. Sc.-Div.	Hor. Force Sc.-Div.	Ther.	Vert. Force Sc.-Div.	Ther.
8 8 20 0	59·5	49·8		53·3	51
25 0	58·9				
30 0	58·5			52·8	

Positions at the usual hours of observation, October 8 and 9.

M. Gött. Time d. h. m. s.	Decl. Sc.-Div.	Hor. Force Sc.-Div.	Ther.	Vert. Force Sc.-Div.	Ther.
8 0 0 0	57·8	54·0	53	54·1	53
1 0 0	58·3	54·2		54·9	
2 0 0	56·8	55·8	52	55·9	51
3 0 0	57·7	55·6		56·0	
4 0 0	54·6	56·1	51	55·1	51
5 0 0	66·0	60·7		53·9	
6 0 0	48·9	62·3	50	41·3	50
7 0 0	62·2	55·1		50·7	
8 0 0	63·9	49·9	51	56·2	51
9 0 0	58·2	50·7		52·8	
10 0 0	60·6	54·9	50	52·0	50
11 0 0	55·6	50·2		55·3	
12 0 0	61·6	48·2	51	57·4	53
13 0 0	59·5	46·2		52·3	
14 0 0	58·4	47·5	56	48·7	56
15 0 0	60·2	48·8		44·1	
16 0 0	62·5	47·9	60	44·3	58
17 0 0	64·0	47·1		44·0	
18 0 0	63·9	48·1	62	42·9	60
19 0 0	64·2	49·6		42·4	
20 0 0	62·0	47·7	63	44·0	61
21 0 0	58·6	48·9		42·2	
22 0 0	59·0	49·5	62	42·0	60
23 0 0	58·6	49·6		44·4	
9 0 0 0	56·9	50·7	61	41·9	61
1 0 0	57·6	50·9		45·6	
2 0 0	58·8	50·9	60	46·5	59

Mean Positions at the same hours during the Month.

d. h. m. s.	Decl. Sc.-Div.	Hor. Force Sc.-Div.	Ther.	Vert. Force Sc.-Div.	Ther.
0 0 0	58·3	52·2	59	45·8	58
1 0 0	57·2	52·1		47·1	
2 0 0	57·2	52·8	58	47·0	57
3 0 0	57·1	52·7		46·5	
4 0 0	58·4	53·0	56	49·7	55
5 0 0	59·6	54·1		49·9	
6 0 0	58·4	55·0	54	49·7	53
7 0 0	59·4	54·8		50·9	
8 0 0	59·4	54·6	54	52·4	53
9 0 0	57·5	54·8		51·6	
10 0 0	56·0	54·6	53	51·6	53
11 0 0	53·7	53·4		51·1	
12 0 0	53·7	51·5	55	50·2	55
13 0 0	55·7	49·5		49·9	
14 0 0	59·4	48·7	58	48·6	57
15 0 0	64·0	49·0		46·7	
16 0 0	67·3	50·0	61	45·0	59
17 0 0	68·5	50·9		44·0	
18 0 0	68·0	51·3	62	43·8	60
19 0 0	66·6	50·8		44·7	
20 0 0	64·3	50·3	63	44·7	60
21 0 0	62·0	50·1		45·3	
22 0 0	60·6	50·8	61	45·0	59
23 0 0	59·2	51·7		45·5	

OCTOBER 8 and 9, 1841.

ANTARCTIC EXPEDITION AT THE BAY OF ISLANDS, NEW ZEALAND.

Decl. 1 Scale Division = 0'·73
H. F. k = ·00018; q =
V. F. k = ; q =

Positions at the usual hours of observation, October 8 and 9.

M. Gött. Time d. h. m. s.	Decl. Sc.-Div.	Hor. Force Sc.-Div.	Ther.	Vert. Force Sc.-Div.	Ther.
8 0 0 0	45·0	73·5	60	58·0	60
1 0 0	45·4	77·7		56·3	
2 0 0	44·9	79·9	57	60·4	59
3 0 0	44·1	81·2		59·7	
4 0 0	44·8	84·2	55	62·1	56
5 0 0	48·6	88·6		61·3	
6 0 0	48·9	94·3	53	63·3	55
7 0 0	51·9	86·3		59·6	
8 0 0	47·3	83·3	53	58·6	55
9 0 0	44·5	80·5		59·0	
10 0 0	43·6	74·8	60	56·8	58
11 0 0	43·6	63·7		53·1	
12 0 0	46·5	55·9	69	50·0	68
13 0 0	48·0	51·9		50·5	
14 0 0	47·2	51·9	73	52·0	72
15 0 0	49·2	51·9		52·5	
16 0 0	49·3	49·6	75	52·6	75
17 0 0	49·7	48·9		52·3	
18 0 0	49·7	49·5	77	52·6	78
19 0 0	49·5	52·5		53·5	
20 0 0	51·7	56·2	72	56·4	72
21 0 0	49·9	62·5		57·7	
22 0 0	48·2	65·4	65	58·9	67
23 0 0	47·7	69·5		59·5	
9 0 0 0(b)	47·6	71·6	61	59·7	63

Mean Positions at the same hours from the 1st to 22nd inclusive.

d. h. m. s.	Decl. Sc.-Div.	Hor. Force Sc.-Div.	Ther.	Vert. Force Sc.-Div.	Ther.
0 0 0	46·8	67·2	64	59·4	65
1 0 0	46·2	69·5		59·2	
2 0 0	46·3	71·0	62	59·3	64
3 0 0	46·1	71·1		58·5	
4 0 0	47·1	71·7	61	58·9	63
5 0 0	47·4	73·2		59·3	
6 0 0	47·5	74·9	60	59·5	62
7 0 0	46·8	75·3		59·4	
8 0 0	45·1	75·0	59	59·2	61
9 0 0	43·3	72·8		58·8	
10 0 0	42·8	68·1	64	56·8	64
11 0 0	43·2	62·7		54·4	
12 0 0	45·6	58·8	69	53·0	70
13 0 0	48·5	57·2		52·8	
14 0 0	50·7	56·1	72	52·5	72
15 0 0	51·4	55·9		52·5	
16 0 0	51·2	55·7	73	53·3	74
17 0 0	50·3	55·5		53·9	
18 0 0	49·6	55·3	73	54·3	74
19 0 0	48·8	55·7		56·6	
20 0 0	49·0	58·1	69	56·6	70
21 0 0	47·8	60·8		57·7	
22 0 0	47·3	63·1	66	58·2	67
23 0 0	47·1	65·0		58·9	

b Saturday midnight at New Zealand.

OCTOBER 24, 25, and 26, 1841.

TORONTO

Decl. 1 Scale Division = 0'·72
H. F. k = ·000076 ; q = ·0002
V. F. k = ·000092 ; q = ·00018

Regular and extra observations.

The V. F. was observed at 1m. 30s. before, and the H. F. 2m. after the times specified. [a]

M. Gött. Time. d. h. m. s.	Decl. Sc.-Div	Hor. Force. Sc.-Div	Ther.	Vert. Force. Sc.-Div	Ther.
24 18 0 0 [b]	150·7	409·7	44	54·9	45
30 0	95·3	361·7		35·4	
35 0	82·9	328·5		27·4	
40 0	83·0	280·0		16·3	
45 0	92·4	306·7		7·3	
50 0	97·5	335·0		9·4	
55 0	96·2	346·6		10·4	
19 0 0	103·2	363·7	44	14·2	46
5 0	117·2	374·5		22·7	
10 0	122·1	373·0		25·2	
15 0	127·4	378·7		31·3	
20 0	134·7	388·2		37·7	
25 0	140·7	393·6		38·4	
30 0	145·1	384·0		38·4	
35 0	152·5	393·9		37·7	
40 0	159·3	390·0		37·1	
45 0	166·6	399·1		36·4	
50 0	172·9	412·0		35·9	
55 0	171·9	414·4		33·5	
20 0 0	167·0	427·1	44	32·9	46
5 0	161·4	424·5		33·6	
10 0	153·6	406·6		34·5	
15 0	146·4	394·2		33·5	
20 0	142·8	387·2		35·5	
25 0	143·0	363·1		36·6	
30 0	140·7	404·2		37·8	
35 0	151·9	422·2		40·2	
40 0	154·3	418·4		42·4	
45 0	156·8	405·4		39·9	
50 0	158·1	398·0		35·3	
55 0	162·6	387·2		30·3	
21 0 0	165·3	392·6	44	29·7	46
5 0	161·9	392·0		32·2	
10 0	155·2	402·5		34·9	
15 0	155·1	405·5		33·2	
20 0	160·4	414·0		32·3	
25 0	162·1	429·2		34·7	
30 0	161·0	440·5		36·7	

OCTOBER 24, 25, and 26, 1841.

M. Gött. Time. d. h. m. s.	Decl. Sc.-Div	Hor. Force. Sc.-Div	Ther.	Vert. Force. Sc.-Div	Ther.
24 21 35 0	157·6	442·0		43·6	
40 0	155·4	454·0		46·4	
45 0	149·1	438·5		48·9	
50 0	150·0	429·5		44·6	
55 0	153·1	438·0		42·6	
22 0 0	156·5	451·0	44	42·7	46
5 0	154·0	453·9		43·8	
10 0	154·5	451·0		41·3	
15 0	155·2	442·3		39·7	
20 0	157·8	444·7		38·3	
25 0	159·8	441·3		44·1	
30 0	160·1	445·5		48·5	
35 0	158·1	450·4		53·4	
40 0	149·9	453·7		57·8	
45 0	138·4	454·2		60·5	
50 0	131·8	446·7		60·3	
55 0	127·2	440·3		58·0	
23 0 0	127·9	437·5	44	50·4	46
5 0	133·0	453·9		53·8	
10 0	133·7	458·0		54·3	
15 0	133·1	445·7		54·2	
20 0	133·8	444·1		50·7	
25 0	136·8	454·5		50·9	
30 0	138·9	459·9		53·0	
35 0	138·6	456·5		52·9	
40 0	138·5	450·5		52·3	
25 0 0 0	127·9	445·9	44	54·2	46
2 0 0	122·6	412·1	44	52·1	46
20 0	127·4	422·9		59·7	
25 0	122·8	412·4		57·4	
30 0	123·7	393·7		57·8	
35 0	124·1	387·4		57·7	
40 0	123·7	401·2		60·7	
45 0	121·8	388·0		61·1	
50 0	120·5	380·0		60·8	
55 0	118·8	396·8		61·4	
3 0 0	117·0	418·0	46	62·9	46
5 0	120·8	438·3		65·5	
10 0	121·0	448·4		67·6	
15 0	116·9	433·1		67·5	
20 0	123·6	435·8		68·3	
25 0	129·4	425·9		69·4	
30 0	129·3	451·8		69·4	
35 0	132·3	459·3		69·5	
40 0	135·1	465·3		69·5	
45 0	136·4	466·1		69·5	

OCTOBER 24, 25, and 26, 1841.

M. Gött. Time. d. h. m. s.	Decl. Sc.-Div	Hor. Force. Sc.-Div	Ther.	Vert. Force. Sc.-Div	Ther.
25 3 50 0	137·5	456·7		68·4	
55 0	144·1	457·3		67·3	
4 0 0	141·1	436·1	47	66·5	47
5 0	138·9	433·3		65·6	
10 0	136·9	429·6		66·6	
15 0	136·8	418·1		65·5	
20 0	133·4	415·3		66·5	
25 0	130·7	411·3		66·2	
30 0	127·5	423·3		66·7	
35 0	130·4	418·2		67·2	
40 0	130·1	424·3		66·8	
45 0	129·2	426·7		70·2	
50 0	129·0	427·7		73·7	
55 0	130·4	429·1		73·7	
5 0 0	120·7	436·2	48	73·4	47
10 0	120·7	428·4		67·0	
15 0	122·8	437·0		73·3	
20 0	121·1	429·2		73·5	
25 0	124·4	429·0		73·4	
30 0	129·2	443·8		73·5	
35 0	127·1	441·3		73·8	
40 0	129·5			73·8	
45 0	128·7	447·9		73·2	
50 0	131·7	438·5		72·5	
55 0	129·0	423·8		69·2	
6 0 0	125·2	426·3	48	73·5	48
5 0	124·3	433·1		69·3	
10 0	126·5	426·1		69·9	
15 0	126·9	429·2		66·2	
20 0	132·5	424·1		72·9	
25 0	134·2	438·7		69·5	
30 0	133·6	413·7		69·5	
35 0	132·3	402·7		73·9	
40 0	132·0	398·7		73·5	
8 0 0	122·9	433·2	48	74·5	48
10 0	127·7	423·2	48	73·8	48
12 0 0	136·4	435·7	47	70·7	48
14 0 0	148·0	440·0	47	68·6	48
16 0 0	146·3	439·0	46	69·1	48
18 0 0	135·1	437·7	46	56·2	47
20 0 0	135·9	442·7	47	49·3	47
22 0 0	141·3	424·5	47	52·1	47
26 0 0 0	125·7	417·6	46	52·0	47
2 0 0	131·6	450·2	46	61·2	47
4 0 0	133·9	446·5	48	67·4	47

---

[a] Commencing immediately after Sunday midnight, at Toronto.

[b] 24 18 0 Faint arch of auroral light in N. At 18h. 15m., Aurora very brilliant; very bright patches, surrounded by a magnificent arch in the N.E., and a large number of streamers, reaching an altitude of 40° to 60°.

30 Aurora much the same.

35 The streamers and patches that rose about N.E., have progressed to the windward, dying away in N.W. H. F. vibrating very much.

45 Arch broken up into a number of patches and banks, intermixed with a great number of streamers; the whole very beautiful. Calm and clear, light cirri in W. disappeared.

19 0 Calm and clear. Slight pulsations in N.W. Streamers as high as 60°; banks very bright.

10 Calm and clear. Aurora growing fainter; streamers just visible.

20 Patches and streamers still fainter.

35 Faint patches only visible.

45 Ditto　　ditto.

20 0 Calm and clear. Bank and streamers in N.W.; nothing to E. of N. but faint light.

15 Bright bank under an arch; arch extending 20° on each side of N.; no streamers.

20 Bank formed like an arch, bright streamers issuing from it; altitude 60°.

30 Arch broken up, very bright bank extending 35° on each side of N., altitude 20°.

d. h. m.
24 20 40 No alteration.

50 Bank again assuming the form of an arch, and more contracted, very bright in centre.

21 5 Calm and clear. Two very bright arches, altitude of upper 40°; lower 20°; lower arch the brightest; streamers from both.

15 Arches extending to N.W., a few streamers only visible.

25 Arches clear and well formed; bright banks and a great number of streamers extending from N.E. to N.W.; light now brighter, and streamers more brilliant than at any other time during the night.

35 Upper arch disappeared; lower arch more faint, extending from N.W. to N.E.; altitude of centre 20°; bright streamers rising from centre of arch.

45 Arch visible but fainter; streamers have disappeared; bright bank inside of arch, reaching down to the horizon.

55 Arch still fainter; a few streamers in centre, and at the N.W. extremity.

22 5 Faint arch; a few streamers from the centre, and patches under the arch.

15 Arch disappearing; a few bright streamers from the centre.

25 Only a faint irregular light at present, in form of banks and patches.

50 Bank brightened up with streamers at both extremities; at 22h. 45m. a faint arch, and at 55m. a faint undefined light alone remained of the Aurora.

23 0 Auroral light scarcely perceptible.

15 All traces of Aurora have disappeared.

October 24, 25, and 26, 1841.

M. Gött. Time. d. h. m. s.	Decl. Sc.-Div^ns	Hor. Force. Sc.-Div^ns	Ther. °	Vert. Force. Sc.-Div^ns	Ther. °
6 0 0	129·6	430·9	49	68·8	48
8 0 0	134·1	438·4	50	67·7	49
10 0 0	141·7	445·4	51	66·6	50
12 0 0	139·0	440·9	51	64·0	51
14 0 0	146·1	443·0	51	60·6	51
16 0 0	138·8	442·2	51	62·5	50
18 0 0	135·8	445·3	51	59·5	51
20 0 0	133·3	444·4	50	57·1	50
22 0 0	134·6	439·5	49	60·4	50

Positions at the usual hours of observation during the Month are given in page 89.

St. Helena { Decl. 1 Scale Division $= 0'·71$
{ H. F. $k = ·00019$; $q = ·00025$
{ V. F.* $k = ·00046$; $q =$

Regular and extra observations.

The V. F. was observed at $2^m. 30^s.$ before, and the H. F. $2^m. 30^s.$ after the times specified.

M. Gött. Time. d. h. m. s.	Decl. Sc.-Div^ns	Hor. Force. Sc.-Div^ns	Ther.	Vert. Force. Sc.-Div^ns	Ther.
24 14 0 0ᵇ	18·6	59·6	62	45·4	61
15 0 0	18·0	59·2	62	46·0	61
16 0 0	18·3	62·4	62	45·8	61
18 0 0	15·2	58·0	61	45·8	61
19 0 0	13·9	55·0	61	45·8	60
5 0	14·0			45·8	
10 0	13·9	56·0		45·8	
15 0	13·4				
20 0	13·1	57·0		47·9	
25 0	13·1				
30 0	13·1	57·1		47·9	
35 0	13·1				
40 0	13·2	57·8		48·1	
45 0	13·1				
50 0	13·7	57·2		48·2	
55 0	13·8				
20 0 0	14·0	57·0	61	48·2	61
5 0	14·0			48·2	
10 0	14·0	56·2		48·2	
15 0	14·0				
20 0	13·2	55·5		48·2	
25 0	13·0				
30 0	12·9	55·2		48·2	
35 0	12·9				
40 0	12·9	55·0		48·1	
45 0	12·3				
50 0	12·0	55·0		48·0	
55 0	12·0				
21 0 0	12·0	53·5	61	48·2	61
5 0	12·0				
10 0	12·3	52·2		48·2	
15 0	12·0				
20 0	12·6	52·0		48·3	
25 0	13·0				
30 0	13·5	51·2		48·4	
35 0	13·9				
40 0	14·0	52·5		48·4	
45 0	14·8				
50 0	14·9	53·7		48·5	
55 0	14·0				
22 0 0	13·9	52·0	61	48·3	61
5 0	13·9				

* New V. F. magnet received from England.

October 24, 25, and 26, 1841.

M. Gött. Time. d. h. m. s.	Decl. Sc.-Div^ns	Hor. Force. Sc.-Div^ns	Ther.	Vert. Force. Sc.-Div^ns	Ther.
24 22 10 0	14·1	51·0		48·3	
15 0	14·9				
20 0	14·4	50·9		48·7	
25 0	14·2				
30 0	13·9	53·0		49·2	
35 0	14·0				
40 0	14·2	53·7		49·3	
45 0	15·0				
50 0	15·5	50·0		49·4	
55 0	16·1				
23 0 0	16·7	52·0	62	49·8	62
5 0	16·0				
10 0	15·9	51·0		49·8	
15 0	15·8				
20 0	16·1	51·0		50·1	
25 0	15·8				
30 0	15·6	51 8		50·1	
35 0	15·4				
40 0	15·2	50·4		50·1	
45 0	14·9				
50 0	15·4	49·7		50·1	
55 0	16·0				
25 0 0 0	16·8	50·2	63	50·6	63
5 0	17·4				
10 0	17·3	50·0		50·7	
15 0	16·8				
20 0	16·8	50·6		51·2	
25 0	17·6				
30 0	18·0	51·0		51·7	
35 0	19·1				
40 0	20·2	51·9		52·1	
45 0	21·0				
50 0	21·4	50·1		52·4	
55 0	20·4				
1 0 0	20·0	47·0	64	52·6	64
5 0	19·8				
10 0	19·0	43·4		52·7	
15 0	18·2				
20 0	18·2	43·0		53·3	
25 0	18·3				
30 0	19·4	43·2		53·6	
35 0	20·1				
40 0	21·1	43·2		53·8	
45 0	21·8				
50 0	21·8	42·4		54·2	
55 0	22·1				
2 0 0	23·1	42·7	66	54·2	65
5 0	22·2				
10 0	21·1	40·0		54·9	
15 0	22·0				
20 0	22·9	39·0		55·1	
25 0	23·8				
30 0	24·0	37·2		55·3	
35 0	23·6				
40 0	22·4	35·2		55·6	
45 0	22·9				
50 0	23·0	35·0		55·0	
55 0	22·1				
3 0 0	21·1	36·9	66	54·9	66
5 0	19·9				
10 0	19·2	39·0		55·4	
15 0	19·0				
20 0	19·0	41·0		55·8	
25 0	18·3				

October 24, 25, and 26, 1841.

M. Gött. Time. d. h. m. s.	Decl. Sc.-Div^ns	Hor. Force. Sc.-Div^ns	Ther.	Vert. Force. Sc.-Div^ns	Ther.
25 3 30 0	18·2	42·0		55·7	
35 0	18·7				
40 0	18·1	43·0		55·9	
45 0	18·0				
50 0	18·0	44·0		56·2	
55 0	19·0				
4 0 0	19·0	44·0	67	56·3	66
5 0	19·5				
10 0	19·2	41·4		56·4	
15 0	19·1				
20 0	19·0	38·0		56·2	
25 0	18·9				
30 0	18·0	37·0		56·8	
35 0	17·9				
40 0	17·9	35·1		56·8	
45 0	17·7				
50 0	17·8	35·0		56·3	
55 0	18·0				
5 0 0	17·0	33·0	67	56·3	67
5 0	16·8				
10 0	16·7	30·8		56·2	
15 0	16·4				
20 0	16·2	32·9		56·4	
25 0	16·1				
30 0	16·0	37·1		56·4	
35 0	16·0				
40 0	16·3	39·0		55·8	
45 0	16·8				
50 0	16·8	37·0		55·7	
55 0	16·1				
6 0 0	16·2	35·9	67	55·8	67
5 0	16·5				
10 0	16·9	34·5		55·8	
15 0	17·0				
20 0	17·0	32·2		55·8	
25 0	16·8				
30 0	16·2	30·0		55·8	
35 0	16·1				
40 0	16·1	28·5		55·5	
45 0	16·8				
50 0	17·0	29·0		55·5	
55 0	17·1				
7 0 0	17·1	29·2		55·5	
5 0	17·1				
10 0	17·0	29·0		54·8	
15 0	17·0				
20 0	17·0	28·0		55·0	
25 0	17·1				
30 0	17·1	28·9		55·0	
35 0	17·0				
40 0	17·0	28·9		55·0	
45 0	17·0				
50 0	16·2	28·0		55·0	
55 0	16·2				
8 0 0	17·0	28·7	66	55·1	66
5 0	17·0				
10 0	17·2	32·0		56·2	
15 0	17·1				
20 0	17·0	35·4		56·0	
25 0	17·0				
30 0	17·1	34·5		56·2	
35 0	17·8				
40 0	17·7	34·3		55·6	
45 0	17·5				

ᵇ Commencing after Sunday midnight at St. Helena.

October 24, 25, and 26, 1841.

M. Gött. Time.				Decl.	Hor. Force.		Vert. Force.	
d.	h.	m.	s.	Sc.-Div.	Sc.-Div.	Ther.	Sc.-Div.	Ther.
25	8	50	0	17·6	34·0			
		55	0	17·5				
	9	0	0	18·0	35·1		55·4	
		5	0	18·1				
		10	0	18·5	36·1		55·1	
		15	0	18·6				
		20	0	18·7	36·5		55·1	
		25	0	18·3				
		30	0	18·1	35·1			
		35	0	18·0				
		40	0	18·0	35·0		55·1	
		45	0	17·9				
		50	0	17·7	34·8		54·5	
		55	0	17·6				
	10	0	0	17·8	34·9	65	54·5	64
		5	0	18·0				
		10	0	18·1	35·4		52·5	
		15	0	18·4				
		20	0	18·8	36·4		53·4	
		25	0	18·8				
		30	0	18·8	37·2		53·4	
		35	0	18·7				
		40	0	18·6	37·8		53·4	
		45	0	18·8				
		50	0	18·8	38·9		53·4	
		55	0	18·8				
	11	0	0	19·0	39·3	65	53·5	64
		5	0	19·2				
		10	0	19·2	39·5		54·0	
		15	0	19·7				
		20	0	19·4	40·6		53·9	
		25	0	19·4				
		30	0	19·6	40·0		53·5	
		35	0	19·0				
		40	0	20·0	41·4		53·8	
		45	0	19·0				
		50	0	20·0	41·6		53·6	
		55	0	20·0				
	12	0	0	20·0	40·6	64	53·9	63
		5	0	20·0				
		10	0	20·0	41·6		53·0	
		15	0	20·0				
		20	0	20·0	43·8		53·8	
		25	0	19·8				
		30	0	20·0	45·6		53·4	
		35	0	20·0				
		40	0	19·8	45·3		53·2	
		45	0	20·0				
		50	0	20·2	45·7		53·2	
		55	0	20·0				
	13	0	0	20·8	49·5	64	53·1	63
		5	0	21·4				
		10	0	21·6	50·4		53·1	
		15	0	21·8				
		20	0	21·0	49·2			
		25	0	21·0				
		30	0	20·8	48·8		52·8	
		35	0	20·4				
		40	0	20·3	48·4		53·0	
		45	0	20·4				
		50	0	20·3	48·0		53·0	
		55	0	20·1				
	14	0	0	20·1	48·0	63	53·0	63
		5	0	20·1				
		10	0	20·1	48·3		52·8	
		15	0	20·1				

October 24, 25, and 26, 1841.

M. Gött. Time.				Decl.	Hor. Force.		Vert. Force.	
d.	h.	m.	s.	Sc.-Div.	Sc.-Div.	Ther.	Sc.-Div.	Ther.
25	14	20	0	20·1	48·7		52·8	
		25	0	20·1				
		30	0	20·1	48·0		52·8	
		35	0	20·1				
		40	0	20·0	47·9		52·8	
		45	0	20·0				
		50	0	20·0	47·0		52·8	
		55	0	20·0				
	15	0	0	20·0	47·0	63	52·7	63
		5	0	20·0				
		10	0	20·0	47·5		52·7	
		15	0	20·0				
		20	0	20·0	47·0		52·6	
		25	0	20·0				
		30	0	20·0	47·3		52·5	
		35	0	20·0				
		40	0	20·0	47·8		52·5	
		45	0	19·1				
		50	0	18·1	47·4		52·4	
		55	0	18·2				
	16	0	0	17·7	47·5	63	52·4	62
		5	0	18·0			52·4	
		10	0	18·0				
		15	0	18·0				
		20	0	18·1	47·9		52·4	
		25	0	18·9				
		30	0	18·9	47·0		52·4	
		35	0	18·9				
		40	0	18·9	47·0		52·4	
		45	0	18·1				
		50	0	18·0	47·0		52·4	
		55	0	18·0				
	17	0	0	18·0	47·0		Vibr?.	
		5	0	18·0				
		10	0	18·0	47·7		51·0	
		15	0	18·0				
		20	0	18·0	47·9		50·9	
		25	0	18·0				
		30	0	18·0	48·0		50·9	
		35	0	18·6				
		40	0	18·9	48·5		50·9	
		45	0	18·9				
		50	0	18·9	48·9		50·8	
		55	0	18·9				
	18	0	0	18·9	49·3	63	50·6	62
		5	0	18·9				
		10	0	18·9	49·2		50·5	
		15	0	19·0				
		20	0	19·1	49·1		50·6	
		25	0	19·1				
		30	0	19·1	49·3		50·7	
		35	0	19·1				
		40	0	19·1	49·9		50·0	
		45	0	19·0				
		50	0	19·0	50·3		50·0	
		55	0	19·0				
	19	0	0	18·9	51·0	63	50·0	62
		5	0	18·9				
		10	0	18·9	51·0		50·0	
		15	0	18·1				
		20	0	18·0	50·0		51·2	
		25	0	17·1				
		30	0	16·9	49·1		50·7	
		35	0	16·0				
		40	0	16 3	48·3		51·4	
		45	0	16·5				

October 24, 25, and 26, 1841.

M. Gött. Time.				Decl.	Hor. Force.		Vert. Force.	
d.	h.	m.	s.	Sc.-Div.	Sc.-Div.	Ther.	Sc.-Div.	Ther.
25	19	50	0	16·0	48·0		51·4	
		55	0	16·0				
	20	0	0	17·0	48·0	63	52·0	62
		30	0	14·0	48·0		51·5	
	22	0	0	13·3	47·9	62	50·8	62
		5	0	13·0				
		10	0	14·2	47·9		51·0	
		15	0	14·1				
		20	0	14·5	47·5		51·0	
		25	0	14·1				
		30	0	14·1	45·5		51·1	
		35	0	13·8				
		40	0	14·0	45·8		51·0	
		45	0	14·5				
		50	0	14·5	46·0		51·3	
		55	0	14·9				
	23	0	0	15·0	47·8	63	51·4	62
		5	0	15·6				
		10	0	15·6	47·5		51·4	
		15	0	15·3				
		20	0	16·1	47·0		51·7	
		25	0	16·4				
		30	0	16·4	46·1		52·0	
		35	0	16·7				
		40	0	16·4	46·5		51·9	
		45	0	16·8				
		50	0	16·8	45·2		52·2	
		55	0	17·1				
26	0	0	0	17·0	44·2	63	52·2	63
		30	0	20·4	49·1		52·4	
	1	0	0	19·6	49·0	64	52·9	63
		30	0	20·2	50·0		53·2	
	2	0	0	21·0	48·2	64	57·7	64
	3	0	0	20·2	43·1	65	56·9	64
		30	0	19·2	42·0		56·7	
	4	0	0	18·7	44·0	65	56·9	
	4	30	0	18·1	41·1		56·9	64
	5	0	0	18·0	34·9	65	56·4	64
	6	0	0	18·9	42·5	65	55·9	64
	8	0	0	18·7	43·2	64	55·6	63
	10	0	0	20·2	49·0	63	52·4	63
	11	0	0	20·2	47·9	63	52·3	63
	12	0	0	20·0	48·9	63	52·4	62
	13	0	0	19·8	50·0	63	52·6	62
	14	0	0	20·0	52·0	62	53·2	62
	15	0	0	19·2	51·0	62	53·1	62
	16	0	0	19·0	50·5	62	53·1	62
	18	0	0	19·0	50·1	62	52·8	61
	19	30	0	16·1	50·9	62	52·6	60
	20	0	0	15·5	51·4	62	53·7	61
	20	30	0	14·8	51·2	62	53·7	61
	22	0	0	16·0	53·0	62	53·9	61
	23	0	0	16·3	53·0	62	52·0	62

The Mean Positions at the usual hours during the Month are given in page 89.

OCTOBER 24, 25, and 26, 1841.

CAPE OF GOOD HOPE
$\begin{cases}\text{Decl. 1 Scale Division} = 0'\cdot75 \\ \text{H. F. } k = \cdot000180;\ q = \cdot0003 \\ \text{V. F.}^a\ k = \qquad ;\ q = \end{cases}$

Positions at the usual hours of observation,
October 24, 25, and 26.

M. Gött. Time.				Decl.	Hor. Force.		Vert. Force.	
d.	h.	m.	s.	Sc.-Div^{ns}.	Sc.-Div^{ns}.	Ther.	Sc.-Div^{ns}.	Ther.
24	12	0	0^b	55·0	74·0	64		
	13	0	0	54·6	74·0			
	14	0	0	53·5	76·0	63		
	15	0	0	52·0	76·0			
	16	0	0	53·0	78·0	63		
	17	0	0	50·0	76·9			
	18	0	0	49·0	75·0	63		
	19	0	0	48·0	71·0			
	20	0	0	51·0	74·4	63		
	21	0	0	51·2	69·6			
	22	0	0	53·0	65·3	63		
	23	0	0	55·4	64·7			
25	0	0	0	55·1	61·8	62		
	1	0	0	58·0	56·6			
	2	0	0	63·1	48·2	62		
	3	0	0	54·4	48·0			
	4	0	0	54·9	57·3	63		
	5	0	0	54·6	44·3			
	6	0	0	50·9	50·9	62		
	7	0	0	51·8	46·6			
	8	0	0	45·0	51·5	62		
	9	0	0	49·4	55·0			
	10	0	0	48·2	54·7	62		
	11	0	0	51·3	57·1			
	12	0	0	51·1	60·2	62		
	13	0	0	55·9	66·1			
	14	0	0	53·9	65·4	62		
	15	0	0	50·9	64·7			
	16	0	0	49·0	64·7	62		
	17	0	0	49·0	64·2			
	18	0	0	49·6	68·0	62		
	19	0	0	47·9	71·0			
	20	0	0	46·7	67·2	62		
	21	0	0	45·6	65·1			
	22	0	0	46·7	62·6	62		
	23	0	0	49·2	59·2			
26	0	0	0	51·8	51·8	63		
	1	0	0	52·6	57·4			
	2	0	0	55·0	57·3	64		
	3	0	0	54·0	53·0			
	4	0	0	52·0	55·3	64		
	5	0	0	48·0	47·1			
	6	0	0	50·4	55·7	64		
	7	0	0	51·1	58·0			
	8	0	0	50·9	58·6	64		
	9	0	0	52·8	65·8			
	10	0	0	53·2	65·0	64		
	11	0	0	52·3	63·5			
	12	0	0	52·5	64·8	63		
	13	0	0	53·7	65·6			
	14	0	0	54·1	67·6	63		
	15	0	0	51·8	66·8			
	16	0	0	51·0	65·9	63		
	17	0	0	50·4	66·2			
	18	0	0	48·7	67·0	63		
	19	0	0	47·4	65·4			

OCTOBER 24, 25, and 26, 1841.

M. Gött. Time.				Decl.	Hor. Force.		Vert. Force.	
d.	h.	m.	s.	Sc.-Div^{ns}.	Sc.-Div^{ns}.	Ther.	Sc.-Div^{ns}.	Ther.
26	20	0	0	47·4	65·7	63		
	21	0	0	48·3	64·2			
	22	0	0	50·8	63·1	63		
	23	0	0	53·7	61·0			

The Mean Positions at the same hours during the Month
are given in page 90.

VAN DIEMEN ISLAND
$\begin{cases}\text{Decl. 1 Scale Division} = 0'\cdot71 \\ \text{H. F. } k = \cdot0003;\ = q \\ \text{V. F. } k = \qquad ;\ = q\end{cases}$

Regular and extra observations.
The V. F. was observed at 2^m. 30^s. before, and the H. F.
2^m. 30^s. after the times specified.

M. Gött. Time.				Decl.	Hor. Force.		Vert. Force.	
d.	h.	m.	s.	Sc.-Div^{ns}.	Sc.-Div^{ns}.	Ther.	Sc.-Div^{ns}.	Ther.
24	3	0	0^c	59·8	54·2		52·8	
	4	0	0	58·7	55·0	55	51·0	54
	5	0	0	61·9	54·8		52·7	
	6	0	0	59·8	55·2	55	51·2	54
	7	0	0	59·2	56·0		50·6	
	8	0	0	58·4	56·7	54	50·4	53
	9	0	0	59·1	56·6		50·0	
	10	0	0	57·8	56·9	54	49·6	53
	11	0	0	52·8	56·1		49·4	
	12	0	0	53·0	54·4	55	46·2	55
	13	0	0	53·7	51·8		48·0	
	14	0	0	57·9	49·7	57	48·3	58
	15	0	0	64·7	50·8		45·7	
	16	0	0	69·9	54·1	60	41·9	60
	17	0	0	75·0	55·0		44·4	
	18	0	0	77·5	55·0	62	45·3	60
	19	0	0	75·2	48·8		54·9	
	20	0	0	76·2	49·0	62	55·7	60
	21	0	0	75·0	49·0		59·1	
	22	0	0	65·3	47·6	61	60·0	59
	23	0	0	59·8	49·9		63·2	
25	0	0	0	58·8	47·5	59	64·9	58
	1	0	0	53·3	44·9		61·3	
	2	0	0	53·0	45·7	58	61·4	57
		15	0	66·6	51·2		43·8	
		22	30	58·7	51·7		34·9	
		30	0	51·6	55·2		27·6	
		37	30	57·0	54·8		23·9	
		45	0	52·1	51·6		21·8	
	3	0	0	42·3	46·6	58	23·5	
		10	0	45·5	46·3			
		15	0	45·4				
		20	0	47·1	47·0		32·0	
		25	0	48·4				
		30	0	50·2	47·3		34·7	
		35	0	51·5				
		40	0	53·1	47·3		36·1	
		45	0	54·8				
		50	0	55·4	47·5		38·0	
	4	0	0	51·9	44·4	58	37·9	
		10	0	43·7	41·7			
		15	0	40·7				
		20	0	37·8	41·0		34·8	59
		25	0	37·0				
		30	0	39·0	41·3		34·3	
	35	·	0	43·5				
		40	0	49·7	43·4		37·8	
		45	·0	54·1				

OCTOBER 24, 25, and 26, 1841.

M. Gött. Time.				Decl.	Hor. Force.		Vert. Force.	
d.	h.	m.	s.	Sc.-Div^{ns}.	Sc.-Div^{ns}.	Ther.	Sc.-Div^{ns}.	Ther.
25	4	50	0	52·8	47·2		37·4	
	55	0		53·8				
	5	0	0	53·5	47·4	58	37·9	
		20	0	52·4	46·8			
		25	0	55·1				
		30	0	54·1	47·2		37·9	58
		35	0	54·8				
		40	0	55·0	47·2		37·9	
		45	0	55·3				
		50	0	54·9			38·4	
	6	0	0	52·6	47·4	56	38·6	56
	7	0	0	52·1	48·8		38·6	
	8	0	0	58·2	46·1	53	50·2	53
		10	0	59·2	45·3			
		15	0	59·9				
		20	0	61·3	46·9		50·9	
		25	0	60·9				
		30	0	61·4	49·8		48·2	
		35	0	61·3				
		40	0	61·7	49·3		46·1	
		45	0	60·3				
		50	0	59·0	49·1		43·9	
		55	0	57·2				
	9	0	0	55·3	49·5	53	43·3	53
		10	0	59·2	49·3	53	46·2	53
	11	0	0	53·1	47·9		45·4	
	12	0	0	53·3	46·2	57	44·8	57
	13	0	0	56·6	41·6		47·3	
	14	0	0	60·3	42·9	61	45·2	60
	15	0	0	64·3	44·7		38·3	
	16	0	0	69·1	45·2	63	40·7	62
	17	0	0	68·4	46·1		41·5	
	18	0	0	70·6	47·2	65	41·0	63
	19	0	0	69·4	47·5		45·9	
	20	0	0	68·3	47·5	65	48·0	63
	21	0	0	65·9	46·0		49·7	
	22	0	0	56·2	48·5	63	41·9	61
	23	0	0	58·5	47·9		41·1	
26	0	0	0	60·5	57·2	62	31·1	59
	1	0	0	56·9	51·7		44·3	
	3	0	0	53·3	49·3		43·1	
	5	0	0	56·8	52·8		39·1	
	6	0	0	52·1	51·8	55	43·7	54
	7	0	0	57·3	52·7		48·2	
	8	0	0	62·1	51·7	54	52·3	53
	9	0	0	59·6	50·8		52·5	
	10	0	0	57·6	51·3	54	51·2	54
	11	0	0	56·1	51·2		51·7	
	12	0	0	55·9	48·9	56	46·2	57
	13	0	0	56·5	44·5		43·3	
	14	0	0	60·5	44·5	61	41·9	61
	15	0	0	63·1	45·8		37·1	
	16	0	0	67·0	46·8	66	35·9	64
	17	0	0	67·4	48·6		33·3	
	18	0	0	65·7	48·8	68	33·2	66
	19	0	0	63·2	49·1		35·2	
	20	0	0	64·0	48·1	68	37·0	66
	21	0	0	60·4	47·3		36·4	
	22	0	0	61·0	48·1	67	35·5	65
	23	0	0	56·6	46·8		35·7	

The Mean Positions at the usual hours of observation
during the Month, are given in page 90.

---

^{a} V. F. magnetometer not in satisfactory adjustment.
^{b} Commencing after midnight of Sunday at the Cape of Good Hope.
^{c} Commencing after midnight of Sunday at Van Diemen Island.

^{d} "A very faint light to the Southward, rendered now more visible by the moon
having set."

NOVEMBER 4, 5, and 6, 1841.

TORONTO*
Decl. 1 Scale Division = 0'·72
H. F. k = ·000076; q = ·0002
V. F. k = ·000092; q = ·00018

Extra observations.
The V. F. was observed at 1ᵐ. 30ˢ. before, and the H. F. 2ᵐ. after the times specified.

M. Gött. Time.				Decl.	Hor. Force.		Vert. Force.	
d.	h.	m.	s.	Sc.-Div⁰ˢ.	Sc.-Div⁰ˢ.	Ther.	Sc.-Div⁰ˢ.	Ther.
6	0	20	0	114·9	422·9	49		
		25	0	113·9	422·1		90·2	50
		30	0	113·2	423·1		90·0	
		35	0	114·6	427·8		91·4	
		40	0	116·7	428·1		92·3	
		45	0	114·9	429·9		92·5	
		50	0	115·2	431·2		92·6	
		55	0	116·3	436·1		92·3	
1	0	0	118·3	435·6	50	94·1	51
		5	0	118·6	428·1		94·3	
		10	0	120·6	428·8		93·0	
		15	0	124·8	435·8		92·8	
		20	0	127·7	429·4		93·3	
		25	0	128·0	434·7		91·3	
		30	0	127·5	443·7		92·9	
		35	0	127·1	449·6		95·1	
		40	0	127·1	446·4		95·3	
		45	0	129·4	448·8		96·1	

Positions at the usual hours of observation, November 4, 5 and 6.

d.	h.	m.	s.	Sc.-Div⁰ˢ.	Sc.-Div⁰ˢ.	Ther.	Sc.-Div⁰ˢ.	Ther.
4	0	0	0	138·3	435·4	53	71·3	54
	2	0	0	138·9	421·2	52	93·4	53
	4	0	0	128·9	410·9	52	96·9	53
	6	0	0	131·3	421·9	53	97·4	53
	8	0	0	132·3	425·9	54	99·1	54
	10	0	0	136·6	438·8	54	98·5	54
	12	0	0	139·3	449·1	53	98·0	54
	14	0	0	139·7	431·8	53	97·8	54
	16	0	0	140·8	428·5	53	96·7	53
	18	0	0	137·0	431·7	52	96·7	53
	20	0	0	138·3	423·4	52	88·5	52
	22	0	0	125·0	397·3	51	85·9	52
5	0	0	0	137·5	458·2	51	88·6	52
	2	0	0	137·3	440·7	51	93·2	52
	4	0	0	133·4	440·0	51	97·7	52
	6	0	0	132·4	429·2	51	99·4	52
	8	0	0	131·5	414·4	51	102·5	52
	10	0	0	130·9	413·6	52	113·2	52
	12	0	0	140·3	433·5	52	105·1	52
	14	0	0	145·6	439·1	51	101·7	51
	16	0	0	138·2	422·1	51	98·0	51

NOVEMBER 4, 5, and 6, 1841.

M. Gött. Time.			Decl.	Hor. Force.		Vert. Force.	
d.	h.	m.	s.	Sc.-Div⁰ˢ.	Sc.-Div⁰ˢ.	Ther.	Sc.-Div⁰ˢ.	Ther.
5	18	0	0	143·9	424·2	51	85·4	51
	20	0	0	140·8	421·9	50	85·4	51
	22	0	0	142·9	442·8	50	92·8	50
6	0	0	0	119·3	410·7	49	90·3	50
	2	0	0	131·0	446·9	50	97·9	51
	4	0	0	142·3	438·4	49	102·3	50
	6	0	0	134·1	432·4	49	103·5	50
	8	0	0	131·0	426·5	49	105·5	50
	10	0	0	134·2	451·2	49	107·0	49
	12	0	0	140·6	430·4	49	107·6	49
	14	0	0	146·0	434·3	49	105·2	49
	16	0	0ᵃ	148·5	448·4	48	107·2	49

Mean Positions at the same hours during the Month.

d.	h.	m.	s.					
	0	0	0	136·9	456·7	47	99·7	49
	2	0	0	140·2	455·7	47	103·2	48
	4	0	0	137·2	447·4	48	103·0	48
	6	0	0	132·4	445·2	49	102·9	49
	8	0	0	132·2	451·9	49	103·7	49
	10	0	0	135·0	456·4	49	103·9	49
	12	0	0	138·7	456·5	49	104·9	49
	14	0	0	140·6	455·2	49	104·2	49
	16	0	0	139·2	453·7	48	101·4	49
	18	0	0	139·3	450·7	48	101·3	49
	20	0	0	139·2	449·3	48	99·9	49
	22	0	0	139·3	454·5	48	102·0	48

ST. HELENA
Decl. 1 Scale Division =0'·711
H. F. k = ·00019; q = ·00025
V. F. k = ·00046; q =

Positions at the usual hours of observation, November 4, 5, and 6.

d.	h.	m.	s.					
4	0	0	0	18·9	44·9	62	54·6	62
	2	0	0	20·9	37·5	62	53·9	62
	3	0	0	22·0	37·0	62	54·1	62
	4	0	0	21·9	37·1	63	53·4	63
	5	0	0	21·0	41·0	63	54·5	63
	6	0	0	20·1	41·0	63	57·4	63
	8	0	0	20·2	40·1	62	56·5	62
	10	0	0	20·0	43·0	62	57·6	62
	11	0	0	20·2	43·2	62	57·4	62
	12	0	0	21·0	44·0	62	57·4	62
	13	0	0	20·8	44·2	62	57·4	61
	14	0	0	19·0	44·0	62	57·2	61
	15	0	0	19·0	44·0	62	57·0	61
	16	0	0	19·0	46·0	62	56·9	61
	18	0	0	19·0	45·0	62	58·3	61
	19	30	0	17·1	47·4	62	58·7	61

NOVEMBER 4, 5, and 6, 1841.

M. Gött. Time.			Decl.	Hor. Force.		Vert. Force.	
d.	h.	m.	s.	Sc.-Div⁰ˢ.	Sc.-Div⁰ˢ.	Ther.	Sc.-Div⁰ˢ.	Ther.
4	20	0	0	17·0	47·0	62	57·7	61
	20	30	0	15·9	47·9	61	57·1	61
	22	0	0	15·9	47·7	61	56·8	61
	23	0	0	18·0	50·1	61	56·8	61
5	0	0	0	20·2	48·4	62	56·8	61
	2	0	0	21·5	44·9	63	57·4	62
	3	0	0	22·4	44·5	63	57·5	62
	4	0	0	23·0	40·1	63	57·5	62
	5	0	0	22·0	45·0	63	57·4	62
	6	0	0	21·4	44·0	63	57·0	62
	8	0	0	20·1	41·0	62	57·5	62
	10	0	0	19·0	34·1	62	57·0	62
	11	0	0	20·0	38·2	62	57·4	62
	12	0	0	20·0	39·1	62	57·4	61
	13	0	0	19·9	42·0	62	57·5	61
	14	0	0	20·0	42·9	62	56·6	61
	15	0	0	20·0	43·0	62	56·0	61
	16	0	0	18·5	42·9	62	55·7	61
	18	0	0	17·1	44·0	61	56·3	61
	19	30	0	18·1	47·8	61	56·2	61
	20	0	0	18·5	47·0	61	56·6	61
	20	30	0	18·0	46·0	61	56·6	61
	22	0	0	18·2	45·9	61	57·0	61
	23	0	0	20·0	48·0	61	57·0	61
6	0	0	0	21·6	44·9	62	57·4	62
	2	0	0	20·0	42·1	63	57·5	63
	3	0	0	21·0	45·0	63	57·6	63
	4	0	0	19·8	46·2	64	57·5	64
	5	0	0	19·9	45·2	64	57·7	63
	6	0	0	20·9	44·0	64	57·7	63
	8	0	0	20·7	39·1	64	58·7	63
	10	0	0	20·3	39·9	63		
	11	0	0	21·1	44·5	63	59·0	61
	12	0	0ᶜ	21·1	41·5	63	57·7	62

Mean Positions at the same hours during the Month.

d.	h.	m.	s.					
	0	0	0	22·5	50·4	64	55·1	64
	2	0	0	23·6	47·3	66	56·2	65
	3	0	0	23·4	45·9	66	56·6	65
	4	0	0	22·0	45·0	67	56·8	66
	5	0	0	21·0	44·4	67	57·2	66
	6	0	0	20·7	43·0	66	57·4	66
	8	0	0	21·8	42·2	66	56·9	65
	10	0	0	22·4	42·0	65	56·2	65
	11	0	0	22·6	42·6	65	56·2	64
	12	0	0	22·4	43·4	65	56·1	64
	13	0	0	22·4	43·7	65	56·0	64
	14	0	0	22·2	44·1	64	55·6	64
	15	0	0	21·9	44·6	64	55·5	64

---

* TORONTO, November, 1841.—*Times of observation at which the Magnetometers were disturbed, but the mean readings were not materially changed.*

d.	h.	
1	10	H. F. moderate shocks.
2	22	V. F. slightly vibrating.
3	0	H. F. moderately vibrating.
	18	Decl. and H. F. moderate vibrations and shocks.
4	22	H. F. moderately vibrating.
5	0	Decl. slight shocks.
	18	H. F. moderate shocks and vibrating ; 10ʰ. Decl. moderate shocks.
8	12	H. F. slight shocks.
10	6	Moderate vibrations and shocks.
11	2	Moderate vibrations and shocks.
12	2	Decl. and H. F. strong shocks.
	4	H. F. moderate vibrations; 6ʰ. H. F. moderate vibrations.
	8	H. F. slight vibrations.
20		H. F. slight shocks.
15	0	Decl. strong shocks; H. F. vibrating much with shocks.
	2	Decl. and H. F. strong shocks.
	4	Decl. and H. F. moderate vibrations and shocks; V. F. slight vibrations.

d.	h.	
15	8	H. F. slight shocks.
16	2	H. F. moderate shocks.
17	16	H. F. moderate shocks.
18	2	Decl. and H. F. moderate shocks.
19	4	H. F. vibrating much.
	8	H. F. considerable vibrations.
	18	Decl. moderate vibrations ; H. F. considerable vibrations.
20	0	Decl. and H. F. strong shocks.
	2	H. F. vibrating much with shocks.
23	2	H. F. moderate vibrations and shocks.
24	22	V. F. slight vibrations.
25	0	H. F. moderate shocks.
26	0	Decl. moderate shocks.
	4	H. F. moderate shocks.
29	10	H. F. slight vibrations.

ᵃ Saturday midnight at Toronto.
ᵇ Vibrating too much for exact observation.
ᶜ Saturday midnight at St. Helena.

NOVEMBER 4, 5, and 6, 1841.

M. Gött. Time. d. h. m. s.	Decl. Sc.-Div.	Hor. Force. Sc.-Div.	Ther.	Vert. Force. Sc.-Div.	Ther.
16 0	21·6	45·1	64	55·4	64
18 0	21·0	46·0	64	55·5	63
19 30	19·4	47·6	64	55·5	63
20 0	18·5	48·5	64	55·3	63
20 30	18·0	49·0	64	55·1	63
22 0	19·1	50·5	63	54·8	63
23 0 0	20·4	51·1	64	54·7	63

CAPE OF GOOD HOPE   { Decl. 1 Scale Division = 0″·75   H. F. k = ·000180 ; q = ·0003   V. F. k = ·000047 ; q =

Positions at the usual hours of observation, November 4, 5, and 6.

M. Gött. Time. d. h. m. s.	Decl. Sc.-Div.	Hor. Force. Sc.-Div.	Ther.	Vert. Force. Sc.-Div.	Ther.
4 0 0 0	54·1	61·6	62	123·9	62
1 0 0	54·0	59·0	62		
2 0 0	55·6	59·0	62	124·2	62
3 0 0	57·1	60·2	62		
4 0 0	55·0	58·8	62	128·4	63
5 0 0	55·0	62·5	63		
6 0 0	54·3	63·9	63	128·5	63
7 0 0	53·9	64·4	63		
8 0 0	52·6	67·3	63	114·0	63
9 0 0	54·0	70·0	63		
10 0 0	52·9	70·0	62	109·9	62
11 0 0	53·9	70·8	62		
12 0 0	54·7	71·9	62	106·9	62
13 0 0	54·5	72·7	62		
14 0 0	52·6	73·0	61	117·6	61
15 0 0	52·0	73·1	61		
16 0 0	51·3	75·1	61	115·1	61
17 0 0	50·6	75·3	61		
18 0 0	46·8	76·3	61	119·1	61
19 0 0	48·0	76·9	60		
20 0 0	49·4	75·1	60	118·4	61
21 0 0	50·4	72·0	61		
22 0 0	53·0	70·3	61	114·9	61
23 0 0	54·5	67·8	61	106·4	61
5 0 0 0	55·9	67·2	62		
1 0 0	57·0	64·6	62	110·1	62
2 0 0	55·8	67·0	62		
3 0 0	54·9	62·7	62	128·6	62
4 0 0	53·9	67·9	62		
5 0 0	53·8	68·0	62	115·4	62
6 0 0	53·4	67·4	62		
7 0 0	50·2	69·6	62	123·7	62
8 0 0	49·2	68·0	62		
9 0 0	49·4	65·8	62	129·7	62
10 0 0	53·7	68·0	62		
11 0 0	54·0	68·6	62	115·1	62
12 0 0	52·9	69·9	62		
13 0 0	53·1	70·2	62	116·9	62
14 0 0	51·9	70·3	62		
15 0 0	54·0	70·5	62	120·0	62
16 0 0	48·2	71·0	62		
17 0 0	48·8	71·0	61	121·6	62
18 0 0	50·3	71·1	61		
19 0 0	54·8	73·3	61	98·8	62
20 0 0	54·0	72·6	61		
21 0 0	54·2	71·8	61	109·3	62

NOVEMBER 4, 5, and 6, 1841.

M. Gött. Time. d. h. m. s.	Decl. Sc.-Div.	Hor. Force. Sc.-Div.	Ther.	Vert. Force. Sc.-Div.	Ther.
5 23 0 0	54·6	70·9	61		
6 0 0 0	56·2	63·9	62	107·8	62
1 0 0	56·0	59·8	62		
2 0 0	53·9	62·0	62	122·3	62
3 0 0	53·8	67·6	63		
4 0 0	54·0	68·2	63	107·2	63
5 0 0	53·9	67·5	63		
6 0 0	54·0	68·8	63	110·3	63
7 0 0	54·0	69·0	63		
8 0 0	51·0	67·1	63	124·6	63
9 0 0	51·3	67·8	62		
10 0 0	52·4	68·0	62	120·2	63
11 0 0[a]	54·4	73·6	63		

Mean Positions at the same hours during the Month.[b]

M. Gött. Time. d. h. m. s.	Decl. Sc.-Div.	Hor. Force. Sc.-Div.	Ther.
0 0 0	53·9	69·7	64
1 0 0	55·0	69·4	64
2 0 0	55·4	68·6	65
3 0 0	54·6	68·5	65
4 0 0	53·5	68·2	65
5 0 0	52·8	68·3	65
6 0 0	52·8	67·8	65
7 0 0	52·9	68·3	65
8 0 0	52·8	68·3	65
9 0 0	53·1	69·1	65
10 0 0	52·7	69·5	65
11 0 0	53·1	70·7	65
12 0 0	52·9	71·4	64
13 0 0	53·1	72·6	64
14 0 0	52·5	72·9	64
15 0 0	51·8	73·2	64
16 0 0	51·3	74·0	64
17 0 0	50·5	75·1	63
18 0 0	48·9	75·8	63
19 0 0	47·5	76·5	63
20 0 0	47·1	75·9	63
21 0 0	47·6	74·3	63
22 0 0	49·7	72·5	64
23 0 0	51·8	71·1	64

VAN DIEMEN ISLAND   { Decl. 1 Scale Division = 0″·71   H. F. k = ·0003 ; q =   V. F. k = q =

Positions at the usual hours of observation, from November 3rd, 12h. to November 6th, 2h.

M. Gött. Time. d. h. m. s.	Decl. Sc.-Div.	Hor. Force. Sc.-Div.	Ther.	Vert. Force. Sc.-Div.	Ther.
3 12 0 0	37·9	64·0	55	41·8	55
13 0 0	40·0	62·9		39·9	
14 0 0	44·5	63·5	57	35·7	57
15 0 0	52·1	65·7		31·2	
16 0 0	57·8	65·2	60	32·0	59
17 0 0	57·4	64·8		31·0	
18 0 0	56·4	69·3	62	34·4	60
19 0 0	62·4	62·4		42·7	
20 0 0	63·0	62·6	62	43·2	60
21 0 0	56·3	66·5		32·2	
22 0 0	52·4	63·9	61	33·3	59
23 0 0	43·3	54·3		42·5	
4 0 0 0	44·8	55·3	60	51·8	57
1 0 0	40·0	58·9		43·9	

NOVEMBER 4, 5, and 6, 1841.

M. Gött. Time. d. h. m. s.	Decl. Sc.-Div.	Hor. Force. Sc.-Div.	Ther.	Vert. Force. Sc.-Div.	Ther.
4 2 0 0	36·0	60·3	58	40·0	56
3 0 0	41·7	61·1		40·0	
4 0 0	41·8	66·9	57	30·9	55
5 0 0	40·6	61·5		39·4	
6 0 0	43·9	62·2	56	41·9	54
7 0 0	44·7	63·0		41·9	
8 0 0	43·8	63·2	54	43·3	53
9 0 0	43·2	63·5		42·8	
10 0 0	43·0	63·5	54	42·5	53
11 0 0	39·5	64·6		39·0	
12 0 0	38·1	64·0	55	36·8	54
13 0 0	44·4	62·2		34·5	
14 0 0	43·6	60·4	58	37·6	58
15 0 0	48·1	60·5		38·1	
16 0 0	52·4	60·1	59	37·5	59
17 0 0	56·1	62·7		33·5	
18 0 0	58·5	62·2	60	35·5	59
19 0 0	57·8	63·0		36·6	
20 0 0	56·2	61·4	60	38·8	59
21 0 0	52·0	61·7		39·3	
22 0 0	46·3	65·3	59	34·2	58
23 0 0	41·5	62·0		35·1	
5 0 0 0	36·9	59·5	58	37·7	56
1 0 0	39·9	60·8		39·7	
2 0 0	36·3	60·2	57	37·6	56
3 0 0	40·1	61·4		35·9	
4 0 0	45·0	68·0	57	27·9	56
5 0 0	43·9	62·9		38·0	
6 0 0	41·5	63·5	56	36·4	54
7 0 0	42·3	63·9		36·9	
8 0 0	42·2	63·8	55	36·0	54
9 0 0	43·4	63·7		36·5	
10 0 0	43·6	63·4	54	36·1	53
11 0 0	46·3	63·5		35·6	
12 0 0	42·0	62·5	56	34·9	54
13 0 0	41·6	62·0		35·4	
14 0 0	44·9	59·8	57	37·4	56
15 0 0	48·8	58·6		42·1	
16 0 0	54·7	58·2	59	42·6	58
17 0 0	58·2	60·2		39·8	
18 0 0	56·8	60·8	60	39·0	59
19 0 0	50·4	63·6		38·0	
20 0 0	52·5	59·6	61	44·9	59
21 0 0	50·9	59·8		42·6	
22 0 0	47·7	60·8	60	38·9	58
23 0 0	47·9	63·0		38·6	
6 0 0 0	52·7	66·1	58	28·4	57
1 0 0	41·6	63·7		36·3	
2 0 0[c]	45·6	63·7	57	36·5	56

Mean Positions at the same hours during the Month.

M. Gött. Time. d. h. m. s.	Decl. Sc.-Div.	Hor. Force. Sc.-Div.	Ther.	Vert. Force. Sc.-Div.	Ther.
0 0 0	44·6	61·0	63	29·4	61
1 0 0	43·3	61·5		29·9	
2 0 0	44·2	63·0	62	30·1	60
3 0 0	44·0	63·2		31·0	
4 0 0	45·1	63·6	60	32·0	59
5 0 0	45·9	63·6		33·9	
6 0 0	46·2	64·0	59	34·1	57
7 0 0	46·3	64·2		35·1	
8 0 0	45·4	64·8	58	35·4	57
9 0 0	43·6	64·7		34·8	

---

[a] Saturday midnight at the Cape of Good Hope.

[b] The observations of the V. F. magnet during the month are not sufficiently regular to admit of mean positions being derived : the mean positions of the Decl. are from the 1st to the 25th inclusive.

[c] Saturday midnight at Van Diemen Island.

November 4, 5, and 6, 1841.

M. Gött. Time.				Decl.	Hor. Force.		Vert. Force.	
d.	h.	m.	s.	Sc.-Divns.	Sc.-Divns.	Ther. °	Sc.-Divns.	Ther. °
	10	0	0	42·5	64·2	57	34·6	57
	11	0	0	40·3	63·0		33·8	
	12	0	0	39·9	60·9	59	33·3	59
	13	0	0	42·3	59·1		32·4	
	14	0	0	45·8	58·5	62	31·3	62
	15	0	0	50·9	58·5		29·7	
	16	0	0	54·8	59·6	65	28·1	63
	17	0	0	56·0	60·7		26·5	
	18	0	0	56·0	61·2	66	26·2	64
	19	0	0	54·0	60·9		26·2	
	20	0	0	52·3	60·8	66	26·8	64
	21	0	0	50·3	60·7		27·1	
	22	0	0	48·4	61·1	65	27·1	63
	23	0	0	46·2	61·0		27·8	

November 18, 19, and 20, 1841.

TORONTO { Decl. 1 Scale Division = 0′·72; H. F. k = ·000076; q = ·0002; V. F. k = ·000092; q = ·00018 }

Extra observations.

The V. F. was observed at 1m. 30s. before, and the H. F. 2m. after the times specified.

M. Gött. Time.				Decl.	Hor. Force.		Vert. Force.	
d.	h.	m.	s.	Sc.-Divns.	Sc.-Divns.	Ther. °	Sc.-Divns.	Ther. °
18	12	20	0*	135·8	432·7	49	151·4	49
		25	0	145·9	435·7		147·4	
		30	0	135·2	431·4		135·6	
		35	0	128·5	445·8		138·9	
		40	0	131·6	448·9		145·8	
		45	0	138·0	470·7		144·1	

November 18, 19, and 20, 1841.

M. Gött. Time.				Decl.	Hor. Force.		Vert. Force.	
d.	h.	m.	s.	Sc.-Divns.	Sc.-Divns.	Ther. °	Sc.-Divns.	Ther. °
18	12	50	0	117·1	459·1		143·1	
		55	0	134·2	477·9		138·3	
	13	0	0	146·2	485·9	48	157·2	49
		5	0	148·3	436·5		147·1	
		10	0	150·5	400·8		133·0	
		15	0	148·0	390·5		124·3	
		20	0	167·8	457·1		135·7	
		25	0	185·2	450·4		131·3	
		30	0	141·0	425·3		100·0	
		35	0	135·1	410·1		112·7	
		40	0	131·4	416·3		119·9	
		45	0	134·1	418·7		123·0	
		50	0	126·3	414·4		122·0	
		55	0	114·6	359·8		128·0	

* An Auroral light in the North was first seen at 12h.; weather calm and clear. The following observations will show the rapid and various changes of its features:—

d. h. m.

18 12 10 Patches and streamers in North to an alt. of about 30°, luminous haze extending to zenith.

25 Double arch of light from N.E. to N.W., alt. about 10°, streamers shooting from it to an alt. of 40°; patches and luminous haze from East to West.

40 Bank of strati rising in North; streamers in N.W., and luminous haze in East and zenith.

13 0 Detached banks and patches of light covering $\frac{1}{4}$ of the sky to Northward; eight or nine distinct groups of bright streamers dispersed about, appearing all to follow a general course Westward.

10 The detached banks and patches of light united, forming one splendid sheet of light. A very bright range of streamers extended from N.E. to N.W., all travelling from the Eastward, and disappearing regularly as they reached the N.W. extremity.

15 A very brilliant double arch of light extending from the N.E. to N.W.; alt. of exterior arch about 20°, interior 10°: ranges of bright streamers between the arches.

22 A very extraordinary bank of light appeared in N.N.E. throwing out brilliant streamers in all directions; streamers tinged with deep red and blue.

25 A very large number of streamers overspreading the whole Northern portion of the sky; banks and patches of light of different shapes and colours appearing and disappearing too rapidly to be accurately described.

30 A remarkably bright streak of light extending across the zenith from North to S.E., remained steady for a few seconds, afterwards the Southern end moved slowly across the sky and disappeared in the West horizon. Bright arch of light in the North with streamers.

35 Streamers disappeared, leaving only a bright bank of light.

40 Bank sinking lower in horizon, occasionally shooting up a bright streamer to the zenith.

43 Bank of light becoming brighter, streamers appearing and disappearing rapidly.

45 A low, well-defined, dark arch stretching from N.E. to N.W., alt. about 5°, bright masses of light (like cirro-cumuli clouds) moving along its border; no streamers visible.

50 Dark arch still visible, large masses of light of a reddish blue colour issuing from behind it, and after assuming various shapes, disappearing suddenly; double arch from North to N.W. and a number of streamers.

14 0 Wind N.N.W., light, clear. Numberless streamers and patches waving and moving rapidly past one another from East to West.

10 Light becoming fainter, but extending over a greater portion of the sky; a few small streamers.

15 Light covering about $\frac{5}{8}$ths of the sky, appearing like haze; streamers moving Westward.

25 Nothing remarkable since last observation.

30 Size of arch rather increased; streamers and patches of light moving Westward.

45 Double arch; alt. of smaller one about 20°, of larger 45°; patches and streamers visible between them; a very large streamer shot up from East extremity of larger arch almost to the zenith, and disappeared in a few seconds.

50 The whole Northern portion of the sky one sheet of pale light, several bright streamers in East; stars visible through the light.

55 Magnificent arch of light across the sky from East to N.W., alt 50°; very brilliant streamers in N.W.

15 0 Arch of light becoming fainter, brightest at the extremities.

5 Arch brighter, the whole North sky brilliantly illuminated, a few faint streamers in N.E.

10 Arch of light still very bright; streamers occasionally appearing and disappearing in East and N.W., those in N.W. moving round and dying away in West.

d. h. m.

18 15 15 A very splendid arch extending from East to W.N.W., breadth about 10°, alt. 55°; a few faint streamers at each extremity.

20 Arch considerably fainter, several bright patches underneath.

25 Bright wavy patches appearing and disappearing in N., arch of light the same as at the last observation.

30 Arch of light fainter, alt. 45°, very bright pulsations in N.E.

35 Brilliant bank covering the whole North sky, brightest at the extremities.

40 Light much fainter in zenith, a few very faint streamers in N.E.

45 A double arch of light from East to N.W.; patches of light appearing and disappearing rapidly in North horizon.

50 Numerous bright streamers shot up suddenly in East, and moved rapidly across the sky to N.W., then suddenly disappeared.

55 A splendid range of wavy light appeared, and after moving rapidly backward and forward between East and N.W. several times, disappeared gradually in N.; lower part of a bright red colour.

16 0 Brilliant streamers covering the whole North sky; waving light rapidly moving backwards and forwards.

5 A broad band of light partly across the sky, about 20° South of the zenith, gradually sinking into the North horizon.

10 Light much fainter; a few very bright patches of light round the North horizon, and a few streamers visible.

15 A double bank of light across the North sky, lower bank considerably the brightest.

20 Streamers and vivid pulsations, covering the whole North sky.

25 Very bright pulsations flashing and quivering over $\frac{3}{4}$ of the sky.

30 A small steady patch of light in zenith; very vivid pulsations; streamers shooting upwards from North and East. The whole North sky in continual motion, broad patches of light traversing backwards and forwards with great velocity.

35 Light fainter, but vivid pulsations still apparent.

40 Bright streamers round the whole of North and N.E. horizon. Auroral light extending directly across the zenith to S.E. by South.

45 Clouds rising in South and S.E.; several remarkably bright streamers in North, moving Westward and suddenly disappearing. Pulsations flashing and quivering without intermission since 16h. 30m.

50 $\frac{3}{4}$ of the sky to Northward covered with vivid pulsations, and remarkably bright streamers rapidly appearing and disappearing.

55 Appearances the same.

17 0 Wind N.W. very light. Partially clouded in North and N.E. cirro-cumuli, strati, and cirro-strati round South horizon (clouding from Eastward), light alternately faint and bright.

5 Bright bank across the whole of North sky, a few bright streamers occasionally.

10 Bank a little fainter, a few faint pulsations in zenith.

15 Bank fainter, and gradually decreasing in size, a few faint streamers visible occasionally.

20 Bank again brighter, a few bright streamers appearing occasionally above it.

25 Fainter, no streamers.

30 Fainter, no streamers; wind Northerly, very light, $\frac{3}{8}$ clouded, cirro-cumuli and cirro-strati dispersed generally over the sky.

35 Light considerably fainter.

40 Moderately bright. Bank of light, alt. 30°, extending from N.W. to N.E.

45 Considerably fainter, partially clouded in North.

50 Rather brighter.

55 Faint bank, alt. 20°.

18 0 Wind North by West, light, $\frac{1}{4}$ clouded, principally round horizon, cirro-cumuli and cirro-strati, remainder clear, fair. Auroral light fainter.

10 Several bright streamers shooting up from the Northward and rapidly disappearing; very bright pulsations across the whole North horizon.

20 Aurora very brilliant; a great number of streamers shooting up to an alt. of 50°.

25 Streamers as before; a number of very vivid pulsations rising from N.E. Streamers passing quickly from East to West.

NOVEMBER 18, 19, and 20, 1841.

M. Gött. Time			Decl.	Hor. Force		Vert. Force	
h.	m.	s.	Sc.-Div.	Sc.-Div.	Ther.	Sc.-Div.	Ther.
18 14	0	0	124·2	424·0	47	119·5	49
	5	0	137·7	459·4		152·9	
	10	0	141·4	484·2		146·4	
	15	0	137·9	468·7		144·8	
	20	0	145·9	483·1		141·1	
	25	0	138·7	459·8		141·2	
	30	0	131·4	423·8		130·1	
	35	0	131·2	438·2		132·5	
	40	0	139·0	414·8		136·7	
	45	0	131·1	384·3		121·9	
	50	0	142·5	413·6		133·6	
	55	0	150·0	446·4		142·4	
15	0	0	155·0	421·2	47	142·4	48
	5	0	149·4	412·1		136·1	
	10	0	149·3	399·1		134·7	
	15	0	158·2	407·9		132·3	
	20	0	153·7	409·8		129·2	
	25	0	153·7	403·9		130·3	
	30	0	147·0	400·4		128·2	
	35	0	148·0	417·2		129·9	
	40	0	146·8	416·6		132·9	
	45	0	143·8	397·9		129·5	
	50	0	160·5	490·2		115·3	
	55	0	191·9	413·8		107·3	
16	0	0	131·3	394·5	45	51·6	47
	5	0	82·6	429·3		69·8	
	10	0	131·6	399·2		116·4	
	15	0ᵃ	104·6	Off Scale	b	95·1	
	20	0	65·9	Off Scale		79·8	
	25	0	95·7	4·5		70·5	
	30	0	41·5	143·9		46·0	
	35	0	52·3	162·9		63·5	
	40	0	102·9	245·8		84·4	
	45	0	119·3	320·9		69·5	
	50	0	95·3	333·8		67·7	
	55	0	122·7	312·6		99·8	
17	0	0	142·0	349·9	45	02·3	47
	5	0	137·1	343·5		93·2	
	10	0	134·7	333·4		95·2	
	15	0	142·1	336·4		96·9	
	20	0	149·0	348·9		101·7	
	25	0	152·9	347·9		101·4	
	30	0	154·9	375·4		100·2	
	35	0	152·0	398·2		102·5	
	40	0	149·3	386·8		106·8	
	45	0	150·8	377·6		106·5	
	50	0	147·5	372·7		106·0	
	55	0	149·7	389·9		105·5	
18	0	0	150·3	387·2	44	108·9	46
	5	0	147·5	384·2		108·6	
	10	0	140·7	362·3		100·5	
	15	0	124·3	343·0		87·4	
	20	0	110·7	313·3		72·1	
	25	0	147·5	416·6		70·8	
	30	0	105·2	376·7		50·2	
	35	0	75·3	332·7		53·3	
	40	0	98·5	344·8		81·1	
	45	0	120·8	388·4		87·5	

NOVEMBER 18, 19, and 20, 1841.

M. Gött. Time			Decl.	Hor. Force		Vert. Force	
d.	h.	s.	Sc.-Div.	Sc.-Div.	Ther.	Sc.-Div.	Ther.
18 18	50	0	124·8	373·5		81·7	
	55	0	131·3	369·8		80·9	
19 0	0	0	139·8	378·2	45	79·7	46
	5	0	134·5	365·7		68·0	
	10	0	136·4	354·2		62·8	
	15	0	134·5	344·8		53·2	
	20	0	124·1	330·3		40·4	
	25	0	110·0	311·6		44·2	
	30	0	106·9	275·8		57·8	
	35	0	124·7	313·2		72·3	
	40	0	134·4	317·1		78·9	
	45	0	143·1	329·1		77·4	
	50	0	151·3	328·2		77·4	
	55	0	155·9	329·7		74·1	
20	5	0	156·6	324·2	44	69·8	46
	10	0	152·5	320·0		68·3	
	15	0	152·0	334·9		67·9	
	20	0	155·0	352·1		67·4	
	25	0	162·8	382·6		72·4	
	30	0	165·7	381·5		74·2	
	35	0	170·3	378·7		74·8	
	40	0	172·1	371·7		73·9	
	45	0	170·6	355·1		72·8	
	50	0	165·8	337·8		69·1	
	55	0	161·8	337·2		64·9	
21	0	0	162·1	357·1	44	72·6	46
	5	0	160·7	361·4		80·1	
	10	0	160·0	360·2		83·0	
	15	0	157·0	372·1		84·1	
	20	0	155·6	377·6		89·6	
	25	0	153·8	390·0		91·0	
	30	0	152·1	404·8		95·3	
	35	0	151·6	415·3		99·8	
	40	0	150·7	417·4		102·3	
	45	0	149·8	423·0		104·0	
	50	0	147·5	420·1		105·3	
	55	0	145·5	421·9		103·5	
22	5	0	145·0	422·0	45	102·0	46
	10	0	147·0	424·9		102·3	
	15	0	147·2	423·8		101·8	
	20	0	149·2	433·9		102·1	

Positions at the usual hours of observation, November 18, 19, and 20·

			Decl.	Hor. Force		Vert. Force	
18	0	0	140·9	468·0	43	105·2	46
2	0	0	145·0	464·2	45	106·2	46
4	0	0	140·6	465·0	46	104·2	46
6	0	0	133·9	466·7	47	104·2	47
8	0	0	125·8	462·2	48	105·8	48
10	0	0	130·2	447·1	50	107·7	48
12	0	0	140·0	447·1	49	136·5	49
14	0	0	124·2	424·0	47	119·5	49
16	0	0	131·3	394·5	45	51·6	47
18	0	0	150·3	387·2	44	108·9	46
20	0	0	158·4	330·9	44	71·1	46
22	0	0	144·6	421·7	45	103·4	46
19 0	0	0	124·9	441·5	45	98·3	46

NOVEMBER 18, 19, and 20, 1841.

M. Gött. Time			Decl.	Hor. Force		Vert. Force	
d.	h.	m.	Sc.-Div.	Sc.-Div.	Ther.	Sc.-Div.	Ther.
19 2	0	0	144·1	451·1	44	107·1	45
4	0	0	134·1	439·2	44	108·9	44
6	0	0	132·2	443·2	44	108·9	45
8	0	0	133·2	443·3	44	109·2	45
10	0	0	136·8	453·4	45	111·7	45
12	0	0	137·0	435·0	45	114·0	45
14	0	0	156·8	421·8	45	116·7	45
16	0	0	143·5	436·4	45	111·6	45
18	0	0	129·8	422·0	45	85·1	46
20	0	0	141·1	434·0	45	101·9	45
22	0	0	137·1	431·0	45	101·5	45
20 0	0	0	126·4	437·1	44	83·8	46
2	0	0	129·7	413·1	44	99·2	45
4	0	0	127·3	450·5	45	106·0	45
6	0	0	129·7	448·9	46	108·8	46
8	0	0	134·6	451·2	47	107·9	47
10	0	0	139·0	456·5	48	107·3	48
12	0	0	157·0	438·9	48	106·9	48
14	0	0	138·0	455·8	48	104·7	48
16	0	0ᶜ	141·3	441·3	48	106·2	48

The Mean Positions at the same hours during the Month are given in page 95.

ST. HELENA   { Decl. 1 Scale Division $= 0'\cdot711$
   H. F. $k = \cdot000186$; $q = \cdot00025$
   V. F. $k = \cdot00046$; $q =$

Extra observations.

The V. F. was observed at 2ᵐ. 30ˢ. before, and the H. F. 2ᵐ. 30ˢ. after the times specified.

d.	h.	m.	Decl.	Hor. Force		Vert. Force	
18 7	0	0	20·9	38·0	66		66
	5	0	20·9				
	10	0	21·0	37·1		56·0	
	15	0	21·5				
	20	0	21·1	36·4		56·2	
	25	0	21·2				
	30	0	21·2	35·9		56·2	
	35	0	21·3				
	40	0	21·4	35·7		56·2	
	45	0	20·9				
	50	0	20·8	35·8		56·1	
	55	0	20·1				
8	5	0	19·4		66		66
	10	0	19·7	35·0		55·6	
	15	0	20·0				
	20	0	20·4	34·6		55·6	
	25	0	21·3				
	30	0	21·5	35·3		55·7	
	35	0	21·7				
	40	0	21·8	36·1		55·7	
	45	0	21·9				
	50	0	22·0	35·9		56·0	
	55	0	22·1				
9	0	0	21·8	34·3	66	56·1	66
	5	0	21·5				
	10	0	21·3	32·9		56·5	

---

d.	h.	m.	
18	18	27	About ⅔ of the sky covered with remarkably bright patches, banks and streamers, very vivid pulsations. Streamers due North of a deep red colour, those to the East and West perfectly white.
		35	Aurora decreasing very much in brilliancy; no pulsations.
		40	Nothing remaining but a bright luminous haze covering about ⅔ of the North sky; clouding rapidly over from S.E.
		45	Totally clouded, except about ¼ in North. Nothing visible of Aurora but a faint light.
19		15	Wind North by West, moderate; densely clouded.

* The readings of the Decl. magnet between 16ʰ. 15ᵐ. and 16ʰ. 20ᵐ. at intervals of 18ˢ. (the time of vibration of the bar), were as follows: 83·2; 53·4; 60·0; 40·0; 20·0; 11·4; 9·4.

b A light held in the prolongation of the scale showed that the reading would have been about −200 if the scale had been continued.

c Saturday midnight at Toronto.

November 18, 19, and 20, 1841.

M. Gött. Time (m. s.)	Decl. (Sc.-Div.)	Hor. Force (Sc.-Div.)	Hor. Ther.	Vert. Force (Sc.-Div.)	Vert. Ther.
18 9 15 0	20·8				
20 0	20·6	31·4		56·1	
25 0	20·3				
30 0	20·1	29·2		55·6	
35 0	19·9				
40 0	19·5	28·6		55·4	
45 0	19·5				
50 0	19·5	27·9		55·8	
55 0	19·3				
10 5 0	19·1		66		65
10 0	18·9	25·2		55·7	
15 0	18·9				
20 0	18·8	24·2		55·7	
25 0	18·2				
30 0	18·1	24·0		55·5	
35 0	18·1				
40 0	18·1	22·7		55·5	
45 0	18·1				
50 0	18·2	21·2		55·4	
55 0	18·5				
11 5 0	18·1		65		65
10 0	18·1	28·0		55·8	
15 0	18·9				
20 0	19·1	30·8		55·8	
25 0	20·8				
30 0	21·5	30·0		56·0	
35 0	21·1				
40 0	21·3	29·2		56·0	
45 0	21·1				
50 0	20·1	29·8		55·9	
55 0	19·0				
12 5 0	18·8		65		65
10 0	18·9	26·0		55·7	
15 0	18·9				
20 0	19·1	24·1		55·6	
25 0	19·8				
30 0	20·0	23·2		55·5	
35 0	20·1				
40 0	20·5	22·8		55·7	
45 0	20·8				
50 0	20·8	23·0		55·6	
55 0	21·1				
13 5 0	21·9		65		65
10 0	21·0	27·0			
15 0	19·5				
20 0	19·5	26·0		55·6	
25 0	19·1				
30 0	19·1	23·9		55 5	
35 0	19.1				
40 0	19·1	23·0		55·5	
45 0	19·1				
50 0	19·1	24·2		55·5	
55 0	19·0				
14 5 0	18·1		65		64
10 0	17·1	26·5		55·5	
15 0	17·0				
20 0	17·1	28·0		55·4	
25 0	17·2				
30 0	17·0	26·9		55·2	
35 0	17·2				
40 0	17·1	27·0		55·3	
45 0	17·3				
50 0	17·9	25·9		55·2	
55 0	17·9				
15 5 0	18·0		65		64
10 0	18·0	26·9		55·8	
15 0	18·2				

November 18, 19, and 20, 1841.

M. Cött. Time (h. d. m.)	Decl. (Sc.-Div.)	Hor. Force (Sc.-Div.)	Hor. Ther.	Vert. Force (Sc.-Div.)	Vert. Ther.
18 15 20 0	18·0	27·1		56·1	
25 0	18·0				
30 0	17·5	28·0		56·1	
35 0	17·0				
40 0	17·0	28·0		55·9	
45 0	17·0				
50 0	17·0	28·9		55·9	
55 0	16·5				
16 5 0	16·8		65		63
10 0	16·2			55·8	
15 0	16·0				
20 0	16·4	29·0		55·8	
25 0	18·0				
30 0	17·5	27·5		56·2	
35 0	17·8				
40 0	18·0	29·9		56·3	
45 0	18·0				
50 0	18·3	32·0		56·3	
55 0	17·6				
17 0 0	17·4	33·4	65		
5 0	17·1				
10 0	17·0	33·7		55·2	
15 0	16·9				
20 0	16·7	33·7		55·2	64
25 0	17·5				
30 0	17·1	32·9		56·0	
35 0	17·2				
40 0	17·1	32·1		56·0	
45 0	17·9				
50 0	18·0	32·1		55·7	
55 0	18·0				
18 5 0	18·1		65		64
10 0	18·1	31·8		56·2	
15 0	17·9				
20 0	18·0	31·1		57·0	
25 0	18·1				
30 0	18·1	31·2		56·6	
35 0	18·0				
40 0	17·4	31·2		55·9	
45 0	18·5				
50 0	16·8	31·0		55·9	
55 0	16·8				
19 0 0	16·8	31·0	65	55·3	64
5 0	16·8				
10 0	16·6	32·0		55·3	
15 0	16·1				
20 0	15·7	32·4		55·8	
25 0	14·8				
30 0	14·6	33·2		56·1	
35 0	14·1				
40 0	14·1	33·3		56·1	
45 0	14·7				
50 0	14·9	33·8		56·8	
55 0	15·1				
20 5 0	14·4		65		64
10 0	14·4	35·0		56·8	
15 0	14·3				
20 0	14·4	35·1		56·5	
25 0	14·4				
30 0	14·2	35·0		56·6	
35 0	14·6				
40 0	14·6	35·1		56·5	
45 0	14·3				
50 0	14·3	37·6		56·5	
55 0	14·1				
21 5 0	15·9		64		64
10 0	15·3	38·2		57·1	
15 0	15·0				

November 18, 19, and 20, 1841.

M. Gütt. Time (d. s.)	Decl. (Sc.-Div.)	Hor. Force (Sc.-Div.)	Hor. Ther.	Vert. Force (Sc.-Div.)	Vert. Ther.
18 21 20 0	15·3	38·0		56·0	
25 0	15·3				
30 0	13·7	37·0		56·5	
35 0	13·9				
40 0	16·5	36·0		56·6	
45 0	15·6				
50 0	15·6	37·0		56·7	
55 0	15·5				
22 5 0	15·7		64		63
10 0	15·9	38·0		56·5	
15 0	16·0				
20 0	15·8	36·9		56·5	
25 0	15·7				
30 0	15·5	36·9		56·0	
35 0	15·7				
40 0	16·1	38·1		55·4	
45 0	16·5				
50 0	16·3	37·4		55·7	
55 0	16·9				
23 5 0	17·8		65		64
10 0	17·2	37·0		55·7	
15 0	18·0				
20 0	18·6	38·1		55·7	
25 0	18·4				
30 0	18·2	37·0		55·1	
35 0	18·1				
40 0	18·1	34·9		55·1	
45 0	18·0				
50 0	18·5	33·0		55·1	
55 0	19·0				
19 0 5 0	20·1		65		64
10 0	20·0	35·1		55·1	
15 0	20·0				
20 0	20·1	36·1		55·1	
25 0	20·2				
30 0	20·3	37·0		55·1	
35 0	20·5				
40 0	20·5	36·2		55·2	
45 0	20·8				
50 0	20·9	36·0		55·1	
55 0	20·9				
1 0 0	20·9	35·9	65	55·1	65
5 0	21·0				
10 0	21·0	35·0		55·1	
15 0	20·9				
20 0	20·9	35·0		55·1	
25 0	21·2				
30 0	21·5	35·7		55·1	
35 0	21·9				
40 0	22·0	35·9		55·1	
45 0	22·1				
50 0	22·4	35·4		55·1	
55 0	22·7				
2 5 0	22·9		65		65
10 0	23·0	35·0		55·5	
15 0	23·1				
20 0	23·3	35·0		55·8	
25 0	23·8				
30 0	23·6	34·0		55·8	
35 0	23·4				
40 0	23·8	34·0		56·0	
45 0	23·8				
50 0	24·0	34·0		56·0	
55 0	24·0				
3 5 0	24·0		66		66
10 0	24·0	34·2		56·0	
15 0	24·0				

NOVEMBER 18, 19, and 20, 1841.

M. Gött. Time. d. h. m. s.	Decl. Sc.-Div^ns.	Hor. Force. Sc.-Div^ns.	Ther.	Vert. Force. Sc.-Div^ns.	Ther.
19 3 20 0	24·0	34·5		56 3	
25 0	24·0				
30 0	23·9	34·1		56·0	
35 0	23·5				
40 0	23·5	34·1		56·0	
45 0	23·6				
50 0	23·9	34·8		56·1	
55 0	23·8				
4 5 0	23·4		67		66
10 0	23·1	34·9		56·2	
15 0	23·1				
20 0	23·0	35·0		56·4	
25 0	23·0				
30 0	23·0	35·0		56 6	
35 0	22·9				
40 0		35·0			
45 0	22·4				
50 0	22·3	35·1		56·8	
55 0	22·1				

Positions at the usual hours of observation, November 18, 19, and 20.

M. Gött. Time. d. h. m. s.	Decl. Sc.-Div^ns.	Hor. Force. Sc.-Div^ns.	Ther.	Vert. Force. Sc.-Div^ns.	Ther.
18 0 0 0	24·0	54·0	65	53·6	65
2 0 0	26·4	51·2	66	55·1	65
3 0 0	25·5	49·1	66	55·5	66
4 0 0	23·3	47·0	67	55·3	66
5 0 0	21·9	44·8	67	56·2	66
6 0 0	21·7	41·9	67	56·7	66
8 0 0	19·6	35·6	66	55·5	66
10 0 0	19·5	26·4	66	55·9	65
11 0 0	18·0	21·0	65	55·4	65
12 0 0	18·1	28·5	65	55·5	65
13 0 0	21·7	25·1	65	55·6	65
14 0 0	18·8	25·9	65	55·2	64
15 0 0	18·0	25·9	65	55·0	64
16 0 0	16·8	27·9	65	55·7	64
18 0 0	18·1	31·2	65	56·2	64
19 30 0	14·6	33·2	65	55·3	64
20 0 0	14·8	34·3	65	56·7	64
20 30 0	14·2	35·0	64	57·0	64
22 0 0	15·7	37·9	64	56·7	63
23 0 0	17·1	38·8	64	55·8	64
19 0 0 0	20·0	35·0	65	55·1	64
2 0 0	22·9	35·2		55·1	65
3 0 0	24·0	34·0	66	56·0	66
4 0 0	23·5	34·9	67	56·2	66
5 0 0	22·0	35·5	67	56·8	67
6 0 0	21·4	34·3	67	57·2	67
7 0 0	22·1	35·1	67	56·1	66
8 0 0	24·5	36·3	66	56·7	66
9 0 0	23·1	38·0	65	56·0	65
10 0 0	23·6	36·0	65	56·0	65
11 0 0	23·0	33·1	65	55·8	65
12 0 0	22·0	34·9	65	55·5	65
13 0 0	23·0	34·0	65	55·5	64
14 0 0	21·5	40·9	65	55·0	64
15 0 0	20·9	40·9	65	54·4	64
16 0 0	20·0	40·0	64	54·7	64
19 30 0	20·9	40·8	64	54·4	64
20 0 0	20·0	41·0	64	55·1	63
20 30 0	19·0	38·8	63	54·4	63
22 0 0	18·9	38·8	63	54·1	63
23 0 0	18·5	38·5	64	53·4	63

NOVEMBER 18, 19, and 20, 1841.

M. Gött. Time. d. h. m. s.	Decl. Sc.-Div^ns.	Hor. Force. Sc.-Div^ns.	Ther.	Vert. Force. Sc.-Div^ns.	Ther.
20 0 0 0	20·2	40·0	64	53·5	64
2 0 0	21·6	34·7	66		
3 0 0	22·1	32·0	66	56·0	65
4 0 0	19·8	33·5	66	55·9	65
5 0 0	21·9	36·0	66	55·9	66
6 0 0	20·8	35·2	66	55·5	65
8 0 0	23·0	37·9	65	55·1	65
10 0 0	23·9	40·0	65	54·1	64
11 0 0	23·8	42·7	65	53·6	64
12 0 0	23·8	44·4	64	53·8	64
13 0 0*	23·0	41·7	64		

The Mean Positions during the Month are given in page 95.

CAPE OF GOOD HOPE { Decl. 1 Scale Division = 0'·75; H. F. k = ·000180; q = ·0003; V. F. k = ·000047; q = }

Positions at the usual hours of observation, November 18, 19, and 20.

M. Gött. Time. d. h. m. s.	Decl. Sc.-Div^ns.	Hor. Force. Sc.-Div^ns.	Ther.	Vert. Force. Sc.-Div^ns.	Ther.
18 0 0	55·5	75·9	63	68·2	63
1 0 0	56·0	76·0	63		
2 0 0	55·9	74·0	64	87·2	63
3 0 0	54·5	73·2	64		
4 0 0	52·9	71·5	64	99·0	64
5 0 0	51·1	70·0	65		
6 0 0	50·9	67·1	65	106·8	64
7 0 0	49·0	62·8	65		
8 0 0	45·0	62·9	65	124·2	64
9 0 0	50·0	62·4	64		
10 0 0	45·2	56·3	64	127·9	64
11 0 0	42·8	54·4	64		
12 0 0	44·5	62·7	64	114·8	64
13 0 0	47·1	62·0	64		
14 0 0	41·8	66·2	64	119·2	64
15 0 0	40·0	67·5	64		
16 0 0	41·1	68·0	63	106·8	63
17 0 0	43·0	71·8	63		
18 0 0	43·3	67·0	63	100·3	63
19 0 0	42·2	63·0	63		
20 0 0	43·9	59·9	63	101·2	63
21 0 0	44·2	59·3	63		
22 0 0	43·2	58·7	63	84·0	63
23 0 0	47·0	57·0	63		
19 0 0 0	51·1	52·2	63	85·6	63
1 0 0	51·6	53·8	63		
2 0 0	53·0	56·8	64	85·8	63
3 0 0	52·0	57·0	64		
4 0 0	51·9	58·0	64	94·0	64
5 0 0	50·8	60·3	64		
6 0 0	50·0	61·7	64	99·4	64
7 0 0	52·4	63·0	64		
8 0 0	51·9	66·5	63	89·9	64
9 0 0	51·0	69·5	63		
10 0 0	50·5	68·5	63	99·6	64
11 0 0	49·0	66·4	63		
12 0 0	48·2	68·0	63	112·8	63
13 ·0 0	49·9	78·2	63		
15 0 0	46·9	73·9	62	94·7	63
16 0 0	45·3	72·7	62	98·4	63
17 0 0	45·0	73·2	62		

NOVEMBER 18, 19, and 20, 1841.

M. Gött. Time. d. h. m. s.	Decl. Sc.-Div^ns.	Hor. Force. Sc.-Div^ns.	Ther.	Vert. Force. Sc.-Div^ns.	Ther.
19 18 0 0	48·3	73·4	62	105·2	62
19 0 0	48·8	74·3	62		
20 0 0	45·9	74·5	62	97·1	62
21 0 0	45·0	69·1	63		
22 0 0	46·0	64·9	63	88·0	63
23 0 0	48·0	60·8	63		
20 0 0 0	48·4	61·1	64	85·4	63
1 0 0	51·2	67·8	64		
2 0 0	53·0	58·6	65	74·9	64
3 0 0	50·6	57·1	65		
4 0 0	49·0	57·2	65	107·6	65
5 0 0	49·5	61·2	66		
6 0 0	49·0	60·5	65	93·9	65
7 0 0	50·5	63 5	65		
8 0 0	51·5	63·2	65	83·8	65
9 0 0	52·0	64·6	65		
10 0 0	52·0	68·1	64	84·7	64
11 0 0^b	52·0	71·4	64		

The Mean Positions at the same hours are given in page 95.

VAN DIEMEN ISLAND { Decl. 1 Scale Division = 0'·71; H. F. k = ·0003; q = ; V. F. k = ; q = }

Extra observations.

The V. F. was observed at 2^m. 30^s. before, and the H. F. 2^m. 30^s. after the times specified.

M. Gött. Time. d. h. m. s.	Decl. Sc.-Div^ns.	Hor. Force. Sc.-Div^ns.	Ther.	Vert. Force. Sc.-Div^ns.	Ther.
18 8 10 0	57·5	68·1	57		55
15 0	55·5				
20 0	53·0	67·2		30·3	
25 0	52·3				
30 0	51·6	66·8		27·8	
35 0	51·1				
40 0	50·4	66·9		27·0	
45 0	49·8				
50 0	49·9	66·9		26·3	
55 0	49·5				
9 5 0	53·7		57		55
10 0	54·0	68·4		28·0	
15 0	55·3				
20 0	55·3	67·5		27·6	
25 0	55·3				
30 0	53·9	65·6		27·8	
35 0	53·1				
40 0	53·6	64·9		28·6	
45 0	53·6				
50 0	53·5	62·3		31·1	
19 2 10 0		61·6	65		63
15 0	49·8				
20 0	48·7	59·4		18·7	
25 0	46·3				
30 0	46·7	63·7		18·7	
35 0	49·5				
40 0	50·4	59·8		17·4	
45 0	44·8				
50 0	39·8	58·1		15·8	
3 5 0	40·2		64		64
10 0	41·1	61·0		16·6	
15 0	41·3				
20 0	40·0	59·5		16·9	
25 0	40·3				
30 0	39·3	58·9		17·5	

* Saturday midnight at St. Helena.   b Saturday midnight at the Cape of Good Hope.

NOVEMBER 18, 19, and 20, 1841.						NOVEMBER 18, 19, and 20, 1841.						DECEMBER 2, 3, and 4, 1841.					

NOVEMBER 18, 19, and 20, 1841.

M. Gött. Time.			Decl.	Hor. Force.		Vert. Force.	
d.	h.	m. s.	Sc.-Div^{ns}.	Sc.-Div^{ns}.	Ther.	Sc.-Div^{ns}.	Ther.
19	3	35 0	38·8				
		40 0	38·9	58·2		18·7	
		45 0	38·6				
		50 0	38·2	57·0		19·8	
		55 0	38·1				
	4	10 0	38·0	55·8	64		64
		15 0^a	38·7				
		20 0	38·6	55·4		24·2	
		25 0	39·1				
		30 0	39·9	55·6		25·5	
		35 0	40·6				
		40 0	41·0	56·0		26·6	
		45 0	42·0				
		50 0	44·4	57·1		27·5	
	5	5 0	46·7				
		10 0	46·7	57·8		28·0	
		15 0^b	46·6				

Positions at the usual hours of observation,
November 18, 19, and 20.

18	0	0 0	44·1	66·0	59	31·6	58
	1	0 0	45·0	64·7		34·1	
	2	0 0	42·4	64·3	58	32·8	57
	3	0 0	44·3	64·4		35·6	
	4	0 0	46·2	65·6	58	35·2	57
	5	0 0	47·2	65·3		35·2	
	6	0 0	48·2	65·3	57	35·1	56
	7	0 0	50·8	65·0		35·2	
	8	0 0	63·7	67·1	57	43·1	55
	9	0 0	50·9	67·4		27·5	
	10	0 0	53·0	61·8	58	33·0	56
	11	0 0	48·4	55·7		35·5	
	12	0 0	46·3	51·1	58	39·6	57
	13	0 0	44·4	50·0		32·9	
	14	0 0	47·4	58·1	61	34·0	60
	15	0 0	54·7	47·9		36·6	
	16	0 0	61·0	51·6	63	35·8	62
	17	0 0	57·2	57·5		34·2	
	18	0 0	62·3	56·3	67	34·4	65
	19	0 0	58·7	57·7		35·9	
	20	0 0	55·2	57·7	68	38·8	64
	21	0 0	60·6	53·2		42·5	
	22	0 0	52·9	52·8	68	35·6	65
	23	0 0	48·2	53·3		32·2	
19	0	0 0	37·4	50·6	67	29·0	64
	1	0 0	37·0	52·6		29·7	

NOVEMBER 18, 19, and 20, 1841.

M. Gött. Time.			Decl.	Hor. Force.		Vert. Force.	
d.	h.	m. s.	Sc.-Div^{ns}.	Sc.-Div^{ns}.	Ther.	Sc.-Div^{ns}.	Ther.
19	2	0 0	50·4	61·3	65	27·6	63
	3	0 0	38·3	61·5		16·6	
	4	0 0	38·5	56·3	64	21·7	64
	5	0 0	47·5	57·4		28·8	
	6	0 0	48·1	57·8	63	29·6	62
	7	0 0	45·6	59·0		27·4	
	8	0 0	44·2	58·9	62	27·3	61
	9	0 0	42·4	58·4		28·1	
	10	0 0	39·9	58·6	64	29·0	63
	11	0 0	41·7	58·1		30·1	
	12	0 0	41·1	54·0	65	27·8	63
	13	0 0	45·9	52·7		26·4	
	14	0 0	53·4	54·6	68	22·2	67
	15	0 0	57·7	58·0		15·8	
	16	0 0	62·5	57·4	70	17·6	69
	17	0 0	63·7	57·9		21·1	
	18	0 0	61·9	55·7	73	23·9	70
	19	0 0	56·6	56·1		20·9	
	20	0 0	54·6	56·0	72	20·6	69
	21	0 0	51·0	54·8		25·6	
	22	0 0	49·4	55·3	68	25·9	67
	23	0 0	48·0	57·3		25·8	
20	0	0 0	41·8	52·2	66	20·4	65
	1	0 0	42·8	56·6		23·9	
	2	0 0^c	44·6	65·0	65	14·4	64

The Mean Positions at the same hours during the Month
are given in page 96.

DECEMBER 2, 3, and 4, 1841.

TORONTO * { Decl. 1 Scale Division = 0'·72
H. F. k = ·000076 ; q = ·0002
V. F. k = ·000092 ; q = ·00018

Positions at the usual hours of observation, from
December 2nd, 12^h, to December 4th, 16^h.

2	12	0 0	56·9	83·5	47	54·4	47
	14	0 0	59·7	59·7	47	57·7	47
	16	0 0	60·8	63·2	47	56·4	47
	18	0 0	64·1	61·1	47	50·0	47
	20	0 0	57·8	55·6	47	46·0	47
	22	0 0	46·2	25·9	46	40·6	47
3	0	0 0	62·0	96·5	46	48·2	47
	2	0 0	48·5	39·3	46	51·5	47
	4	0 0	42·9	66·6	46	50·0	46

DECEMBER 2, 3, and 4, 1841.

M. Gött. Time.			Decl.	Hor. Force.		Vert. Force.	
d.	h.	m. s.	Sc.-Div^{ns}.	Sc.-Div^{ns}.	Ther.	Sc.-Div^{ns}.	Ther.
3	6	0 0	47·5	46·9	46	56·1	47
	8	0 0	50·2	69·1	47	57·3	47
	10	0 0	56·4	65·1	47	55·6	47
	12	0 0	66·9	49·5	47	57·7	47
	14	0 0	61·2	44·0	48	57·1	48
	16	0 0	50·0	49·4	48	49·9	48
	18	0 0	60·2	57·8	49	53·2	49
	20	0 0	53·9	60·5	49	42·3	48
	22	0 0	58·0	63·8	49	51·0	48
4	0	0 0	51·9	64·7	48	49·7	49
	2	0 0	58·8	78·8	47	51·5	48
	4	0 0	54·0	50·4	47	51·9	48
	6	0 0	52·5	53·2	47	53·3	47
	8	0 0	53·0	58·9	46	56·0	47
	10	0 0	55·0	69·4	46	56·0	47
	12	0 0	61·7	59·2	46	58·7	46
	14	0 0	58·3	72·4	46	57·3	47
	16	0 0	58·9	76·5	46	56·6	46

Mean Positions at the same hours during the Month.

0	0	0 0	58·5	93·0	41	58·9	42
2	0	0 0	56·9	88·1	41	60·0	42
4	0	0 0	56·9	82·7	41	59·1	41
6	0	0 0	52·0	73·2	42	59·6	42
8	0	0 0	50·8	80·3	43	60·4	42
10	0	0 0	54·5	84·1	43	60·1	43
12	0	0 0	59·8	81·5	43	61·0	43
14	0	0 0	60·2	82·1	43	59·9	43
16	0	0 0	58·6	82·9	42	58·9	43
18	0	0 0	58·1	84·9	42	59·4	42
20	0	0 0	56·5	84·6	42	58·0	42
22	0	0 0	57·6	86·1	41	58·7	42

ST. HELENA.^d

CAPE OF { Decl. 1 Scale Division = 0'·75
GOOD HOPE { H. F. k = ·000180 ; q = ·0003
V. F. ^e

Positions at the usual hours of observation, from
December 2nd, 12^h, to December 4th, 11^h.

2	12	0 0	52·5	75·1	67		
	13	0 0	50·2	79·8	67		
	14	0 0	49·7	75·5	66		
	15	0 0	48·8	75·2	66		

---

^a No Aurora visible.
^b During these observations the Decl. bar had a constant slight vertical motion.
^c Saturday midnight at Van Diemen Island.

* TORONTO, December, 1841.—*Times of observation at which the Magnetometers were
disturbed, but the mean readings were not materially changed.*

Dec. 4 4^h H. F. vibrating moderately.
6 10 H. F. slight vibrations and strong shocks.
12 H. F. moderate shocks.
7 22 Decl. and H. F. moderate shocks.
8 4 Decl. and H. F. moderate shocks.
14 H. F. vibrating slightly.
9 0 Decl. and H. F. moderate shocks.
2 Decl. and H. F. moderate shocks.
12 18 H. F. slightly vibrating.
20 H. F. vibrating slightly.
22 V. F. vibrating slightly ; H. F. moderately.
13 6 H. F. much disturbed by vibrations.
14 6 H. F. considerable shocks.
10 H. F. vibrating slightly.
16 12 Decl. and H. F. moderate vibrations and shocks.
18 Decl. and V. F. vibrating very much.

Dec. 16 20 Decl. and V. F. slight vibrations ; H. F. moderate vibrations.
22 Decl. slightly vibrating ; H. F. moderately.
17 0 Decl. vibrating very much ; H. F. slightly.
2 Decl. and H. F. vibrating very much ; V. F. slightly.
4 Decl. and H. F. vibrating very much ; V. F. slightly.
6 Decl. and H. F. vibrating very much ; V. F. slightly.
8 Decl. and H. F. moderate vibrations and shocks.
10 Decl. and H. F. slight shocks.
12 Decl. moderate vibrations.
16 Decl. vibrating slightly ; H. F. much with shocks.
18 Decl. vibrating slightly ; H. F. very much.
20 Decl. slight vibrations ; H. F. very much with shocks.
19 18 H. F. vibrating much.
20 Decl. and H. F. vibrating very much with shocks.
22 Decl. and H. F. slight vibrations and shocks.
20 0 Decl. and H. F. moderate vibrations and shocks.
4 H. F. moderate shocks.
12 H. F. slight shocks.
16 Decl. slight shocks.

^d The magnetometers at St. Helena were employed in temperature experiments.
^e The V. F. magnet was not in satisfactory adjustment.

DECEMBER 2, 3, and 4, 1841.

M. Gött. Time.			Decl.	Hor. Force.		Vert. Force.	
d.	h.	m.	s.	Sc.-Div.	Sc.-Div.	Ther.	Sc.-Div.	Ther.
2	16	0	0	47·9	75·6	66		
	17	0	0	47·0	78·5	66		
	18	0	0	46·1	79·0	66		
	19	0	0	45·8	80·0	66		
	20	0	0	43·4	83·1	66		
	21	0	0	43·1	78·9	66		
	22	0	0	42·0	73·0	67		
	23	0	0	44·9	65·9	67		
3	0	0	0	47·4	69·4	68		
	1	0	0	48·8	69·0	68		
	2	0	0	50·4	61·0	69		
	3	0	0	49·0	49·0	69		
	4	0	0	45·9	56·6	69		
	5	0	0	46·0	54·1	69		
	6	0	0	40·2	53·1	69		
	7	0	0	46·0	61·2	69		
	8	0	0	44·9	59·8	68		
	9	0	0	44·7	60·9	68		
	10	0	0	44·5	64·0	68		
	11	0	0	45·1	68·2	68		
	12	0	0	43·4	69·0	67		
	13	0	0	52·5	68·8	67		
	14	0	0	42·4	67·0	67		
	15	0	0	41·6	68·5	67		
	16	0	0	40·8	68·9	66		
	17	0	0	41·0	71·6	66		
	18	0	0	39·7	72·9	66		
	19	0	0	39·1	73·5	66		
	20	0	0	37·2	72·3	66		
	21	0	0	38·1	71·2	66		
	22	0	0	40·0	68·7	67		
	23	0	0	42·0	66·5	67		
4	0	0	0	43·1	64·2	68		
	1	0	0	45·0	60·8	68		
	2	0	0	45·0	65·1	68		
	3	0	0	56·4	67·0	68		
	4	0	0	44·8	63·8	68		
	5	0	0	42·8	60·2	69		
	6	0	0	42·4	63·2	69		
	7	0	0	44·0	64·8	68		
	8	0	0	39·8	66·9	68		
	9	0	0	45·0	66·5	67		
	10	0	0	44·2	67·8	67		
	11	0	0	48·5	72·1	67		

Mean Positions at the same hours during the Month.

	h.	m.	s.	Decl.	Hor.	Ther.		
	0	0	0	45·0	74·6	66		
	1	0	0	46·6	73·6	67		
	2	0	0	46·5	71·6	67		
	3	0	0	46·5	70·8	68		
	4	0	0	45·0	70·9	68		
	5	0	0	45·0	69·8	68		
	6	0	0	44·4	68·3	68		
	7	0	0	44·8	68·6	68		
	8	0	0	44·7	68·5	68		
	9	0	0	44·6	69·4	67		
	10	0	0	44·7	71·0	67		
	11	0	0	44·8	72·2	67		
	12	0	0	45·6	73·6	66		
	13	0	0	44·1	74·1	66		
	14	0	0	44·1	73·9	66		
	15	0	0	43·6	74·8	66		

DECEMBER 2, 3, and 4, 1841.

M. Gött. Time.			Decl.	Hor. Force.		Vert. Force.	
d.	h.	m.	s.	Sc.-Div.	Sc.-Div.	Ther.	Sc.-Div.	Ther.
	16	0	0	43·1	75·4	66		
	17	0	0	42·3	76·9	66		
	18	0	0	41·1	78·5	65		
	19	0	0	40·1	79·5	65		
	20	0	0	39·7	78·9	65		
	21	0	0	40·0	78·3	65		
	22	0	0	41·8	76·3	66		
	23	0	0	43·6	75·1	66		

VAN DIEMEN ISLAND { Decl. 1 Scale Division = 0'·71
H. F. k = ·0003; q =
V. F. k = ; q =

Extra observations.

The V. F. was observed at 2ᵐ. 30ˢ. before, and the H. F. 2ᵐ. 30ˢ. after the times specified.

d.	h.	m.	s.	Decl.	Hor.	Ther.	Vert.	Ther.
3	2	15	0	48·9		60		59
		20	0	48·9	50·5		63·2	
		25	0	47·3				
		30	0	48·4	53·5		62·1	
		35	0	50·0				
		40	0	52·6	51·9		58·0	
		45	0	52·7				
		50	0	50·7			57·9	
	3	10	0	55·0	43·9	60		59
		15	0	55·0				
		20	0	56·5	44·9		70·9	
		25	0	57·9				
		30	0	58·2	45·0		71·6	
		35	0	58·4				
		40	0	58·1	44·9		71·6	
		45	0	58·2				
		50	0	58·3	45·0		71·6	
		55	0	57·8				
	4	5	0	58·0		60		60
		10	0	58·0	45·0		71·7	
		15	0	58·0				
		20	0	58·1	44·9		71·9	
		25	0	57·3				
		30	0	57·7	46·7		71·6	
		35	0	58·3				
		40	0	58·6	48·1		66·6	
		45	0	59·3				
		50	0	59·7	49·0		66·6	
		55	0	60·1				
	5	5	0	58·4		60		60
		10	0	58·4	48·9		65·2	
		15	0	58·4				
		20	0	58·3	48·0		65·2	
		25	0	59·7				
		30	0	59·7	47·6		68·7	
		35	0	61·2				
		40	0	63·3	47·7		70·0	
		45	0	65·3				
		50	0	67·7			71·6	
		55	0	68·9				
	6	10	0	72·6	47·1	60		59
		15	0	72·6				
		20	0	72·7	46·7		71·7	
		25	0	71·5				

DECEMBER 2, 3, and 4. 1841.

Positions at the usual hours of observation, from December 2nd, 12h., to December 4th, 2h.[a]

M. Gött. Time.			Decl.	Hor. Force.		Vert. Force.	
d.	h.	m.	s.	Sc.-Div.	Sc.-Div.	Ther.	Sc.-Div.	Ther.
2	12	0	0	62·1	48·8	56	80·9	56
	13	0	0	64·8	44·2		83·1	
	14	0	0	73·0	45·5	58	79·0	58
	15	0	0	78·5	47·1		77·4	
	16	0	0	85·0	50·5	60	76·0	59
	17	0	0	88·8	55·6		73·1	
	18	0	0	88·8	52·1	61	77·1	60
	19	0	0	84·5	52·3		74·7	
	20	0	0	82·3	52·7	62	72·7	61
	21	0	0	81·3	50·8		76·6	
	22	0	0	78·0	50·4	61	76·3	61
	23	0	0	69·2	45·9		77·1	
3	0	0	0	72·4	47·6	61	79·0	60
	1	0	0	73·2	47·9		78·6	
	2	0	0	65·5	44·4	60	73·8	59
	3	0	0	53·2	45·4		63·6	
	4	0	0	58·0	45·3	60	71·6	60
	5	0	0	58·9	48·3		65·2	
	6	0	0	70·4	47·1	60	71·7	59
	7	0	0	70·1	47·2		71·0	
	8	0	0	70·1	48·4	58	73·2	57
	9	0	0	66·2	48·2		77·5	
	10	0	0	63·8	45·8	57	77·4	56
	11	0	0	64·2	45·1		78·9	
	12	0	0	64·2	43·5	58	79·4	57
	13	0	0	67·3	43·6		79·4	
	14	0	0	73·4	45·3	61	75·3	59
	15	0	0	80·1	46·2		73·2	
	16	0	0	83·6	48·3	62	72·4	61
	17	0	0	84·2	48·6		72·2	
	18	0	0	83·7	49·5	64	72·6	62
	19	0	0	82·6	47·3		73·8	
	20	0	0	77·3	47·8	64	71·8	62
	21	0	0	76·6	47·8		76·2	
	22	0	0	76·2	48·8	64	74·3	61
	23	0	0	74·1	51·7		68·6	
4	0	0	0	70·7	49·1	62	76·0	60
	1	0	0	68·4	46·4		76·4	
	2	0	0	71·1	48·0	61	75·5	59

Mean Positions at the same hours during the Month.

	h.	m.	s.	Decl.	Hor.	Ther.	Vert.	Ther.
	0	0	0	75·2	47·9	65	69·0	63
	1	0	0	74·0	48·3		69·8	
	2	0	0	73·1	48·2	63	70·3	62
	3	0	0	72·0	48·5		70·6	
	4	0	0	73·1	48·9	62	71·8	61
	5	0	0	72·4	49·2		72·6	
	6	0	0	73·3	49·6	60	74·4	59
	7	0	0	72·9	49·7		74·6	
	8	0	0	73·0	49·6	60	75·5	58
	9	0	0	71·5	49·8		75·7	
	10	0	0	69·2	49·3	59	75·0	58
	11	0	0	68·0	48·1		74·7	
	12	0	0	67·9	46·7	61	73·8	60
	13	0	0	70·3	45·3		72·7	
	14	0	0	74·4	45·2	63	71·7	63
	15	0	0	78·6	45·7		69·9	
	16	0	0	82·0	47·0	66	68·1	64
	17	0	0	84·0	48·1		66·7	
	18	0	0	83·7	48·7	67	66·0	65

[a] Saturday midnight at Van Diemen Island.

DECEMBER 2, 3, and 4, 1841.

M. Gött. Time.			Decl.	Hor. Force.		Vert. Force.	
d.	h.	m.	s.	Sc.-Div	Sc.-Div	Ther.	Sc.-Div	Ther.

d.	h.	m.	s.	Sc.-Div	Sc.-Div	Ther.	Sc.-Div	Ther.
	19	0	0	82·0	48·3		65·5	
	20	0	0	80·0	47·7	68	65·5	66
	21	0	0	78·3	47·2		66·4	
	22	0	0	76·6	47·3	67	67·0	65
	23	0	0	75·9	47·8		67·6	

DECEMBER 8, 1841.

TORONTO { Decl. 1 Scale Division = 0′·72
H. F. $k = ·000076$; $q = ·0002$
V. F. $k = ·000092$; $q = ·00018$

Positions at the usual hours of observation,
December 8.

	h.	m.	s.					
8	0	0	0	68·4	79·4	43	54·6	44
	2	0	0	58·2	82·0	42	57·2	43
	4	0	0	41·2	23·7	44	56·5	44
	6	0	0	42·8	35·2	45	60·7	45
	8	0	0	44·8	58·1	46	62·2	45
	10	0	0	51·8	72·2	46	61·0	46
	12	0	0	59·8	73·0	46	58·4	46
	14	0	0	61·3	67·0	47	55·2	47
	16	0	0	58·0	68·0	47	54·5	47
	18	0	0	56·0	70·0	47	55·5	47
	20	0	0	58·5	69·9	47	54·4	47
	22	0	0	57·1	70·4	48	53·3	48

The Mean Positions at the same hours during the Month,
are given in page 101.

ST. HELENA { Decl. 1 Scale Division = 0′·71
H. F. $k = ·000189$; $q = ·00025$
V. F. $k = ·00047$; $q =$

Extra observations.

The V. F. was observed at 2^m. 30^s. before the times specified.

	h.	m.	s.					
8	4	10	0	*	39·4	68		
		15	0		39·5			
		20	0		39·1		55·1	67
		25	0		39·2			
		30	0		39·0		55·0	
		35	0		38·1			
		40	0		37·6		54·3	
		45	0		37·2			
		50	0		36·5		54·0	
		55	0		36·2			
	5	0	0		36·0	68	53·5	67
		5	0		35·9			
		10	0		35·9		53·7	
		15	0		35·9			
		20	0		35·2		53·7	
		25	0		35·2			
		30	0		35·1		53·3	
		35	0		35·1			
		40	0		34·6		54·6	
		45	0		34·8			
		50	0		34·9		54·6	
		55	0		34·8			
	6	0	0		34·8	68	54·2	67
		5	0		34·4			
		10	0		34·2		54·4	
		15	0		33·9			

DECEMBER 8, 1841.

M. Gött. Time.			Decl.	Hor. Force.		Vert. Force.	
d.	h.	m.	s.	Sc.-Div	Sc.-Div	Ther.	Sc.-Div	Ther.

d.	h.	m.	s.					
8	6	20	0		34·0		54·3	
		25	0		33·8			
		30	0		34·0			
		35	0		34·0			
		40	0		34·7			
		45	0		34·8			
		50	0		34·9			
		55	0		34·9			
	7	0	0		35·0	68		
		5	0		35·0			
		10	0		34·9		53·6	
		15	0		34·9			
		20	0		35·0		52·1	
		25	0		35·0			
		30	0		35·0		51·2	
		35	0		35·2			
		40	0		35·3		51·2	
		45	0		35·8			
		50	0		36·1		51·2	
		55	0		36·2			

Positions at the usual hours of observation,
December 8.

	h.	m.	s.					
8	0	0	0	*	55·2	67	53·1	66
	2	0	0		49·7	67	53·0	67
	3	0	0		44·6	68		67
	4	0	0		38·4	68	54·9	67
	5	0	0		36·0	67	53·5	67
	6	0	0		34·8	68	54·2	67
	8	0	0		36·0	68	52·0	67
	10	0	0		39·6	67	52·2	66
	11	0	0		44·0	67	51·9	66
	12	0	0		46·1	67	51·5	66
	13	0	0		49·0	67	51·2	65
	14	0	0		50·7	66	51·1	65
	15	0	0		52·3	66	51·1	65
	1f	0	0		52·5	66	50·8	65
	18	0	0		52·1	66	50·8	65
	19	30	0		52·2	66	50·0	65
	20	0	0		52·9	66	50·0	65
	20	30	0		53·2	66	48·2	65
	22	0	0		56·0	66	48·1	65
	23	0	0		57·0	66	48·1	65

The Mean Positions of the H. F. magnetometers at the same
hours during the Month, are given under date December 14.

ISLAND OF
ASCENSION[b] { Decl. 1. Scale Division = 2′·98
H. F. $k = ·0003$; $q =$

Positions at the usual hours of observation,
December 8.

	h.	m.	s.					
8	0	0	0	48·3	47·9	82		
	1	0	0	47·2	51·0	82		
	2	0	0	46·8	46·2	82		
	3	0	0	46·3	39·9	82		
	4	0	0	46·0	37·5	82		
	5	0	0	45·0	35·7	82		
	6	0	0	46·0	35·4	82		
	7	0	0	46·2	36·0	81		
	8	0	0	47·1	36·7	80		
	9	0	0	47·7	37·8	81		

DECEMBER 8, 1841.

M. Gött. Time.			Decl.	Hor. Force.		Vert. Force.	
d.	h.	m.	s.	Sc.-Div	Sc.-Div	Ther.	Sc.-Div	Ther.

d.	h.	m.	s.					
8	10	0	0	47·0	39·6	81		
	11	0	0	48·2	40·3	81		
	12	0	0	48·9	43·8	81		
	15	0	0	49·0	48·4	79		
	16	0	0	48·9	48·6	79		
	17	0	0	49·0	48·1	79		
	18	0	0	49·0	48·0	79		
	19	0	0	49·2	47·5	79		
	20	0	0	48·8	48·6	79		
	21	0	0	48·3	50·0	79		
	22	0	0	48·3	50·2	81		
	23	0	0	48·0	49·7	80		

Mean Positions at the same hours from the 5th to the
12th December inclusive.

	h.	m.	s.					
	0	0	0	48·6	57·7	81		
	1	0	0	48·2	56·9	81		
	2	0	0	46·4	54·9	81		
	3	0	0	47·7	52·3	81		
	4	0	0	46·7	47·3	82		
	5	0	0	46·6	45·0	82		
	6	0	0	47·0	45·6	82		
	7	0	0	47·4	45·8	81		
	8	0	0	47·7	46·0	81		
	9	0	0	48·1	46·7	80		
	10	0	0	48·8	47·0	80		
	11	0	0	48·7	47·8	80		
	12	0	0	48·6	47·6	80		
	13	0	0	48·6	48·3	80		
	14	0	0	49·1	49·0	80		
	15	0	0	49·6	48·8	80		
	16	0	0	49·4	49·0	80		
	17	0	0	49·5	49·0	80		
	18	0	0	49·3	49·0	80		
	19	0	0	49·3	49·2	80		
	20	0	0	49·1	50·6	79		
	21	0	0	49·0	51·6	79		
	22	0	0	49·0	53·0	65		
	23	0	0	48·8	57·3	80		

CAPE OF
GOOD HOPE { Decl. 1 Scale Division = 0′·75
H. F. $k = ·00018$; $q =$
V. F. $k = ·000035$; $q =$

Positions at the usual hours of observation,
December 8.

	h.	m.	s.					
8	0	0	0	45·0	74·7	64	138·8	64
	1	0	0	46·8	74·8	65		
	2	0	0	46·7	68·2	65	146·3	65
	3	0	0	45·4	61·2	65		
	4	0	0	43·1	57·8	66	182·6	65
	5	0	0	42·9	54·1	66		
	6	0	0	44·5	53·0	66	172·0	65
	7	0	0	44·4	55·8	66		
	8	0	0	44·5	57·7	66	156·2	65
	9	0	0	44·4	62·0	65		
	10	0	0	43·9	66·0	65	148·6	65
	11	0	0	45·0	70·2	65		
	12	0	0	45·5	73·0	64	126·3	64
	13	0	0	44·3	74·7	64		

* The Decl. magnet was not in adjustment at St. Helena, on December 8th.

[b] Observations made by Captain William Allen, R.N., Commanding H.M. Steamer
Wilberforce, with Weber's transportable magnetometer.

DECEMBER 8, 1841.

M. Gött. Time. d. h. m. s.	Decl. Sc.-Div^ns	Hor. Force. Sc.-Div^ns.	Ther.	Vert. Force. Sc.-Div^ns.	Ther.
8 14 0 0	42·8	75·6	64	135·8	64
15 0 0	42·0	78·3	64		
16 0 0	41·1	78·5	64	142·7	63
17 0 0	39·8	78·4	63		
18 0 0	38·8	78·2	63	155·4	63
19 0 0	37·7	77·6	63		
20 0 0	38·0	76·7	63	155·3	63
21 0 0	39·3	77·7	63		
22 0 0	42·1	77·6	63	107·0	63
23 0 0	45·1	77·6	64		

The Mean Positions at the same hours during the Month are given in page 102.

VAN DIEMEN ISLAND — Decl. 1 Scale Division = 0'·71 ; H. F. k = ·0003 ; q = ; V. F. k = ; q =

Extra observations.

The V. F. was observed at 2^m. 30^s. before, and the H. F. 2^m. 30^s. after the times specified.

M. Gött. Time. d. h. m. s.	Decl.	Hor. Force.	Ther.	Vert. Force.	Ther.
8 2 15 0	61·0		67		65
20 0	60·6	41·5		65·2	
25 0	61·8				
30 0	63·7	42·2		65·7	
35 0	63·3				
40 0	62·4	43·8		64·1	
45 0	65·0				
50 0	64·7	42·5		62·6	
55 0	61·2				
3 10 0	64·5	45·6	67	58·0	66
15 0	65·7				
20 0	69·3	45·5		58·4	
25 0	75·7				
30 0	78·7	46·9		62·2	
35 0	77·8				
40 0	70·3	45·5		59·8	
45 0	67·5				
50 0	66·0	43·1		55·2	
4 10 0	65·2	39·2	67		66
15 0	62·6				
20 0	60·1	37·4		59·9	
25 0	56·6				
30 0	55·4	39·3		59·9	
35 0	54·5				
40 0	53·2	40·3		58·8	
45 0	52·8				
50 0	52·5	42·3		56·6	
5 15 0	52·3		66		66
20 0	52·1	43·3		52·2	
25 0	51·2				
35 0	50·6				
40 0	51·3	42·1		53·4	
45 0	52·0				
50 0	52·2	42·3		55·2	
6 10 0	54·8	41·7	66		65
15 0	55·0				
20 0	56·4	41·1		59·6	
25 0	57·6				
30 0	58·5	41·7		61·5	
35 0	60·0				
40 0	60·5	42·5		62·0	
45 0	60·4				

DECEMBER 8, 1841.

M. Gött. Time. d. h. m. s.	Decl. Sc.-Div^ns	Hor. Force. Sc.-Div^ns.	Ther.	Vert. Force. Sc.-Div^ns.	Ther.
8 6 50 0	60·9	43·1		62·0	
7 10 0	60·3	44·0	65		65
15 0	59·7				
20 0	60·8	44·2		60·7	
25 0	62·4				
30 0	62·5	44·2		62·7	
35 0	62·8				
40 0	63·6	44·3		62·7	
45 0	62·9				
50 0	63·1	44·1		62·7	
55 0	62·9				

Positions at the usual hours of observation, December 8.

d. h. m. s.	Decl.	Hor. Force.	Ther.	Vert. Force.	Ther.
8 0 0 0	67·4	43·3	69	67·8	66
1 0 0	65·7	46·0		63·0	
2 0 0	60·6	42·3	67	64·5	65
3 0 0	57·7	45·1		58·8	
4 0 0	65·2	41·4	67	57·1	66
5 0 0	52·5	43·3		54·3	
6 0 0	53·0	42·1	66	56·1	65
7 0 0	61·7	43·8		61·6	
8 0 0	63·4	44·1	65	63·7	64
9 0 0	66·1	44·7		65·0	
10 0 0	68·3	42·4	66	67·8	64
11 0 0	64·4	40·5		69·8	
12 0 0	63·9	40·5	66	67·9	64
13 0 0	68·0	39·8		65·9	
14 0 0	72·6	40·8	66	66·7	64
15 0 0	79·0	43·6		65·8	
16 0 0	82·4	44·9	66	66·0	65
17 0 0	85·4	47·2		66·0	
18 0 0	84·7	48·0	66	67·4	65
19 0 0	82·3	47·2		67·4	
20 0 0	76·7	47·3	67	64·6	65
21 0 0	76·0	46·7		64·4	
22 0 0	75·0	47·6	66	68·3	64
23 0 0	74·8	46·9		70·0	

The Mean Positions at the same hours during the Month are given in page 102.

DECEMBER 14, 1841.

TORONTO — Decl. 1 Scale Division = 0'·72 ; H. F. k = ·000076 ; q = ·0002 ; V. F. k = ·000092 ; q = ·00018

Extra observations.

The V. F. was observed at 1^m. 30^s. before, and the H. F. 2^m. after the times specified.

d. h. m. s.	Decl.	Hor. Force.	Ther.	Vert. Force.	Ther.
14 12 10 0	63·1	19·5	49	62·2	49
15 0	59·3	27·1		63·1	
20 0	52·7	35·6		61·9	
25 0	49·9	41·0		61·4	
30 0	49·3	42·0		59·0	
35 0	51·5	42·7		57·9	
40 0	55·6	45·1		56·1	
45 0	57·1	43·9		54·4	
50 0	59·1	44·4		53·5	
55 0	60·3	44·7		52·3	
13 0 0	59·1	42·8	51	52·2	51

DECEMBER 14, 1841.

M. Gött. Time. d. h. m. s.	Decl. Sc.-Div^ns	Hor. Force. Sc.-Div^ns.	Ther.	Vert. Force. Sc.-Div^ns.	Ther.
14 13 5 0	58·3	40·4		51·7	
10 0	58·8	38·9		51·7	
15 0	58·7	38·2		51·3	
20 0	58·3	37·7		51·4	
25 0	58·9	36·8		51·6	
30 0	60·2	35·4		51·7	
35 0	60·0	43·1		51·6	
40 0	60·5	45·0		52·1	
45 0	61·3	49·5		52·4	
50 0	59·9	48·8		50·5	
55 0	62·3	46·6		49·5	

Positions at the usual hours of observation, December 14.

d. h. m. s.	Decl.	Hor. Force.	Ther.	Vert. Force.	Ther.
14 0 0 0	58·9	80·1	47	51·7	48
2 0 0	59·4	72·3	48	51·0	48
4 0 0	59·8	80·2	48	50·2	48
6 0 0	53·4	66·5	49	49·6	49
8 0 0	44·1	47·9	50	49·8	49
10 0 0	49·7	49·1	50	55·1	50
12 0 0	89·3	09·2	49	67·9	49
14 0 0	65·0	42·2	51	48·2	50
16 0 0	57·4	46·1	50	46·3	50
18 0 0	56·7	60·7	50	49·3	50
20 0 0	55·3	58·0	50	51·4	50
22 0 0	56·9	61·7	49	51·4	50

The Mean Positions at the same hours during the Month are given in page 101.

ST. HELENA — Decl. 1 Scale Division = 0'·71 ; H. F. k = ·000189 ; q = ·00025 ; V. F. k = ·00047 ; q =

Positions at the usual hours of observation, December 14.

d. h. m. s.	Decl.	Hor. Force.	Ther.	Vert. Force.	Ther.
14 0 0 0	19·0	61·5	67	47·6	66
2 0 0	20·8	60·1	68	48·2	67
3 0 0	20·0	59·0	68	48·4	67
4 0 0	19·9	56·9	68	47·6	67
5 0 0	18·1	53·1	68	47·5	67
6 0 0	17·8	51·5	68	48·0	67
8 0 0	17·1	42·0	67	47·2	67
10 0 0	17·0	42·7	67	47·9	66
11 0 0	15·1	39·1	67	47·9	66
12 0 0	19·9	49·5	67	48·9	66
13 0 0	16·1	45·9	66	48·2	66
14 0 0	15·7	47·0	66	48·2	65
15 0 0	15·1	48·1	66	48·2	66
16 0 0	14·0	49·2	66	48·2	65
18 0 0	16·5	50·0	66	48·2	65
19 30 0	16·8	51·1	66	47·9	65
20 0 0	15·0	51·8	66	47·5	65
20 30 0	14·6	54·1	66	47·6	65
22 0 0	15·1	55·7	66	46·5	65
23 0 0	16·9	58·0	66	46·6	66

Mean Positions at the same hours during the Month. *

d. h. m. s.	Decl.	Hor. Force.	Ther.
0 0 0	19·6	60·1	66
2 0 0	20·0	56·5	67
3 0 0	19·3	55·5	68
4 0 0	18·4	54·0	68
5 0 0	18·0	52·2	68
6 0 0	18·5	50·7	68

* The Mean Positions of the Decl. magnet are from the 10th to 31st December, inclusive; those of the H. F. magnet from the 4th to the 31st inclusive.

DECEMBER 14, 1841.

d. h. m. s. (M. Gött. Time)	Decl. Sc.-Div	Hor. Force Sc.-Div	Ther.	Vert. Force Sc.-Div	Ther.
8 0 0	19·8	49·6	67		
10 0 0	20·6	50·0	67		
11 0 0	20·5	50·4	66		
12 0 0	20·6	51·9	66		
13 0 0	19·9	51·7	66		
14 0 0	19·7	52·1	66		
15 0 0	19·6	52·8	66		
16 0 0	19·1	53·3	66		
18 0 0	18·9	53·8	65		
19 30 0	18·2	55·0	65		
20 0 0	17·2	55·7	65		
20 30 0	16·3	56·4	65		
22 0 0	16·8	58·5	65		
23 0 0	18·3	60·0	66		

CAPE OF GOOD HOPE { Decl. 1 Scale Division = 0·75
H. F. k = ·00018; q = ·0003
V. F. k = ·000035; q =

Positions at the usual hours of observation, December 14.

d. h. m. s.	Decl. Sc.-Div	Hor. Force Sc.-Div	Ther.	Vert. Force Sc.-Div	Ther.
14 0 0 0	45·0	78·2	65	117·8	65
1 0 0	46·2	79·2	66		
2 0 0	47·0	76·0	66	109·5	66
3 0 0	47·1	76·0	66		
4 0 0	46·4	75·7	66	119·2	66
5 0 0	45·1	72·1	67		
6 0 0	43·0	70·3	67	141·5	66
7 0 0	42·1	66·7	66		
8 0 0	39·1	62·4	66	167·5	66
9 0 0	33·4	59·4	66		
10 0 0	38·0	68·6	66	141·5	65
11 0 0	36·8	68·8	65		
12 0 0	43·1	80·0	65	89·6	65
13 0 0	39·9	70·9	65		
14 0 0	39·0	72·0	65	136·1	64
15 0 0	38·0	72·2	65		
16 0 0	37·3	74·1	64	135·4	64
17 0 0	37·3	74·9	64		
18 0 0	37·5	76·7	64	135·8	64
19 0 0	37·6	77·5	64		
20 0 0	37·4	76·5	64	132·8	64
21 0 0	39·0	75·5	65		
22 0 0	41·9	70·7	65	115·1	65
23 0 0	41·9	69·0	66		

The Mean Positions at the same hours during the Month, are given in page 102.

VAN DIEMEN ISLAND { Decl. 1 Scale Division = 0'·71
H. F. k = ·0003; q =
V. F. k = ; q =

Positions at the usual hours of observation, December 14.

d. h. m. s.	Decl. Sc.-Div	Hor. Force Sc.-Div	Ther.	Vert. Force Sc.-Div	Ther.
14 0 0 0	76·1	48·7	67	65·2	65
1 0 0	75·9	50·0		64·9	
2 0 0	71·4	49·4	65	65·5	64
3 0 0	69·7	49·5	63	67·0	
4 0 0	75·1	49·8	63	71·3	62
5 0 0	76·1	51·6		71·4	
6 0 0	77·6	50·5	61	74·8	60
7 0 0	76·6	52·0		74·8	
8 0 0	82·0	51·4	59	76·7	57
9 0 0	83·8	50·1		80·0	
10 0 0	71·5	48·7	60	71·9	58
11 0 0	67·4	43·2		68·5	

DECEMBER 14, 1841.

d. h. m. s. (M. Gött. Time)	Decl. Sc.-Div	Hor. Force Sc.-Div	Ther.	Vert. Force Sc.-Div	Ther.
14 12 0 0	69·5	43·9	64	61·8	64
13 0 0	67·8	43·0		56·6	
14 0 0	71·6	48·7	68	54·7	68
15 0 0	78·3	46·0		56·0	
16 0 0	83·3	47·5	72	54·7	70
17 0 0	82·2	47·2		54·7	
18 0 0	79·8	47·0	75	53·0	72
19 0 0	78·7	44·4		54·2	
20 0 0	78·4	42·7	76	55·7	73
21 0 0	78·0	41·7		56·8	
22 0 0	76·5	41·0	74	59·0	72
23 0 0	76·5	41·8		59·0	

The Mean Positions at the same hours during the Month, are given in page 102.

DECEMBER 30 and 31, 1841.

TORONTO { Decl. 1 Scale Division = 0'·72
H. F. k = ·000076; q = ·0002
V. F. k = ·00009; q = ·00018

Positions at the usual hours of observation, December 30 and 31.

d. h. m. s.	Decl. Sc.-Div	Hor. Force Sc.-Div	Ther.	Vert. Force Sc.-Div	Ther.
30 0 0 0	59·3	112·8	40	58·2	41
2 0 0	41·4	83·3	40	57·9	41
4 0 0	53·0	109·4	41	54·6	40
6 0 0	36·8	42·5	43	58·5	41
8 0 0	41·5	55·7	44	61·9	43
10 0 0	49·5	66·9	44	62·6	43
12 0 0	56·0	47·2	44	65·2	43
14 0 0	59·9	63·9	44	62·4	43
16 0 0	60·9	71·5	43	56·8	43
18 0 0	58·4	66·0	43	56·0	43
20 0 0	57·1	70·3	44	52·0	43
22 0 0	60·6	73·0	43	54·1	43
31 0 0 0	54·1	85·9	42	55·3	43
2 0 0	54·7	82·5	41	57·7	42
4 0 0	55·0	81·5	41	59·2	41
6 0 0	55·0	57·5	41	60·7	41
8 0 0	54·4	80·2	40	62·3	41
10 0 0	58·7	81·4	40	63·6	40
12 0 0	59·0	93·1	39	62·3	40
14 0 0	61·3	95·2	38	62·9	40
16 0 0	60·2	92·8	38	61·0	40
18 0 0	61·0	93·7	38	58·9	39
20 0 0	60·0	88·9	37	59·1	39
22 0 0	50·2	80·9	37	53·5	39

The Mean Positions at the same hours during the Month, are given in page 101.

ST. HELENA { Decl. 1 Scale Division = 0'·71
H. F. k = ·00019; q = ·00025
V. F. k = ·00047; q =

Extra observations.
The V. F. was observed at 2m. 30s. before, and the H. F. 2m. 30s. after the times specified.

d. h. m. s.	Decl. Sc.-Div	Hor. Force Sc.-Div	Ther.	Vert. Force Sc.-Div	Ther.
30 5 0 0	19·5	48·1	67		
5 0	18·9				
10 0	19·0	46·8		36·0	67
15 0	19·0				
20 0	19·0	45·1		36·1	
25 0	19·0				

DECEMBER 30 and 31, 1841.

d. h. m. s. (M. Gött. Time)	Decl. Sc.-Div	Hor. Force Sc.-Div	Ther.	Vert. Force Sc.-Div	Ther.
30 5 30 0	19·0	44·0		31·6	
35 0	18·3				
40 0	18·0	42·0		36·2	
45 0	18·0				
50 0	17·7	39·9		36·2	
55 0	17·5				
6 0 0	17·9	37·9	67	36·2	67
5 0	18·2				
10 0	18·2	37·2		36·2	
15 0	17·9				
20 0	17·9	37·9		36·2	
25 0	17·9				
30 0	18·0	39·9		36·2	
35 0	18·0				
40 0	18·0	40·5		36·0	
45 0	18·1				
50 0	18·1	40·5		36·0	
55 0	18·1				
7 0 0	18·5	40·0	67	36·0	67
5 0	18·9				
10 0	18·9	38·5		36·4	
15 0	18·9				
20 0	19·0	36·5		36·4	
25 0	19·1				
30 0	19·0	35·0		36·4	
35 0	19·2				
40 0	18·1	34·0		36·5	
45 0	18·0				
50 0	18·1	33·1		36·4	
55 0	18·7				
8 0 0	19·0	34·0	67	36·2	66
5 0	19·1				
10 0	19·1	35·9		35·7	
15 0	19·1				
20 0	19·1	37·0		35·5	
25 0	19·2				
30 0	19·2	38·8		35·5	
35 0	19·0				
40 0	18·9	39·1		35·5	
45 0	18·8				
50 0	18·8	40·1		35·7	
55 0	18·9				
9 0 0	19·1	41·6	67	35·6	66
5 0	19·2				
10 0	19·4	43·2		36·8	
15 0	19·2				
20 0	19·2	44·0		36·6	
25 0	19·4				
30 0	19·8	44·0		35·8	
35 0	19·9				
40 0	20·0	44·0		36·7	
45 0	20·0				
50 0	20·0	43·0		36·7	
55 0	20·0				
10 0 0	20·0	43·0	67	36·6	66
5 0	20·0				
10 0	19·9	43·0		37·4	
15 0	20·0				
20 0	20·0	44·0		37·1	
25 0	20·0				
30 0	20·0	44·0		36·3	
35 0	20·0				
40 0	20·1	43·9		Vibr.	
45 0	20·1				
50 0	20·1	44·0		35·3	
55 0	20·0				
11 0 0	20·0	45·0	66	35·3	66

P

DECEMBER 30 and 31, 1841.

M. Gött. Time. d. h. m. s.	Decl. Sc.-Div^ns.	Hor. Force. Sc.-Div^ns.	Ther.	Vert. Force. Sc.-Div^ns.	Ther.
30 11 5 0	19·9				
10 0	19·8	45·5		35·3	
15 0	19·7				
20 0	19·0	46·0		35·4	
25 0	19·0				
30 0	19·0	46·5		35·4	
35 0	19·0				
40 0	18·9	46·2		35·2	
45 0	18·9				
50 0	18·6	45·5		34·8	
55 0	18·4				
12 0 0	18·6	45·7	66	35·1	66
5 0	18·8				
10 0	18·6	45·9		35·5	
15 0	18·1				
20 0	18·1	46·0		35·2	
25 0	18·0				
30 0	18·0	46·0		34·8	
35 0	18·0				
40 0	17·9	45·9		34·6	
45 0	17·9				
50 0	17·9	46·1		34·5	
55 0	17·9				
13 0 0	17·8	46·9	66	34·5	66
5 0	17·8				
10 0	17·8	46·1		35·6	
15 0	17·0				
20 0	16·6	45·2		34·6	
25 0	16·7				
30 0	16·9	46·0		35·8	
35 0	16·7				
40 0	17·1	46·2		35·8	
45 0	17·1				
50 0	17·6	46·8		36·8	
55 0	17·8				
14 0 0	17·9	47·0	66	36·8	65
5 0	18·0				
10 0	18·0	47·0		36·8	
15 0	18·1				
20 0	18·1	47·1		35·8	
25 0	18·1				
30 0	18·1	47·7		Vibr^n.	
35 0	18·1				
40 0	18·2	48·0		35·9	
45 0	18·2				
50 0	18·2	48·0		35·9	
55 0	18·2				

Positions at the usual hours of observation, December 30 and 31.

	Decl.	Hor. Force.	Ther.	Vert. Force.	Ther.
30 0 0	19·9	72·7	65	32·8	65
2 0 0	24·0	60·0	66	34·3	65
3 0 0	22·0	57·0	67	34·3	66
4 0 0	18·2	56·3	67	34·7	67
5 0 0	19·5	48·1	67	36·0	67
6 0 0	17·9	37·9	67	36·2	67
8 0 0	19·0	34·0	67	36·2	66
10 0 0	20·0	43·0	67	36·6	66
11 0 0	20·0	45·0	66	35·3	66
12 0 0	18·6	45·7	66	35·1	66
13 0 0	17·9	46·9	66	34·5	66
14 0 0	17·9	47·0	66	36·8	65

DECEMBER 30 and 31, 1841.

M. Gött. Time. d. h. m. s.	Decl. Sc.-Div^ns.	Hor. Force. Sc.-Div^ns.	Ther.	Vert. Force. Sc.-Div^ns.	Ther.
30 15 0 0	18·3	48·6	66	35·9	65
16 0 0	18·6	49·9	66	35·5	65
18 0 0	17·4	50·6	65	36·4	65
19 30 0	17·8	51·5	65	35·7	65
20 0 0	17·4	51·6	65	36·1	65
20 30 0	16·0	51·8	65	35·9	64
22 0 0	14·2	54·1	65	35·3	65
23 0 0	16·2	53·2	65	35·8	65
31 0 0 0	17·9	53·8	65	36·5	65
2 0 0	16·9	50·9	66	37·8	66
3 0 0	17·5	51·0	66	37·9	67
4 0 0	18·0	51·9	68	37·6	67
5 0 0	18·6	51·0	67	37·1	67
6 0 0	19·0	48·5	68	37·5	67
8 0 0	18·4	50·9	67	36·3	67
10 0 0	20·0	53·9	67	35·7	66
11 0 0	19·7	53·9	67	36·3	66
12 0 0	19·6	53·2	66	35·6	66
13 0 0	19·7	52·9	66	35·5	66
14 0 0	19·8	53·4	66	35·3	65
15 0 0	19·9	54·3	66		
18 0 0	19·8	55·2	65	34·9	65
18 0 0	18·3	57·2	65	34·8	64
19 30 0	17·4	56·9			
20 0 0	15·0	57·9	65	34·6	64
20 30 0	18·8	58·8			
22 0 0	11·9	59·5	65	Vibr^n.	
23 0 0	12·8	58·8	65	33·9	65

The Mean Positions at the same hours during the Month are given in page 104.

ISLAND OF ASCENSION * { Decl. 1 Scale Division = 3'·00 H. F. k = ·0003; q =

Positions at the usual hours of observation, December 30.

	Decl.	Hor. Force.	Ther.
30 0 0	52·5	82·4	82
1 0 0	53·5	83·4	82
2 0 0	51·5	78·5	82
3 0 0	52·0	74·5	82
4 0 0	52·6	74·0	82
5 0 0	52·6	72·1	82
6 0 0	53·2	67·6	81
7 0 0	55·0	68·9	80
8 0 0	56·8	66·0	80
9 0 0	58·8	70·0	80
10 0	58·6	69·9	79
11	59·1	69·3	79
12	59·0	70·0	79
13 0 0	58·3	70·0	79

Mean Positions at the same hours from December 24 to December 30, inclusive.

	Decl.	Hor. Force.	Ther.
0 0 0	55·3	76·6	82
1 0 0	55·1	76·8	82
2 0 0	54·1	74·0	83
3 0 0	53·5	74·2	82
4 0 0	53·3	75·1	82
5 0 0	53·2	74·0	82
6 0 0	53·4	72·3	82
7 0 0	54·6	72·3	81

DECEMBER 30 and 31, 1841.

M. Gött. Time. h. m. s.	Decl. Sc.-Div^ns.	Hor. Force. Sc.-Div^ns.	Ther.	Vert. Force. Sc.-Div^ns.	Ther.
8 0 0	55·5	72·2	81		
9 0 0	56·6	73·0	81		
10 0 0	57·1	73·0	81		
11 0 0	57·6	74·4	81		
12 0 0	57·5	73·2	81		
13 0 0	57·7	73·4	81		

CAPE OF GOOD HOPE { Decl. 1 Scale Division = 0'·75 H. F. k = ·00018; q = ·0003 V. F. k = ·000035; q =

Positions at the usual hours of observation, December 30 and 31.

	Decl.	Hor. Force.	Ther.	Vert. Force.	Ther.
30 0 0 0	50·8	84·9	68	49·9	68
1 0 0	53·7	81·5	68		
2 0 0	53·0	72·3	69	102·1	68
3 0 0	50·2	69·2	69		
4 0 0	49·2	73·0	69	103·7	69
5 0 0	46·4	66·5	69		
6 0 0	46·2	51·5	69	152·6	69
7 0 0	44·0	55·0	69		
8 0 0	43·4	52·0	69	155·6	68
9 0 0	41·0	63·1	68		
10 0 0	42·4	63·6	68	119·4	68
11 0 0	43·0	66·2	68		
12 0 0	40·8	68·1	68	119·5	68
13 0 0	40·5	68·5	68		
14 0 0	41·8	66·3	68	113·2	67
15 0 0	42·0	67·1	68		
16 0 0	42·5	69·0	67	106·9	67
17 0 0	42·1	70·0	67		
18 0 0	41·1	70·2	67	111·7	67
19 0 0	40·0	72·1	67		
20 0 0	39·0	72·7	67	124·2	67
21 0 0	38·0	74·0	67		
22 0 0	39·0	72·9	67	115·8	67
23 0 0	39·9	73·4	67		
31 0 0 0	42·2	75·2	67	94·3	66
1 0 0	43·0	72·1	68		
2 0 0	42·9	71·4	68	97·8	68
3 0 0	44·0	71·2	68		
4 0 0	43·5	71·0	69	96·8	68
5 0 0	43·0	68·9	69		
6 0 0	42·0	65·5	70	109·1	69
7 0 0	42·7	66·0	70		
8 0 0	43·5	67·4	70	98·7	69
9 0 0	44·2	69·2	69		
10 0 0	44·5	73·1	69	103·6	68
11 0 0	44·2	72·7	68		
12 0 0	44·5	71·8	68	100·6	68
13 0 0	44·0	71·8	67	105·1	67
14 0 0	44·0	71·8	67	107·7	67
15 0 0	44·1	72·8	67	102·1	67
16 0 0	44·6	74·9	67	100·2	67
17 0 0	43·8	76·3	66	100·8	66
18 0 0	42·6	79·3	66	100·5	66
19 0 0	41·7	79·3	66	106·9	66
20 0 0	39·8	80·8	66	107·4	66
21 0 0	37·0	79·9	66	104·5	66
22 0 0	39·1	74·2	66	84·8	67
23 0 0	43·1	70·3	67	75·7	67

The Mean Positions at the same hours during the Month are given in page 102.

* Observations made by Captain William Allen, R.N., Commanding H.M. Steamer Wilberforce, with Weber's transportable magnetometer.

DECEMBER 30 and 31, 1841.

VAN DIEMEN ISLAND {Decl. 1 Scale Division = 0'·71 ; H. F. k = ·0003 ; q = ; V. F. k = ; q =

Extra observations.

The V. F. was observed at 2m. 30s. before, and the H. F. 2m. 30s. after the times specified.*

M. Gött. Time.		Decl.	Hor. Force.		Vert. Force.	
m.	s.	Sc.-Div^ns	Sc.-Div^ns	Ther.	Sc.-Div^ns	Ther.
30 2 15	0	51·2				
20	0	50·4	49·3	63	61·8	62
25	0	50·3				
30	0	51·5	47·9		62·2	
35	0	53·0				
40	0	53·5	46·0		64·8	
45	0	53·2				
50	0	52·6	45·8		66·2	
3 10	0	57·0	46·9	64		
15	0	58·3				
20	0	58·6	47·1		69·0	63
25	0	59·5				
30	0	59·4	48·4		69·0	
35	0	59·4				
40	0	59·0	46·4		68·4	
45	0	59·0				
50	0	58·7	45·8		67·2	
55	0	58·9				

Positions at the usual hours of observation, December 30 and 31.

			Decl.	Hor. Force.		Vert. Force.	
30 0	0	0	69·8	47·3	64	67·3	62
1	0	0	68·9	47·0		68·8	
2	0	0	67·4	47·2	63	63·6	62

DECEMBER 30 and 31, 1841.

M. Gött. Time.				Decl.	Hor. Force.		Vert. Force.	
d.	h.	m.	s.	Sc.-Div^ns	Sc.-Div^ns	Ther.	Sc.-Div^ns	Ther.
30	3	0	0	54·6	46·3		67·3	
	4	0	0	58·5	45·5	64	68·1	63
	5	0	0	59·5	45·3		67·2	
	6	0	6	58·0	45·9	63	65·4	62
	7	0	0	58·0	47·0		65·9	
	8	0	0	58·4	42·5	62	70·1	60
	9	0	0	61·1	42·5		73·7	
	10	0	0	58·4	43·5	61	71·7	59
	11	0	0	56·2	43·5		72·5	
	12	0	0^b	60·4	43·1	60	76·6	59
31	2	0	0	90·0	52·5	62	68·8	61
	3	0	0	88·3	53·4		70·7	
	4	0	0	89·5	53·9	63	71·8	62
	5	0	0	88·6	54·5		72·2	
	6	0	0	92·3	56·7	60	72·7	60
	7	0	0	94·0	56·6		72·4	
	8	0	0	90·4	54·8	60	75·1	59
	9	0	0	87·6	54·2		74·4	
	10	0	0	85·8	54·4	60	73·5	59
	11	0	0	85·7	52·3		69·6	
	12	0	0		52·1	63		
	15	0	0	73·4	45·6		63·5	61
	16	0	0	78·5	48·0	72	60·9	70
	17	0	0	80·9	45·4		61·9	
	18	0	0	81·6	51·8	74	57·2	71
	19	0	0	81·4	53·2		59·6	
	20	0	0	80·0	52·8	73	59·2	70
	21	0	0	79·1	54·4		59·2	
	22	0	0	77·7	54·9	71	58·1	69
	23	0	0	69·2	51·3		61·4	

DECEMBER 30 and 31, 1841.

The Mean Positions corresponding to the position of the H. F. and V. F., namely, from 30^d. 0^h. to 30^d. 12^h., are given in page 102; the Mean Positions corresponding to the observations of the three instruments from 31^d. 2^h. to 31^d. 23^h., are those of the Month of January 1842, and are as follows:—

M. Gött. Time.				Decl.	Hor. Force.		Vert. Force.	
d.	h.	m.	s.	Sc.-Div^ns	Sc.-Div^ns	Ther.	Sc.-Div^ns	Ther.
0	0	0		75·7	54·2	68	62·3	66
1	0	0		74·6	54·4		63·6	
2	0	0		74·2	44·1	66	64·8	64
3	0	0		73·9	54·5		66·0	
4	0	0		73·6	55·2	64	69·0	63
5	0	0		73·6	55·5		67·8	
6	0	0		73·7	56·1	63	68·1	62
7	0	0		73·8	55·9		69·1	
8	0	0		72·3	56·3	62	70·8	60
9	0	0		70·9	56·3		70·3	
10	0	0		68·7	55·7	61	70·1	61
11	0	0		67·9	55·0		69·2	
12	0	0		68·2	53·3	63	68·1	62
13	0	0		70·7	52·2		67·9	
14	0	0		74·1	51·9	66	66·3	65
15	0	0		78·4	51·7		64·5	
16	0	0		81·7	52·6	68	62·7	66
17	0	0		82·3	53·1		61·1	
18	0	0		81·5	53·5	70	59·6	68
19	0	0		80·3	53·5		58·7	
20	0	0		79·1	53·5	71	58·8	68
21	0	0		77·7	53·2		59·4	
22	0	0		76·6	52·9	70	60·1	68
23	0	0		76·3	53·9		61·3	

* The connexion of the Declination series appears to have been broken on the 28th of December, consequently the observations of that magnet from 30^d. 0^h. to 30^d. 12^h., are not comparable with the mean positions given in page 102.

b The connexion of the series was broken at 30^d. 12^h. for the purpose of making the usual monthly determinations of the magnetic moments of the bars.